S0-BMA-422

Applied Functions of a Complex Variable

Applied Functions of a Complex Variable

A. Kyrala

Professor of Physics
Arizona State University
and
Visiting Professor of Mathematics
American University of Beirut

Wiley-Interscience

a division of John Wiley & Sons, Inc.
New York · London · Sydney · Toronto

Library of Congress Catalog Card Number: 74-176285

ISBN 0-471-51129-3

Printed in the United States of America.

10 9 8 7 6 5 4 3 2 1

To Lawrence Benali Kyrala

May his enthusiasm and vigor
be matched by an insatiable desire
for understanding the universe

Preface

This short monograph has been written with the dual purpose of providing a senior/graduate level text suitable for engineering and scientific curricula and of providing a rapid introduction or review for home study by professional scientists and engineers.

Since it is not directed toward pure mathematicians, the basis adopted here has been *maximum didactic simplicity* rather than maximum generality. It should be remembered that those primarily interested in applications of mathematics wish to acquire a rapid but secure intuitive grasp of the subject which will permit them to apply basic principles with effectiveness and confidence. Their basic concern is *not* methodology or the logical structure of the argument, although no one can hope to become an accomplished applied mathematician without a reasonable amount of attention to these aspects. Scientists and engineers are not usually interested in presentations which devote 90% of the space to enlarging the class of admissible functions by 1%. This often happens if one makes maximum generality the primary objective. To the applied mathematician the inclusion of a few more "pathological" functions never to be encountered in practice is not worth the effort. This should not be construed as an attack against the principle of maximum generality which has a useful and aesthetic place in the literature of pure mathematics, but for the applied group the cost of every argument in time and effort must be considered.

To convey to the reader not only a grasp of the content and meaning of the principal theorems of the subject, but also a clear picture of their interrelationships with one another and with applications, numerous worked examples and unsolved problems have been included. The purpose of this is to connect the abstract with the concrete and also to ensure that the book may be used independently of lectures.

Generally in the book the mapping approach to complex variables has been emphasized. The conceptual connection between the Schwarz–Christoffel mapping and the Riemann mapping theorem, the method of stationary phase as a functional inversion, and the simple approach to Cauchy integrals and sectionally holomorphic functions all represent deviations from the usual treatments of those subjects.

The material has been tested in actual presentation for several years and modified on the basis of this experience. The present form of the text therefore is that which the author has found to be most readily assimilable by students in the applied fields.

At this time many mathematics departments are giving courses for engineers and scientists in functions of a complex variable which are really collections of unmotivated theorems. Some go so far as to spend a whole semester on the topological aspects of Cauchy's theorem. While this kind of approach might be well justified for presentation to certain groups of pure mathematics majors, it is fraudulent to pretend that it fulfills the needs of scientists or engineers.

On the other hand, there are many results known to pure analysts which could be applied but have not been because of what might be called a "semantic" gap between the pure and applied groups. It is hoped that this book will help to fill that gap and make more of the beautiful and useful aspects of the subject accessible to the applied group of readers.

The mathematical sources most frequently consulted in preparing the present work were the following:

1. E. T. Copson, *Theory of Functions of a Complex Variable.* Oxford: Oxford University Press, 1935.
2. E. C. Titchmarsh, *Theory of Functions.* Oxford: Oxford University Press, 1939.
3. L. Bieberbach, *Lehrbuch der Funktionentheorie.* Leipzig: Teubner, 1931.
4. N. I. Muskhelishvili, *Some Basic Problems of the Mathematical Theory of Elasticity.* Groningen: Noordhoff, 1953.

The author wishes to express his indebtedness to and admiration for these works.

Portions of the first two chapters of this volume were written in collaboration with the author's wife, Judith Wood Kyrala, to whom grateful thanks are due.

Many thanks are also due to the Provost, Dr. E. Terry Prothro, of the American University of Beirut and to the Mathematics Department of that institution for the opportunity to test the text material internationally for two years after its previous use in the United States.

Finally, the secretarial assistance of Mrs. Kathryn Rogers, Miss Mona Jabbour, and Miss Juanita Moore are gratefully acknowledged.

Tempe, Arizona
December 1971

A. KYRALA

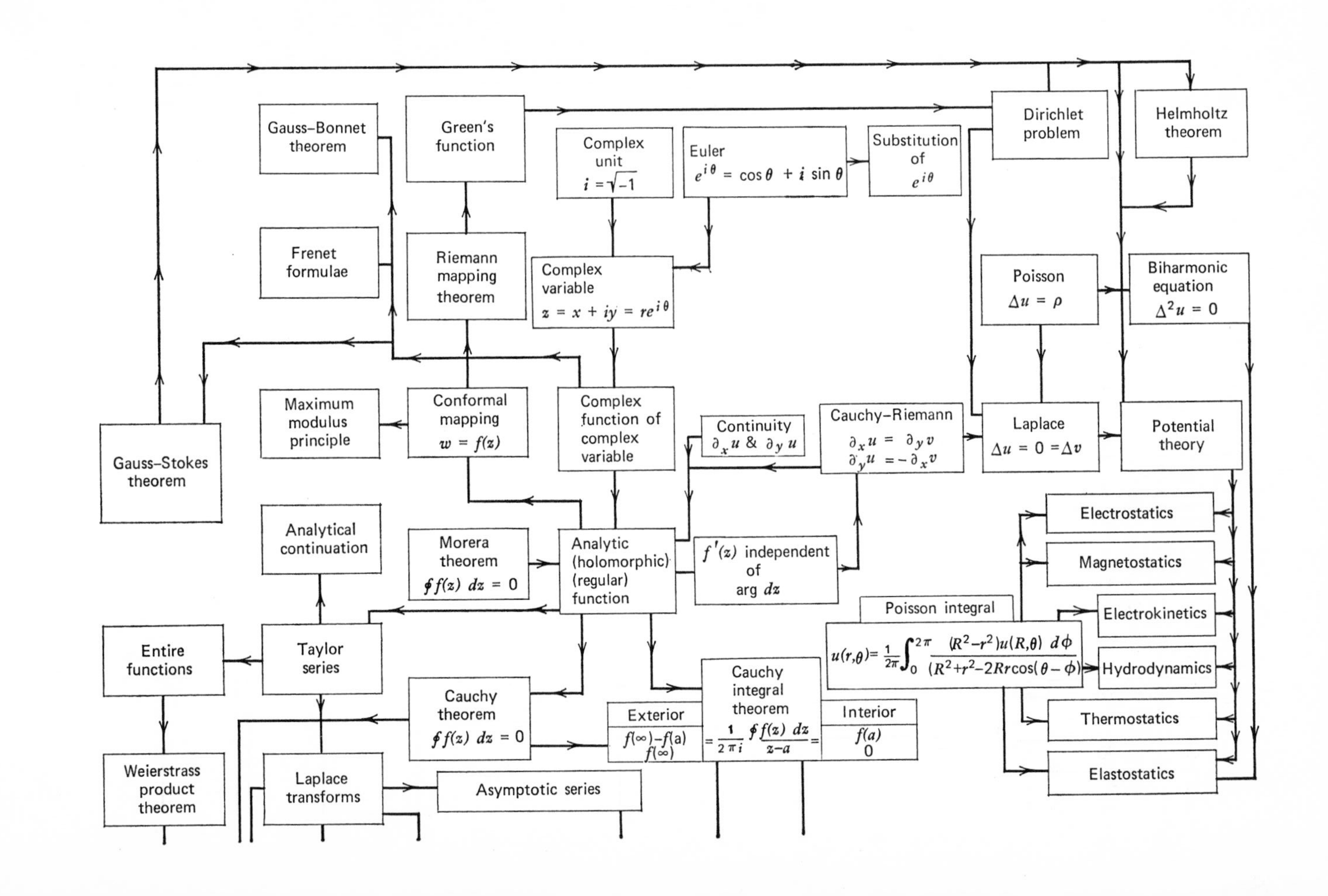

Gauss–Bonnet theorem
Green's function
Complex unit $i = \sqrt{-1}$
Euler $e^{i\theta} = \cos\theta + i \sin\theta$
Substitution of $e^{i\theta}$
Dirichlet problem
Helmholtz theorem
Frenet formulae
Riemann mapping theorem
Complex variable $z = x + iy = re^{i\theta}$
Poisson $\Delta u = \rho$
Biharmonic equation $\Delta^2 u = 0$
Gauss–Stokes theorem
Maximum modulus principle
Conformal mapping $w = f(z)$
Complex function of complex variable
Continuity $\partial_x u$ & $\partial_y u$
Cauchy–Riemann $\partial_x u = \partial_y v$ $\partial_y u = -\partial_x v$
Laplace $\Delta u = 0 = \Delta v$
Potential theory
Analytical continuation
Morera theorem $\oint f(z)\, dz = 0$
Analytic (holomorphic) (regular) function
$f'(z)$ independent of arg dz
Electrostatics
Magnetostatics
Electrokinetics
Hydrodynamics
Thermostatics
Elastostatics
Poisson integral
$u(r,\theta) = \frac{1}{2\pi}\int_0^{2\pi} \frac{(R^2-r^2)u(R,\theta)\, d\phi}{(R^2+r^2-2Rr\cos(\theta-\phi))}$
Entire functions
Taylor series
Cauchy theorem $\oint f(z)\, dz = 0$
Exterior
$\frac{f(\infty)-f(a)}{f(\infty)}$
Cauchy integral theorem
$= \frac{1}{2\pi i}\oint \frac{f(z)\, dz}{z-a} =$
Interior
$f(a)$
0
Weierstrass product theorem
Laplace transforms
Asymptotic series

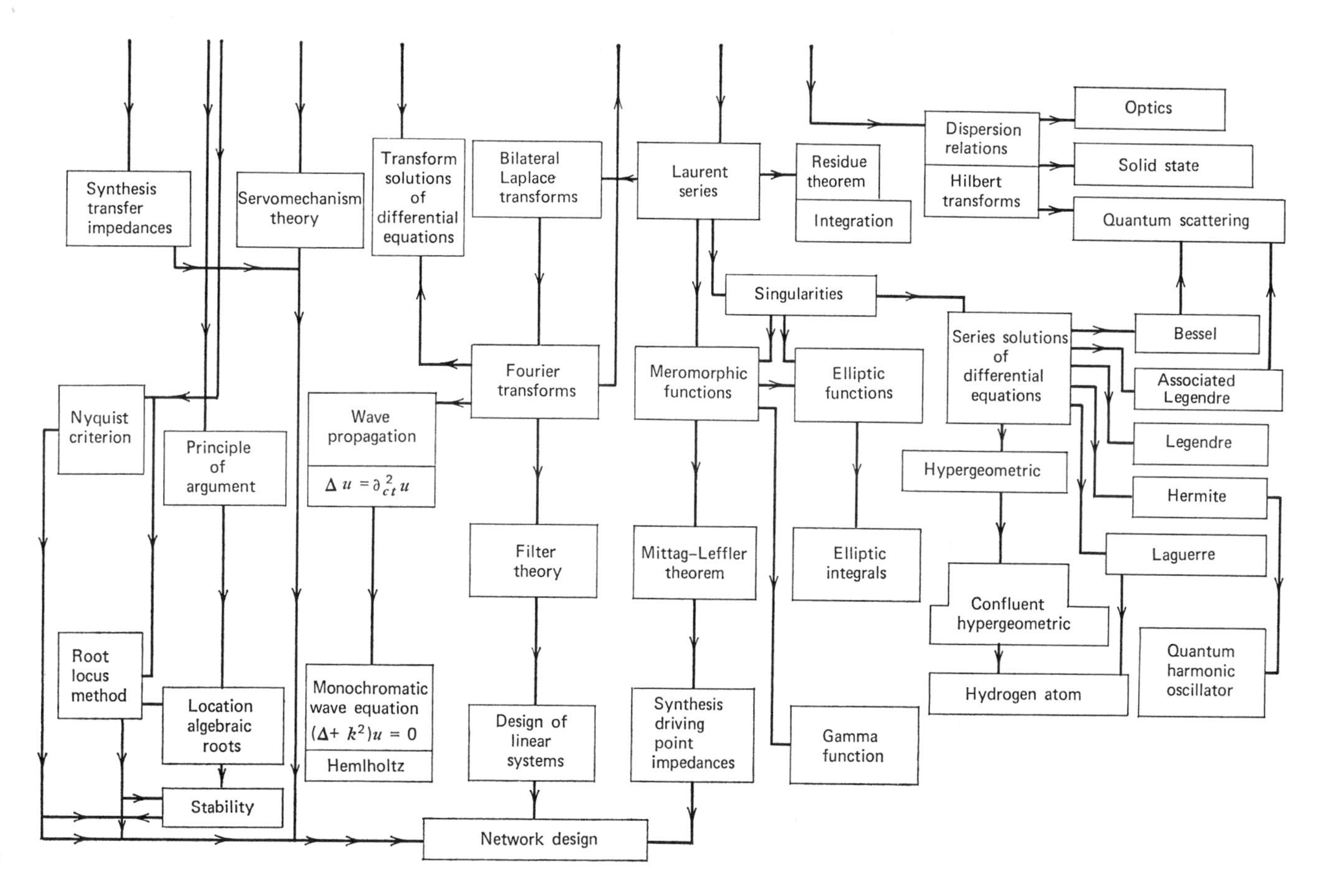
Synthesis transfer impedances
Servomechanism theory
Transform solutions of differential equations
Bilateral Laplace transforms
Laurent series
Residue theorem
Integration
Dispersion relations
Hilbert transforms
Optics
Solid state
Quantum scattering
Singularities
Series solutions of differential equations
Bessel
Associated Legendre
Legendre
Hermite
Laguerre
Nyquist criterion
Principle of argument
Wave propagation
$\Delta u = \partial_{ct}^2 u$
Fourier transforms
Meromorphic functions
Elliptic functions
Hypergeometric
Filter theory
Mittag-Leffler theorem
Elliptic integrals
Confluent hypergeometric
Quantum harmonic oscillator
Root locus method
Location algebraic roots
Monochromatic wave equation
$(\Delta + k^2)u = 0$
Hemlholtz
Design of linear systems
Synthesis driving point impedances
Gamma function
Hydrogen atom
Stability
Network design

Contents

CHAPTER 1: COMPLEX NUMBERS AND DIRECT APPLICATIONS

CHAPTER 2: FUNCTIONS OF A COMPLEX VARIABLE

CHAPTER 3: INFINITE SERIES

CHAPTER 4: CAUCHY'S THEOREM

CHAPTER 5: CAUCHY INTEGRAL THEOREM

CHAPTER 6: LAURENT SERIES AND RESIDUE THEOREM

CHAPTER 7: SINGULARITIES AND ANALYTICAL CONTINUATION

CHAPTER 8: CONFORMAL MAPPING

CHAPTER 9: LAPLACE AND FOURIER TRANSFORMS

CHAPTER 10: INFINITE PRODUCT AND RATIONAL FRACTION EXPANSIONS

CHAPTER 11: DISPERSION RELATIONS

CHAPTER 12: ELLIPTIC FUNCTIONS AND INTEGRALS

CHAPTER 13: DIFFERENTIAL EQUATIONS AND SPECIAL FUNCTIONS

Special Notations Used in the Text

d_z means d/dz

∂_x means $\partial/\partial x$

∂ means $\partial_x + i\partial_y$

$d_z f$ means df/dz

$\partial_x u$ means $\partial u/\partial x$

For $z = x + iy$

$\operatorname{Re}(z^*w)$ means $\vec{z} \cdot \vec{w}$

$\operatorname{Im}(z^*w)$ means $|\vec{z} \times \vec{w}|$

$|z|^2 = z^*z = x^2 + y^2$

$\arg z = \arctan(y/x)$

CHAPTER 1:

Complex Numbers and Direct Applications

1.1 THE COMPLEX NUMBER SYSTEM

It is well known that the equation

$$x^2 - 1 = (x - 1)(x + 1) = 0 \tag{1.1}$$

has solutions $x = \pm 1$. Clearly there are no *real* solutions of the equation

$$x^2 + 1 = 0 \tag{1.2}$$

since the square of any real number is nonnegative and cannot yield zero when added to unity.

Thus if there is a solution to (1.2), it cannot be sought among the real numbers but must be a *new kind of number* with somewhat different properties. Instead of seeking an x satisfying (1.2) from a set of known (real) numbers we may use an identity similar to (1.2) to *define* a unit (denoted by i) of these new numbers which we shall call *imaginary numbers*. Thus by definition,

$$i^2 \equiv -1 \tag{1.3}$$

which defines i to within an algebraic sign. As in the case of (1.1) we may write (1.2) as

$$x^2 + 1 = (x - i)(x + i) = 0 \tag{1.4}$$

with solutions $x = \pm i$. If b is any real number, then ib is a corresponding imaginary number. It might be supposed that one could use these imaginary numbers much as one had used the real numbers. While this would be quite so for addition and subtraction it would not be true for multiplication, since the product of two such imaginary numbers is not again an imaginary number but is a real number. The property of *closure* (of obtaining the same kind of number as the result of addition, subtraction, multiplication, or division except by zero) was an important feature of the real number system, and although it does not hold for imaginary numbers it does hold for the system of *complex numbers* z defined as an additive combination of any real number x with any imaginary number iy:

$$z = x + iy \tag{1.5}$$

Here x is called the *real part* of z, denoted by $x = \operatorname{Re} z$, and y is called the *imaginary part* of z, denoted by $y = \operatorname{Im} z$. A complex number closely related to (1.5) is the *complex conjugate* of z denoted by z^*:

$$z^* = x - iy \tag{1.6}$$

The sum and difference of two complex numbers $z_1 = x_1 + iy_1$ and $z_2 = x_2 + iy_2$ are given by the complex numbers

$$(z_1 + z_2) = (x_1 + x_2) + i(y_1 + y_2) \tag{1.7}$$

$$(z_1 - z_2) = (x_1 - x_2) + i(y_1 - y_2) \tag{1.8}$$

It is convenient to represent the complex numbers by points in a two-dimensional plane called the *complex plane*, with the real part x corresponding to the abscissal distance x from the *imaginary axis* (consisting of all real multiples of i) and the imaginary part y corresponding to the ordinate distance y from the *real axis* (consisting of all real numbers). With these conventions it is readily discerned that (1.7) and (1.8) are equivalent to the laws of addition and subtraction of two-dimensional vectors $\vec{z}_1$ and $\vec{z}_2$:

$$\vec{z}_1 = (x_1, y_1) \tag{1.9}$$

$$\vec{z}_2 = (x_2, y_2) \tag{1.10}$$

so that

$$\vec{z}_1 \pm \vec{z}_2 = (x_1 \pm x_2, y_1 \pm y_2) \tag{1.11}$$

and we may also represent the complex number z by an arrow directed from the origin (zero) to the point z in the complex plane. Also the scalar product $\vec{z}_1 \cdot \vec{z}_2 = x_1x_2 + y_1y_2 = \operatorname{Re}(z_1^*z_2)$ and the absolute value of the vector product $|\vec{z}_1 \times \vec{z}_2| = x_1y_2 - x_2y_1 = \operatorname{Im}(z_1^*z_2)$. Throughout the text such real and imaginary parts of products will be used rather than the vectorial notation for scalar and vector products. Using the (Argand) diagram (Fig. 1.1) it may be seen that

$$x = r\cos\theta \qquad y = r\sin\theta \tag{1.12}$$

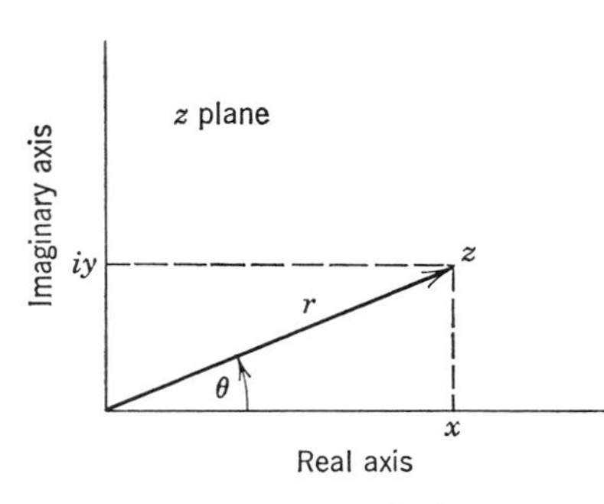

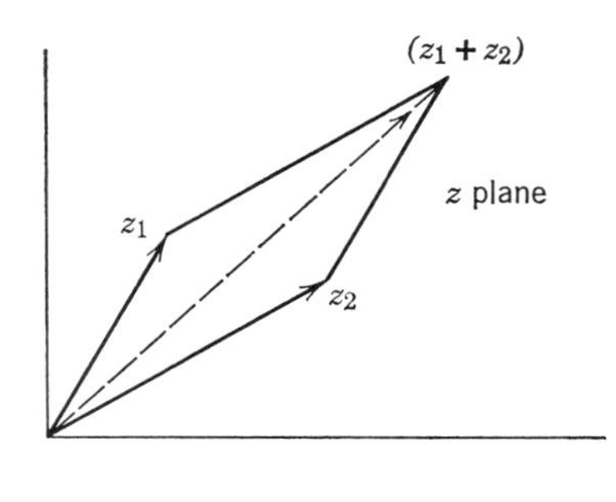

Figure 1.1. Geometrical representation of a complex number $z = x + iy$ and addition law.

Equation (1.5) may now be written in terms of (r,θ):

$$z = x + iy = r(\cos\theta + i\sin\theta) \tag{1.13}$$

It is now convenient to introduce the *modulus* or *magnitude*, $|z|$, of a complex number,

$$|z| = \sqrt{x^2 + y^2} = r \tag{1.14}$$

and the *argument* or *phase*, $\arg z$, of a complex number,

$$\arg z = \arctan(y/x) = \theta \tag{1.15}$$

Note that $w = (\cos\theta + i\sin\theta)$ is unity for $\theta = 0$ and satisfies the differential equation

$$-\sin\theta + i\cos\theta = d_\theta w = \frac{dw}{d\theta} = iw = i(\cos\theta + i\sin\theta) \tag{1.16}$$

The solution of (1.16) may alternatively be seen to be the exponential $e^{i\theta}$, since (1.16) has a *unique* solution reducing to unity for $\theta = 0$, and only an exponential function is proportional to its derivative and reduces to unity at the origin. This uniqueness allows us to conclude that

$$e^{i\theta} \equiv \cos\theta + i\sin\theta \tag{1.17}$$

which is called *Euler's formula.* Thus a more concise form of (1.13) is

$$z = re^{i\theta} \tag{1.18}$$

Since two complex numbers z_1 and z_2 are equal if and only if their real and imaginary parts are equal, they are also equal if and only if their magnitudes are equal and if their arguments are equal to within an integral multiple of 2π. This can easily be shown using (1.18):

$$z_1 = r_1 e^{i\theta_1} \qquad z_2 = r_2 e^{i\theta_2} \tag{1.19}$$

If

$$z_1 = z_2 \tag{1.20}$$

then

$$\frac{r_1}{r_2} = e^{i(\theta_2-\theta_1)} \tag{1.21}$$

$$\frac{r_1}{r_2} = \cos(\theta_2 - \theta_1) + i\sin(\theta_2 - \theta_1) \tag{1.22}$$

But r_1/r_2 is a nonnegative real number, so the imaginary term on the right in (1.22) must be zero and

$$\sin(\theta_2 - \theta_1) = 0 \tag{1.23}$$

$$\theta_2 = \theta_1 + 2n\pi \qquad n = \text{integer} \tag{1.24}$$

This causes $\cos(\theta_2 - \theta_1)$ to be unity, so we obtain the second condition that

$$r_1 = r_2 \qquad \text{or} \qquad |z_1| = |z_2| \tag{1.25}$$

The magnitude of any complex number is a nonnegative number and so the usual concept of logarithms may be used such that

$$e^{\log|z|} = |z| \tag{1.26}$$

It has been shown previously that

$$z = |z|\, e^{i \arg z} \tag{1.27}$$

Therefore,

$$z = e^{\log|z| + i \arg z} = e^{\alpha} \tag{1.28}$$

Let us suppose that there is another solution, α_1, to (1.28). Then

$$z = e^{\alpha} = e^{\alpha_1} \tag{1.29}$$

and using Euler's formula,

$$e^{\alpha_1 - \alpha} = 1 = \cos(\alpha_1 - \alpha) + i\sin(\alpha_1 - \alpha) \tag{1.30}$$

The imaginary part of (1.30) must be zero; therefore,

$$\alpha_1 - \alpha = 2n\pi \qquad n = \text{integer} \tag{1.31}$$

Thus,

$$\alpha = \log|z| + i \arg z + 2n\pi i \tag{1.32}$$

A number α such that $e^{\alpha} = z$ is called a *logarithm* of z and it will be denoted by

$$\alpha = \log z = \log|z| + i \arg z + 2n\pi i \tag{1.33}$$

If $n = 0$ and $-\pi < \arg z < \pi$, then α actually is the *principal logarithm* of z, but it will often be called the logarithm of z. Note that the logarithm of a complex number is not a single-valued function, but rather, is multi-valued depending upon the value of n in (1.33).

1.2 APPLICATIONS TO KINEMATICS

Kinematics is the study of the possible motions of mechanical systems. The concept of complex numbers is a convenient method for describing the motion of a particle in a plane. We start out with the expression for the position of the particle in the complex plane:

$$z = re^{i\theta} \tag{1.34}$$

Differentiating this with respect to time, we get

$$\dot{z} = \dot{r}e^{i\theta} + ire^{i\theta}\dot{\theta} = v \tag{1.35}$$

where v is the complex velocity of the particle, $e^{i\theta}$ corresponds to a unit vector in the radial direction, and $ie^{i\theta}$ corresponds to a unit vector orthogonal to the radial direction. In mechanics the expression for the radial velocity v_r of a particle in a plane is

$$v_r = \dot{r} \tag{1.36}$$

and for the tangential velocity v_θ it is

$$v_\theta = r\dot{\theta} \tag{1.37}$$

so we may express (1.35) as

$$v = (v_r + iv_\theta)e^{i\theta} \tag{1.38}$$

The complex acceleration may similarly be found by differentiating (1.35) to obtain

$$\ddot{z} = (\ddot{r} - r\dot{\theta}^2)e^{i\theta} + ie^{i\theta}(2\dot{r}\dot{\theta} + r\ddot{\theta}) = a \tag{1.39}$$

The radial component of acceleration for the motion of a particle in a plane is

$$a_r = \ddot{r} - r\dot{\theta}^2 \tag{1.40}$$

The tangential component is

$$a_\theta = r\ddot{\theta} + 2\dot{r}\dot{\theta} \tag{1.41}$$

so we may write

$$a = (a_r + ia_\theta)e^{i\theta} \tag{1.42}$$

The term $2\dot{r}\dot{\theta}$ is called the Coriolis acceleration, $r\ddot{\theta}$ is the tangential acceleration, $r\dot{\theta}^2$ is the centripetal acceleration, and $\ddot{r}$ is the radial acceleration.

Another example is the conservation of angular momentum for a particle moving under the influence of a central force. If the (complex) location of the particle is z and its mass is the real positive number m while the (complex) force acting on the particle is F, Newton's second law yields

$$F = m\ddot{z}$$

The angular momentum (directed orthogonally to the z plane) has magnitude

$$L = m \operatorname{Im}(z^*\dot{z})$$

The time derivative of L is

$$\dot{L} = m \operatorname{Im}(\dot{z}^*\dot{z}) + m \operatorname{Im}(z^*\ddot{z})$$

Since $\dot{z}^*\dot{z} = |\dot{z}|^2$ is real,

$$\dot{L} = \operatorname{Im}(z^*F)$$

For a central force the ratio of F to z must be a real number, say λ, so that

$$\dot{L} = \lambda \operatorname{Im}(z^*z) = 0$$

since $z^*z = |z|^2$ is real. Since the angular momentum L does not change with time, it is said to be *conserved*.

1.3 APPLICATIONS TO INTEGRALS AND SUMS

Complex numbers may also be used to conveniently evaluate many types of otherwise more difficult integrals and sums with a great saving of labor. For instance, consider the sum of the geometric progression

$$\sum_{k=0}^{n-1} z^k = \frac{z^n - 1}{z - 1} \tag{1.43}$$

Taking

$$z = e^{i\theta} \tag{1.44}$$

we obtain

$$\sum_{k=0}^{n-1} e^{ik\theta} = \frac{e^{in\theta} - 1}{e^{i\theta} - 1} = \frac{e^{in\theta/2}(e^{in\theta/2} - e^{-in\theta/2})}{e^{i\theta/2}(e^{i\theta/2} - e^{-i\theta/2})} \tag{1.45}$$

Invoking Euler's formula in the form

$$\frac{e^{in\theta/2} - e^{-in\theta/2}}{2i} = \sin\frac{n\theta}{2} \tag{1.46}$$

we obtain

$$\sum_{k=0}^{n-1} e^{ik\theta} = e^{i(n-1)\theta/2}\left[\frac{\sin(n\theta/2)}{\sin(\theta/2)}\right] \tag{1.47}$$

which may also be written as

$$\sum_{k=0}^{n-1} e^{ik\theta} = \left[\frac{\sin(n\theta/2)}{\sin(\theta/2)}\right][\cos\{(n-1)\theta/2\} + i\sin\{(n-1)\theta/2\}] \tag{1.48}$$

But notice that the left side of (1.48) may also be written as

$$\sum_{k=0}^{n-1} e^{ik\theta} = \sum_{k=0}^{n-1}\cos k\theta + i\sum_{k=0}^{n-1}\sin k\theta \tag{1.49}$$

Equating the real and imaginary parts of each side of (1.48) separately yields the following useful *real* results:

$$\sum_{k=0}^{n-1}\cos k\theta = \cos[(n-1)\theta/2]\frac{\sin(n\theta/2)}{\sin(\theta/2)} \tag{1.50}$$

and

$$\sum_{k=0}^{n-1}\sin k\theta = \sin[(n-1)\theta/2]\frac{\sin(n\theta/2)}{\sin(\theta/2)} \tag{1.51}$$

Another example is the evaluation of

$$\int e^{(a+ib)x}\,dx = \frac{e^{(a+ib)x}}{(a+ib)} \tag{1.52}$$

Multiplying the top and bottom of the right side of (1.52) by $(a - ib)$, we obtain

$$\int e^{(a+ib)x}\,dx = e^{(a+ib)x}\left\{\frac{(a-ib)}{a^2+b^2}\right\} \tag{1.53}$$

Equation (1.53) may also be written as

$$\int e^{(a+ib)x}\,dx = (\cos bx + i\sin bx)e^{ax}\left\{\frac{a-ib}{a^2+b^2}\right\} \tag{1.54}$$

Combining real and imaginary terms, we get

$$\int e^{(a+ib)x}\,dx = \frac{(a\cos bx + b\sin bx)e^{ax}}{(a^2+b^2)} + i\,\frac{(a\sin bx - b\cos bx)e^{ax}}{(a^2+b^2)} \tag{1.55}$$

Note that by using Euler's formula again we may write

$$\int e^{(a+ib)x}\,dx = \int e^{ax}\cos bx\,dx + i\int e^{ax}\sin bx\,dx \tag{1.56}$$

Equating the real and imaginary components on each side of (1.55) we then get

$$\int e^{ax}\cos bx\,dx = \frac{(a\cos bx + b\sin bx)e^{ax}}{(a^2+b^2)} \tag{1.57}$$

and

$$\int e^{ax} \sin bx\, dx = \frac{(a \sin bx - b \cos bx)e^{ax}}{(a^2 + b^2)} \tag{1.58}$$

Euler's formula is also useful for determining the nth roots of a given complex number $ae^{i\alpha}$. Thus if the complex number z is sought for which

$$z^n = ae^{i\alpha} \qquad a > 0$$

one may write $z = re^{i\theta}$ and from

$$r^n e^{in\theta} = ae^{i\alpha}$$

it can be concluded that

$$r = a^{1/n}$$

and

$$n\theta = \alpha + 2k\pi$$

since

$$e^{in\theta} = e^{i\alpha}e^{2k\pi i} = e^{i\alpha}(e^{2i\pi})^k = e^{i\alpha}$$

Hence,

$$\theta = \frac{\alpha + 2k\pi}{n}$$

which yields distinct values of θ for each integer value of k from 0 to $(n - 1)$ inclusive. For $k = n$ and above, subtraction of an appropriate multiple of 2π yields one of the values of θ already obtained.

The basic advantage of the complex formulation in the above situations derives from the conciseness of the Euler formula and the attendent simplicity of differentiating or integrating exponential rather than trigonometric functions. Whenever the latter occur prominently in calculations, consideration should be given to the possible advantages of substituting via the Euler formula.

● EXAMPLES 1

1. Show that $z^* = re^{-i\theta}$ if $z = re^{i\theta}$.

Ans.

$$\begin{aligned} z &= x + iy = re^{i\theta} \\ z^* &= x - iy = x + i(-y) \\ &= [r\cos\theta + i(-\sin\theta)] \\ &= [r\cos(-\theta) + i\sin(-\theta)] \\ &= re^{-i\theta} \end{aligned}$$

2. Show that $|z^*| = r = |z|$ and that $-\arg z^* = \arg z$.

Ans.
$$\begin{aligned} |z^*| &= |re^{-i\theta}| = |r|\,|e^{-i\theta}| \\ &= r\,|\cos\theta - i\sin\theta| \\ &= r(\cos^2\theta + \sin^2\theta)^{1/2} = r = |z| \end{aligned}$$
$$\arg z^* = -\theta = -\arg z$$

3. Show that $(z + z^*)/2 = x$ and that $(z - z^*)/2i = y$ if $z = x + iy$.

Ans.
$$\frac{z + z^*}{2} = \frac{(x + iy) + (x - iy)}{2} = \frac{2x}{2} = x$$
$$\frac{z - z^*}{2i} = \frac{(x + iy) - (x - iy)}{2i} = \frac{2iy}{2i} = y$$

4. Show that $\theta = (1/2i)\log(z/z^*)$.

Ans.
$$\frac{z}{z^*} = \frac{re^{i\theta}}{re^{-i\theta}} = e^{2i\theta}$$
$$\log\left(\frac{z}{z^*}\right) = 2i\theta$$
$$\theta = \frac{1}{2i}\log\left(\frac{z}{z^*}\right)$$

5. If $z^2 = i$, find z.

Ans. $z^2 = e^{i[(\pi/2)+2n\pi]}$ implies $z = e^{i[(\pi/4)+n\pi]}$ so

$$z = e^{i(\pi/4)},\ e^{i(5\pi/4)}$$

6. Express $\tanh(1 + \pi i)$ in terms of real numbers only.

Ans.
$$\tanh(1 + \pi i) = \frac{ee^{i\pi} - e^{-1}e^{-i\pi}}{ee^{i\pi} + e^{-1}e^{-i\pi}} = \frac{-e + (1/e)}{-e - (1/e)} = \frac{e^2 - 1}{e^2 + 1}$$

7. Find ω such that $1 + \omega + \omega^2 = 0$.

Ans.
$$\omega = -\frac{1}{2} \pm \frac{i\sqrt{3}}{2}$$

8. Show that $|\cos z| \geq \sinh|y|$ with $z = x + iy$.

Ans.
$$\begin{aligned} 2\,|\cos z| = |e^{iz} + e^{-iz}| = |e^{ix}e^{-y} + e^{-ix}e^{y}| &\geq e^{|y|} - e^{-|y|} \\ &= 2\sinh|y| \end{aligned}$$

9. Find the complex number symmetrical to $(2 + 3i)$ with respect to the line $\arg z = \alpha$.

Ans. Rotate $(2 + 3i)$ through negative angle α, take the conjugate, and then rotate through positive angle α. Thus the complex number sought is $(2 - 3i)e^{2i\alpha}$.

10. Show that $z = pz_1 + (1 - p)z_2$ with p real lies on the line through z_1 and z_2.

Ans. $\text{Im}[(z - z_1)^*(z_2 - z)] = p(1 - p)\text{Im}[(z_2^* - z_1^*)(z_2 - z_1)] = 0$

11. Show that for the complex numbers z and w, $|z + w| \leq |z| + |w|$ and $|z - w| \geq ||z| - |w||$.

Ans. The first relation follows from the geometrical result that the sum of the lengths of two sides of a triangle cannot be exceeded by the length of the third side. The second follows if we take $|z|$ to be the larger of $|z|$ and $|w|$. Then $|w| + |z - w| \geq |w + (z - w)| = |z|$.

12. Show that

$$\text{Re}(z_1^* z_2) = \text{Re}(z_1 z_2^*) = |z_1|\,|z_2| \cos[\arg(z_1/z_2)]$$
$$\text{Im}(z_1^* z_2) = -\text{Im}(z_1 z_2^*) = |z_1|\,|z_2| \sin[\arg(z_2/z_1)]$$

Ans. Let $z_1 = r_1 e^{i\theta_1}$ and $z_2 = r_2 e^{i\theta_2}$. Then

$$\text{Re}(z_1^* z_2) = \text{Re}(z_1 z_2^*) = r_1 r_2 \cos(\theta_1 - \theta_2)$$
$$\text{Im}(z_1^* z_2) = -\text{Im}(z_1 z_2^*) = r_1 r_2 \sin(\theta_2 - \theta_1)$$

which is equivalent to the result stated.

13. For each star in our galaxy (the Milky Way) which moves clockwise in a circular orbit at constant (in time) angular velocity about the galactic center determine (Bottlinger's formula) the radial velocity (of such a star) as seen from the Solar System (i.e., from the sun) in terms of the angular velocities of the sun and of the star, the distance of the galactic center from the sun and the angular difference (galactic longitude) in the directions of the star and galactic center as seen from the sun. Assume the orbits to be coplanar and concentric, and assume that the angular velocities of the stars may differ but each is constant in time.

Ans. Take the galactic center as origin in the galactic plane (z-plane). Let

$z_\odot$ = location of sun

z_* = location of star

$\omega_\odot$ = angular velocity of sun

ω_* = angular velocity of star

ϕ = galactic longitude of star

$z = z_* - z_\odot$

v_r = radial velocity of star

Because the motion is circular and uniform

$$z_* \sim e^{-i\omega_* t} \qquad \text{and} \qquad \dot{z}_* = -i\omega_* z_*$$

$$z_\odot \sim e^{-i\omega_\odot t} \qquad \text{and} \qquad \dot{z}_\odot = -i\omega_\odot z_\odot$$

Hence,

$$\dot{z} = \dot{z}_* - \dot{z}_\odot = i(\omega_\odot z_\odot - \omega_* z_*)$$

$$z^*\dot{z} = i(\omega_* z_\odot^* z_* + \omega_\odot z_*^* z_\odot) - i(\omega_* z_*^* z_* + \omega_\odot z_\odot^* z_\odot)$$

Since the last term is imaginary,

$$\text{Re}(z^*\dot{z}) = -\text{Im}(\omega_* z_\odot^* z_* + \omega_\odot z_*^* z_\odot)$$

Also, $\text{Im}(z_\odot^* z_*) = -\text{Im}(z_*^* z_\odot)$, so

$$\text{Re}(z^*\dot{z}) = (\omega_* - \omega_\odot)\text{Im}(z_*^* z_\odot)$$

Since $z_\odot^* z_\odot$ is real, this may be written as

$$\text{Re}(z^*\dot{z}) = (\omega_* - \omega_\odot)\text{Im}[(z_*^* - z_\odot^*)z_\odot]$$

or

$$\text{Re}(z^*\dot{z}) = (\omega_* - \omega_\odot)\text{Im}(z^* z_\odot)$$

so that

$$v_r = (\omega_* - \omega_\odot)\, |z_\odot| \sin\phi$$

Bottlinger's formula.

14. Show that for stars much closer to the sun than the galactic center the radial velocity (as observed from the sun) is proportional to the product of the $\sin 2\phi$ and the sun–star distance $r = |z|$. Assume the angular velocities depend only on $|z_*|$.

Ans. In the notation of Example 13, for $|z| \ll |z_\odot|$ one has

$$z_*^* z_* = z_\odot^* z_\odot + z^* z_\odot + z z_\odot^* + z^* z \approx z_\odot^* z_\odot + 2\,\text{Re}(z^* z_\odot)$$

Since $r_* = |z_*| \approx |z_\odot| = r_\odot$,

$$|z_*|^2 - |z_\odot|^2 = (|z_*| + |z_\odot|)(|z_*| - |z_\odot|) \approx 2\,|z_\odot|\,(|z_*| - |z_\odot|)$$

it follows that

$$r_\odot(r_* - r_\odot) \approx \text{Re}(z^* z_\odot) = -r r_\odot \cos\phi$$

and with

$$(\omega_* - \omega_\odot) \approx (r_* - r_\odot)\left.\frac{d\omega_*}{dr_*}\right|_\odot = -(r\cos\phi)\left.\frac{d\omega_*}{dr_*}\right|_\odot$$

and the derivative evaluated at the location of the sun the Bottlinger formula implies (Oort formula I)

$$v_r = Ar \sin 2\phi$$

with (Oort's first constant)

$$A = -\left(\frac{r_\odot}{2}\right)\frac{d\omega_*}{dr_*}\bigg|_\odot$$

15. Prove that the transverse velocity, v_T of a star (as seen from the sun) is given (in the notation of Example 13) by (Oort formula II)

$$v_T = \omega_* - Ar \cos 2\phi - A$$

where the star is supposed to be close to the sun.

Ans. The transverse velocity v_T may be obtained from

$$\text{Im}(z^*\dot{z}) = -i[z^*\dot{z} - \text{Re}(z^*\dot{z})]$$

which according to Example 13 is identical with

$$\begin{aligned}\text{Im}(z^*\dot{z}) &= -i[z^*\dot{z} - (\omega_* - \omega_\odot)\text{Im}(z^*z_\odot)] \\ &= -iz^*\dot{z} + (\omega_* - \omega_\odot)[z^*z_\odot - (\text{Re}\, z^*z_\odot)] \\ &= z^*(\omega_\odot z_\odot - \omega_* z_*) + (\omega_* - \omega_\odot)z^*z_\odot \\ &\quad - (\omega_* - \omega_\odot)\text{Re}(z^*z_\odot)\end{aligned}$$

which may be written as

$$\begin{aligned}\text{Im}(z^*\dot{z}) &= \omega_* z^*(z_\odot - z_*) - (\omega_* - \omega_\odot)\text{Re}(z^*z_\odot) \\ &= -[(\omega_* z^* z) + (\omega_* - \omega_\odot)\text{Re}(z^*z_\odot)] \\ rv_T &= \omega_* r^2 - (\omega_* - \omega_\odot)r_\odot r \cos\phi\end{aligned}$$

and for stars close to the sun

$$\omega_* - \omega_\odot \approx \frac{2A}{r_\odot} r \cos\phi$$

so that

$$v_T = \omega_* r - 2Ar \cos^2\phi$$

or

$$v_T = [\omega_* - A(1 + \cos 2\phi)]r$$

which is the Oort formula for transverse velocity of stars close to the sun.

If for stars near the sun (v_r/r) is observed and plotted as a function of ϕ, the amplitude of the oscillation yields A. If (v_T/r) is observed and plotted, the (dc component) value of

$B = (A - \omega_*)$ (called Oort's second constant) is obtained. Since A is then already known, $\omega_* \approx \omega_\odot$ is determined. This corresponds to a rotational period of approximately 250 million years. By plotting $(\omega_* - \omega_\odot)$ (calculated from the Bottlinger formula) versus r_*, one can determine the derivative of angular velocity with respect to stellar distance from the galactic center at the location of the sun. Since A is already known, this determines $r_\odot$, which turns out to be about 10 kiloparsecs (a parsec is about 3.26 light-years).

● PROBLEMS 1

1. Evaluate $\int_0^a t^n e^{xt} \begin{pmatrix} \cos \\ \sin \end{pmatrix} yt\, dt$ using complex number analysis.

   [*Hint:* If $I(a,z) = \int_0^a e^{zt}\, dt = (e^{az} - 1)/z$, then

$$\partial_z^n I(a,z) = \int_0^a t^n e^{zt}\, dt$$

$$= \frac{(-1)^n n!\,(e^{az} - 1)}{z^{n+1}} + \sum_{k=1}^{n} (-1)^{n-k} \frac{n!}{k!} \frac{a^k e^{az}}{z^{n-k+1}}$$

   where $z = x + iy$.]

2. Show that

$$\sum_{n=0}^{\infty} \frac{\sin n\theta}{n!} = [\sin(\sin\theta)]e^{\cos\theta}$$

   and

$$\sum_{n=0}^{\infty} \frac{\cos n\theta}{n!} = [\cos(\sin\theta)]e^{\cos\theta}$$

   [*Hint:* $e^z = \sum_{n=0}^{\infty} (z^n/n!)$]

3. Verify and plot all complex numbers graphically:
   (a) $(2,-7)\cdot(3,-4) = (-5,-29)$
   (b) $(2,3) + (5,6) = (7,9)$
   (c) $(4,1) - (6,0) = (-2,1)$
   (d) $\dfrac{(6,1)}{(2,1)} = \dfrac{(6,1)(2,-1)}{(4+1,0)} = \dfrac{(6,1)(2,-1)}{(5,0)} = \dfrac{(13,-4)}{(5,0)} = \left(\dfrac{13}{5}, -\dfrac{4}{5}\right)$

4. Plot the cube roots of unity in the complex plane. Do the same for all the fourth and fifth roots of unity in separate diagrams. What is significant about the placement of the roots on the diagrams?

5. Express $[(1 - i)/(1 + i)]^{1+i}$ in the forms $x + iy$ and $re^{i\theta}$.
6. Find the roots of $z^4 + 1 = 0$.
7. Find the roots of $z^6 + 7z^3 - 8 = 0$.
8. The complex number z has $\operatorname{Re} z > 0 < \operatorname{Im} z$ and $|z| < 1$. Find the complex number w which together with z forms a pair symmetrical with respect to the following:
   (a) the point $(2 + 3i)$
   (b) the line $(1 - i)z + (1 + i)z^* = 4$
9. Describe the region of the z plane for which the following hold:
   (a) $1 < \operatorname{Im} z < 3$
   (b) $|z - 1| \geq 2\,|z + 1|$
10. Describe the region of the z plane for which the following hold:
   (a) $3 < \operatorname{Re}(z^2) < 5$
   (b) $\alpha^* z + \alpha z^* + 2c \geq 0$ with c real
11. Show that $\arg[(z - z_1)/(z_3 - z_1)]/[(z - z_2)/(z_3 - z_2)] = 0$ is the equation of a circle.
12. If t is (real) time, describe the motion of the point $z \cos^2 t + w \sin^2 t$ in the complex plane; z and w are fixed complex numbers.
13. Find the roots of $z^5 - 1 = 0$.
14. Find the center and radius of the circle $|z - a| = k\,|z - b|$.
15. Find the vectorial (complex) velocity and acceleration of a moving point at location z:
   (a) $z = t + it^2$
   (b) $z = t^3 + ie^{-t}$
16. Find the complex velocity and acceleration for $z = te^{it^2}$.
17. Find the radial, centripetal, tangential, and Coriolis acceleration in Problem 16.
18. Find the radial, centripetal, tangential, and Coriolis acceleration if $z = \cos t\, e^{i \sin t}$, and describe the trajectory.
19. Sum the finite Fourier series $\sum_{n=1}^{N} n \sin n\theta$.
20. Sum the infinite Fourier series $\sum_{n=1}^{\infty} \sin n\theta/n(n + 1)$.
21. If $\int_{-\infty}^{\infty} e^{-z^2t^2}\, dt = \sqrt{\pi}/z$ for $|z| > 0$ and $|\arg z| < \pi/4$, find the value of $\int_{-\infty}^{\infty} e^{-(x^2-y^2)t^2} \cos(2xyt^2)\, dt$.
22. Find the value of $\int_{-\infty}^{\infty} t^4 e^{-(x^2-y^2)t^2} \sin(2xyt^2)\, dt$.
23. Sum the infinite series $\sum_{n=0}^{\infty} e^{-nx} \cos ny$.
24. Sum the infinite series $\sum_{n=0}^{\infty} n^3 e^{-nx} \sin ny$.
25. Evaluate $\int_0^a t^3 e^{xt} \cos yt\, dt$ by differentiation of $\int_0^a e^{xt} \cos yt\, dt$ with respect to x.
26. Show that $\sum_{n=0}^{\infty} \sin n\theta/n! = [\sin(\sin \theta)]e^{\cos \theta}$.
27. Show that $k^* z + kz^* + 2c = 0$ with c real is the equation of a straight line with slope $(\operatorname{Im} k^*/\operatorname{Re} k)$, and perpendicular to the direction of k.
28. Show that the angle between the diagonals of a parallelogram with sides a and b is $\arccos[\operatorname{Re}[(a^* + b^*)(a - b)]/|a + b|\,|a - b|]$.

29. Integrate $\int d\theta/(a + b\cos\theta + c\sin\theta)$ for $a^2 < b^2 + c^2$ by setting $z = e^{i\theta}$.

30. Prove the *Schwarz inequality* $\sum_k |a_k|^2 \sum_k |b_k|^2 \geq |\sum_k a_k^* b_k|^2$ by considering $I(z) = \sum_k (a_k^* - z^* b_k^*)(a_k - z b_k)$, which is never negative for any complex z.
[*Hint:* Choose $z = \sum_k a_k b_k^* / \sum_k b_k^* b_k$.]
Also show that equality holds only for a_k proportional to b_k for all k.

31. Show that the general equation for a conic section may be expressed in the form

$$A^* z^2 + 2Hz^* z + A(z^*)^2 + 2F^* z + 2Fz^* + K = 0.$$

with $H = H^*$ and $K = K^*$ real and A and F complex.

32. Show that the three nondegenerate distinct types of curve correspond to

$H^2 > \lvert A\rvert^2$	ellipse
$H^2 = \lvert A\rvert^2$	parabola
$H^2 < \lvert A\rvert^2$	hyperbola

33. Show that the equation of the conic section passing through the five points z_1, z_2, z_3, z_4, z_5 is given by

$$0 = \begin{vmatrix} z^2 & z^* z & (z^*)^2 & z & z^* & 1 \\ z_1^2 & z_1^* z_1 & (z_1^*)^2 & z_1 & z_1^* & 1 \\ z_2^2 & z_2^* z_2 & (z_2^*)^2 & z_2 & z_2^* & 1 \\ z_3^2 & z_3^* z_3 & (z_3^*)^2 & z_3 & z_3^* & 1 \\ z_4^2 & z_4^* z_4 & (z_4^*)^2 & z_4 & z_4^* & 1 \\ z_5^2 & z_5^* z_5 & (z_5^*)^2 & z_5 & z_5^* & 1 \end{vmatrix}$$

34. Show that if there exist real numbers $M_1 = M_1^*$ and $M_2 = M_2^*$ and complex numbers Λ_1 and Λ_2 such that

$$A = \Lambda_1 \Lambda_2$$
$$H = \mathrm{Re}(\Lambda_1^* \Lambda_2)$$
$$K = M_1 M_2$$

the conic in Problem 31 is degenerate, forming a pair of straight lines.

35. Show that the distance of the straight line

$$a^* z + a z^* = 2b = 2b^*$$

from the origin ($z = 0$) is $b/\sqrt{a^* a}$.

36. Show that the distance of the point c from the line $a^* z + a z^* = b = b^*$ is given by $(a^* c + a c^* - 2b)/2\sqrt{a^* a}$.

37. Show that the necessary and sufficient condition for two triangles with vertices at z_1, z_2, z_3 and w_1, w_2, w_3 to be similar of the same sense is

$$\begin{vmatrix} 1 & z_1 & w_1 \\ 1 & z_2 & w_2 \\ 1 & z_3 & w_3 \end{vmatrix} = 0$$

38. Show that the area of the triangle with vertices at z_1, z_2, z_3 is

$$A = \frac{i}{4} \begin{vmatrix} 1 & z_1 & z_1^* \\ 1 & z_2 & z_2^* \\ 1 & z_3 & z_3^* \end{vmatrix}$$

39. Show that the straight line passing through z_1 and z_2 has the equation

$$\begin{vmatrix} 1 & z & z^* \\ 1 & z_1 & z_1^* \\ 1 & z_2 & z_2^* \end{vmatrix} = 0$$

40. Show that

$$|z| = \varepsilon(\operatorname{Re} z + k)$$

is the equation of a conic section with origin at focus, eccentricity ε, and directrix $\operatorname{Re} z = -k$.

CHAPTER 2:

Functions of a Complex Variable

2.1 REAL AND IMAGINARY PARTS OF COMPLEX FUNCTIONS

If for each value of a complex number z, there corresponds a value of another complex variable w, then w is said to be a *function* of z:

$$w = f(z) = f(x + iy) \tag{2.1}$$

A function is said to be *single-valued* if it has just *one* value corresponding to each value of z. Since w is also a complex number, it may also be written in terms of its real and imaginary components, u and v, respectively:

$$w = f(z) = u + iv = u(x,y) + iv(x,y) \tag{2.2}$$

where u, v are real numbers which may be expressed as functions of x and y.

For example, if

$$f(z) = z(1 + z) = (x + iy)(1 + x + iy) \tag{2.3}$$

then

$$\begin{aligned} u &= x(1 + x) - y^2 \\ v &= y(1 + 2x) \end{aligned} \tag{2.4}$$

and w may be written in terms of x, y as

$$w = [x(1 + x) - y^2] + i[y(1 + 2x)] \tag{2.5}$$

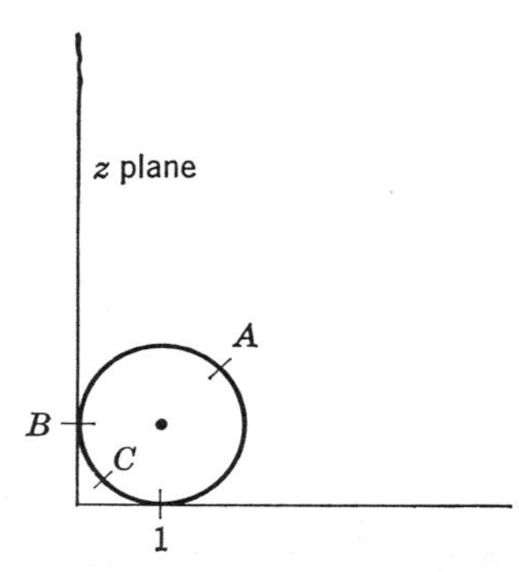

Figure 2.1. Unit circle $|z - (1 + i)| = 1$ centered at $z = 1 + i$.

2.2 COMPLEX FUNCTIONS AS MAPPINGS

Since a complex function such as w is a function of a complex number z, difficulties in the geometrical representation of the function w are encountered, because a *plane* is required to represent *each* variable z and w. Therefore some method of plotting four variables x, y and u, v is needed.

The usual method used is to plot a set of points in the z plane and then to plot their corresponding points in the w plane. For example, take the points on the unit circle in the z plane (see Fig. 2.1). If $f(z)$ is given by

$$w = f(z) = z^2 \tag{2.6}$$

then the image of the circle $|z - (1 + i)| = 1$ [locus of all points at unit distance from $(1 + i)$], as in Figure 2.2, with the points A, B, C going into A', B', C'.

In general then, a point in the z plane has a corresponding or *image* point in the w plane. When such a double graph has been made, one is said to

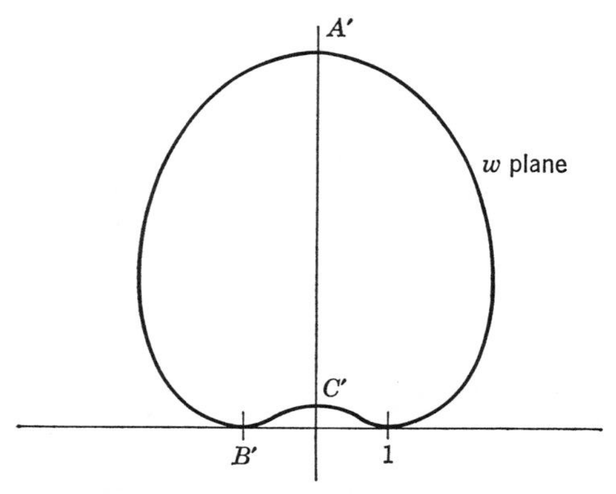

Figure 2.2. Image of unit circle in $w = z^2$ plane.

have *mapped* or *transformed* the points in the z plane into corresponding points in the w plane by means of the transforming function $f(z)$.

2.3 ANALYTIC FUNCTIONS AND DIFFERENTIABILITY

A function $f(z)$ is *regular* or *holomorphic* in a region R of the z plane if $f(z)$ is both single-valued and uniquely differentiable at all points within the region. If a function is both single-valued and differentiable at a point z, then $f(z)$ is said to be *analytic* at that point. In most of the technical literature the term analytic is used as a synonym for regular or holomorphic. Points where $f(z)$ is not analytic are called the *singularities* or *singular points* of $f(z)$.

A function is differentiable at a point z if

$$\lim_{\delta\to 0}\left\{\frac{1}{\delta}\left[f(z+\delta)-f(z)\right]\right\}=\frac{df}{dz} \tag{2.7}$$

exists and is independent of the manner in which $\delta = \varepsilon + i\eta$ approaches zero. For example, let

$$f(z) = z^2 \tag{2.8}$$

Then

$$\begin{aligned} d_z f = \frac{df}{dz} &= \lim_{\delta\to 0}\left\{\frac{1}{\delta}\left[(z+\delta)^2 - z^2\right]\right\} \\ &= \lim_{\delta\to 0}\left\{\frac{\delta}{\delta}(2z+\delta)\right\} \\ &= \lim_{\delta\to 0}\{2z+\delta\} = 2z \end{aligned} \tag{2.9}$$

Then $f(z)$ is analytic everywhere in the finite z plane since $2z$ exists everywhere in that plane and z^2 is single-valued in the z plane.

But notice the difference when

$$f(z) = |z|^2 \tag{2.10}$$

Then

$$\begin{aligned} \frac{df}{dz} &= \lim_{\delta\to 0}\left\{\frac{1}{\delta}\left[|z+\delta|^2 - |z|^2\right]\right\} \\ &= \lim_{\delta\to 0}\left\{\frac{1}{\delta}\left[(z+\delta)(z^*+\delta^*) - zz^*\right]\right\} \\ &= \lim_{\delta\to 0}\left[\frac{1}{\delta}(\delta z^* + \delta^* z + \delta^*\delta)\right] \\ &= \lim_{\delta\to 0}\left\{z^* + z\frac{\delta^*}{\delta} + \delta^*\right\} \end{aligned} \tag{2.11}$$

At $z = 0$, $f'(z) = 0$ and $|z|^2$ is analytic, but at other points (2.11) depends upon the manner in which $\delta \to 0$. If δ approaches zero through real values, then $\delta = \varepsilon$ and

$$\frac{df}{dz} = \lim_{\varepsilon \to 0} \left\{ z^* + z\frac{\varepsilon}{\varepsilon} + \varepsilon \right\} = z^* + z \tag{2.12}$$

and if δ approaches zero through imaginary values, then $\delta = i\eta$ and

$$\begin{aligned} \frac{df}{dz} &= \lim_{i\eta \to 0} \left\{ z^* + z\frac{-\eta i}{\eta i} + (-\eta i) \right\} \\ &= z^* - z \end{aligned} \tag{2.13}$$

Since (2.13) is not the same as (2.12), $f(z) = |z|^2$ is *not* analytic in general.

2.4 CAUCHY–RIEMANN CONDITIONS

A necessary condition for a function to be analytic is that the partial derivatives $\partial_x u$, $\partial_y u$, $\partial_x v$, $\partial_y v$ exist and satisfy the *Cauchy–Riemann conditions* that

$$\partial_x u = \partial_y v \qquad \partial_y u = -\partial_x v \tag{2.14}$$

This can be seen to be true since $f'(z)$ exists and the ratio

$$\frac{f(z + \delta) - f(z)}{\delta} \tag{2.15}$$

must thus tend to a definite limit as $\delta \to 0$ in *any* manner, where $\delta = \varepsilon + i\eta$.

We first let $\eta = 0$, or, in other words, we approach the point z parallel to the real axis. Then

$$\begin{aligned} f'(z) &= \lim_{\varepsilon \to 0} \frac{u(x + \varepsilon, y) + iv(x + \varepsilon, y) - u(x,y) - iv(x,y)}{\varepsilon} \\ &= \lim_{\varepsilon \to 0} \frac{u(x + \varepsilon, y) - u(x,y)}{\varepsilon} + i \lim_{\varepsilon \to 0} \frac{v(x + \varepsilon, y) - v(x,y)}{\varepsilon} \\ &= \partial_x u + i\partial_x v \end{aligned} \tag{2.16}$$

Now we let $\varepsilon = 0$, and approach z along the imaginary axis so that

$$\begin{aligned} f'(z) &= \lim_{\eta \to 0} \frac{u(x,y + \eta) + iv(x,y + \eta) - u(x,y) - iv(x,y)}{i\eta} \\ &= \lim_{\eta \to 0} \frac{-iu(x,y + \eta) + v(x,y + \eta) + iu(x,y) - v(x,y)}{\eta} \\ &= \lim_{\eta \to 0} \frac{\{v(x,y + \eta) - v(x,y)\}}{\eta} - i \lim_{\eta \to 0} \frac{u(x,y + \eta) - u(x,y)}{\eta} \end{aligned}$$

$$d_z f = f'(z) = \partial_y v - i\partial_y u \tag{2.17}$$

But $f'(z)$ must be unique, so (2.17) and (2.16) must be equal. Therefore we see that the conditions (2.14) must be satisfied. This implies that the real and imaginary parts of an *analytic* function are not independent. However, the Cauchy–Riemann conditions alone are *not sufficient* for analyticity. For example, consider

$$f(z) = \begin{cases} \left(\dfrac{x^3 - y^3}{x^2 + y^2}\right) + i\left(\dfrac{x^3 + y^3}{x^2 + y^2}\right) & \text{for } |z| \neq 0 \\ 0 & \text{for } |z| = 0 \end{cases} \tag{2.18}$$

$$\begin{aligned} \partial_x u(0,0) &= 1 & \partial_x v(0,0) &= 1 \\ \partial_y u(0,0) &= -1 & \partial_y v(0,0) &= 1 \end{aligned} \tag{2.19}$$

At $z = 0$ the Cauchy–Riemann conditions are satisfied, but using (2.16) and (2.19),

$$f'(0) = \partial_x u + i\partial_x v = 1 + i \tag{2.20}$$

and approaching z along the line $x = y$,

$$f'(0) = \lim_{z\to 0} \frac{f(z) - f(0)}{2 - 0} = \lim_{z\to 0} \frac{ix}{x + ix} = \frac{1 + i}{2} \tag{2.21}$$

Since (2.20) and (2.21) are not equal, $f(z)$ is *not* analytic at $z = 0$ even though the Cauchy–Riemann equations are satisfied there. A function $f(z)$ *is* analytic, however, if the four partial derivatives *exist*, are *continuous*, and satisfy the Cauchy–Riemann conditions (2.14).

The sufficiency proof is based on the *mean value theorem* for functions of two variables. Let $\varepsilon = \varepsilon_x + i\varepsilon_y$. Then

$$f(z + \varepsilon) - f(z) = \{u(z + \varepsilon) - u(z)\} + i\{v(z + \varepsilon) - v(z)\} \tag{2.22}$$

By the mean value theorem, then

$$f(z + \varepsilon) - f(z) = (\varepsilon_x \partial_x u + \varepsilon_y \partial_y u) + i(\varepsilon_x \partial_x v + \varepsilon_y \partial_y v) \tag{2.23}$$

where the partial derivatives are supposed to be continuous and are evaluated at an appropriate point of the small rectangle of base ε_x and altitude ε_y.

Using (2.14) one obtains

$$f(z + \varepsilon) - f(z) = (\varepsilon_x + i\varepsilon_y)(\partial_x u + i\partial_x v) \tag{2.24}$$

from (2.23). Thus as $|\varepsilon| \to 0$, the unique limit

$$f'(z) = \lim_{\varepsilon\to 0} \frac{f(z + \varepsilon) - f(z)}{\varepsilon} = \partial_x u + i\partial_x v \tag{2.25}$$

exists ($\partial_x u$ and $\partial_x v$ are now evaluated at z).

If $f(z)$ is analytic, it may also be shown that u and v are *harmonic functions*, i.e., they satisfy the equations (Laplace)

$$\begin{aligned}\Delta u &= (\partial_x^2 + \partial_y^2)u = 0 \\ \Delta v &= (\partial_x^2 + \partial_y^2)v = 0\end{aligned} \tag{2.26}$$

Since by (2.14) we have

$$\partial_x(\partial_x u) = \partial_x(\partial_y v) \tag{2.27}$$

and

$$\partial_y(\partial_y u) = -\partial_y(\partial_x v)$$

then

$$\Delta u = \partial_x(\partial_y v) - \partial_y(\partial_x v) = 0.$$

Similarly, $\Delta v = 0$ may be concluded.

● EXAMPLES 2

1. Is e^{e^z} an analytic function of z?

   Ans. Yes. An analytic function of an analytic function is analytic, since if $f(z)$ and $g(z)$ are analytic, then

   $$d_z f[g(z)] = f'[g(z)]g'(z)$$

   exists and is unique. Thus it suffices to show that e^z is analytic. But

   $$e^z = e^x(\cos y + i \sin y)$$

   so

   $$u = e^x \cos y$$
   $$v = e^x \sin y$$

   and

   $$\partial_x u = e^x \cos y = \partial_y v$$
   $$\partial_y u = -e^x \sin y = -\partial_x v$$

   all of which are continuous.

2. Show that the Cauchy–Riemann equations may be replaced by $\partial_{z^*} f = 0$ if f is considered a function of z and z^*.

   Ans. Since $z = x + iy$, $z^* = x - iy$ and $2x = z + z^*$, $2iy = z - z^*$, f may be considered a function of z and z^* (instead of x and y). Then

   $$\partial_{z^*} f = \partial_x f \partial_{z^*} x + \partial_y f \partial_{z^*} y$$
   $$2\partial_{z^*} f = (\partial_x u + i\partial_x v) + i(\partial_y u + i\partial_y v)$$

   or

   $$2\partial_{z^*} f = (\partial_x u - \partial_y v) + i(\partial_y u + \partial_x v) = 0$$

3. Find the Cauchy–Riemann equations in polar coordinates (r,θ).

Ans.
$$f = u(r,\theta) + iv(r,\theta)$$
$$f'(z) = e^{-i\theta}\partial_r f = e^{-i\theta}(\partial_r u + i\partial_r v)$$

Also,
$$rf'(z) = -ie^{-i\theta}\partial_\theta f = e^{-i\theta}(\partial_\theta v - i\partial_\theta u)$$

Hence,
$$r\partial_r u = \partial_\theta v \qquad \text{and} \qquad r\partial_r v = -\partial_\theta u$$

4. If $|f(z)|$ is constant everywhere in a region where $f(z)$ is analytic, show that $f(z)$ is also constant there.
Ans. Let $f^*f = c^2 \neq 0$; then $\partial_{z^*}(f^*f) = 0$ and
$$f^*\partial_{z^*}f + f\partial_{z^*}f^* = 0$$
But f is analytic, so $\partial_{z^*}f = 0$ and $f\partial_{z^*}f^* = 0$. The conjugate of this yields $ff^*\partial_z f = 0$, so that $\partial_z f = 0$ and f is a constant.

PROBLEMS 2

1. Show that $\arg f(z)$ is a harmonic function if $f(z)$ is analytic and nonvanishing in a region.
2. Show that $f(z) = 2xy^2/(x^2 + y^4)$ for $z \neq 0$, with $f(0) = 0$ not analytic at $z = 0$, by considering paths for which $x = y^2$.
3. Show that if $f(z)$ is analytic, $|f(z)|^2$ is not harmonic unless $f(z)$ is a constant.
4. Show that the lines along which $u =$ constant cut those along which $v =$ constant orthogonally (perpendicularly) wherever $f'(z) \neq 0$.
5. Verify that $\cosh z$ and $\sinh z$ are analytic functions for all finite points z.
6. Show that $|z|$ is not analytic.
7. Show that if $f(z)$ is analytic, then $\phi(z) = [f(z) - f(a)]/[z - a]$ is analytic, including at $z = a$.
8. Show that $\arcsin z$ is analytic.
9. Show that $\arg z$ is not analytic.
10. If the real part of an analytic function is a function of its imaginary part, find the derivative of the analytic function.
11. Show that if $f = f(z,z^*)$, then
$$\Delta f = (\partial_x^2 + \partial_y^2)f = 4\partial_{z^*}\partial_z f$$

12. Show that the general solution of Laplace's equation, $\Delta f = 0$, is given by

$$f(z,z^*) = g(z) + h(z^*)$$

where g and h are arbitrary (differentiable) functions.

13. By taking $y = ict$ in Laplace's equation show that the general solution of the wave equation $(\partial_x^2 - \partial_{ct}^2)f = 0$ is given by

$$f = g(x - ct) + h(x + ct)$$

(i.e., a wave to the right and a wave to the left traveling at phase velocity c).

14. Show that the spherical wave equation

$$\frac{1}{r^2}\partial_r(r^2\partial_r\phi) = \partial_{ct}^2\phi$$

may be rewritten as

$$\partial_r^2(r\phi) = \partial_{ct}^2(r\phi)$$

with general solution

$$\phi = \frac{g(r - ct)}{r} + \frac{h(r + ct)}{r}$$

(i.e., exploding and imploding waves).

15. Find the general solution of the biharmonic equation $\Delta^2 f = 0$ by expressing Δ^2 in terms of ∂_z and ∂_{z^*}.

CHAPTER 3: Infinite Series

3.1 GEOMETRICAL SERIES

The finite geometrical series

$$S_n(z) = 1 + z + z^2 + \cdots + z^{n-1} \tag{3.1}$$

may be easily summed if $z \neq 1$ by subtracting $zS_n(z)$ from (3.1) and dividing by $(1 - z)$ with the result

$$S_n(z) = \frac{1 - z^n}{1 - z} \tag{3.2}$$

If $z = 1$, the sum of (3.1) is $S_n(1) = n$. If $|z| < 1$,

$$\left| \frac{1}{1 - z} - S_n(z) \right| = \frac{|z|^n}{|1 - z|} \leq \frac{|z|^n}{1 - |z|} \tag{3.3}$$

approaches zero as $n \to \infty$ so that for $|z| < 1$ the sum of an infinite number of terms exists and is equal to $S_\infty(z)$:

$$S_\infty(z) = \frac{1}{1 - z} \tag{3.4}$$

One calls the region $|z| < 1$ for which $S_\infty(z)$ exists the *circle of convergence* of the infinite geometrical series.

3.2 POWER SERIES AND RADIUS OF CONVERGENCE

Series of positive powers of z are of frequent occurrence in analysis. The partial sum (of a finite number of terms) is

$$S_n(z) = \sum_{k=0}^{n-1} a_k z^k \tag{3.5}$$

If there are values of z for which the limit $S_\infty(z)$ exists, then

$$|S_\infty(z) - S_n(z)| = \left| \sum_{k=n}^{\infty} a_k z^k \right| \le \sum_{k=n}^{\infty} |a_k|\, |z|^k \tag{3.6}$$

If R is the limit of $|a_n|^{-1/n}$ as $n \to \infty$, then $|a_n|^{-1/n} > R/2$ for large n and

$$|S_\infty(z) - S_n(z)| \le \sum_{k=n}^{\infty} (|a_k|^{1/k}\, |z|)^k = \sum_{k=n}^{\infty} \left(\frac{|z|}{|a_k|^{-1/k}} \right)^k$$

$$|S_\infty(z) - S_n(z)| \le 2 \sum_{k=n}^{\infty} \left| \frac{z}{R} \right|^k = 2 \left| \frac{z}{R} \right|^n \left(\frac{1}{1 - |z/R|} \right)$$

if $|z| < R$ so that

$$|S_\infty(z) - S_n(z)| \to 0 \qquad \text{as} \quad n \to \infty$$

The number

$$R = \lim_{n \to \infty} |a_n|^{-1/n} \tag{3.7}$$

is the *radius of* the circle of *convergence* of $\sum_{k=0}^{\infty} a_k z^k$. In case it exists, one has for sufficiently large n

$$|a_n| \approx \frac{1}{R^n} \tag{3.8}$$

so that

$$\left| \frac{a_n}{a_{n+1}} \right| \approx R \tag{3.9}$$

and one also has

$$R = \lim_{n \to \infty} \left| \frac{a_n}{a_{n+1}} \right| \tag{3.10}$$

The convergence criteria corresponding to (3.7) and (3.10) for the complex series $\sum_n a_n$,

$$\lim_{n \to \infty} |a_n|^{1/n} < 1 \tag{3.11}$$

and

$$\lim_{n \to \infty} \left| \frac{a_{n+1}}{a_n} \right| < 1 \tag{3.12}$$

are called the Cauchy nth-root test and the D'Alembert ratio test, respectively.

3.3 ABSOLUTE AND UNIFORM CONVERGENCE

There are several interesting theorems concerning the convergence of series of analytic functions. It should be noted that a series of analytic functions

$$\sum_{k=1}^{\infty} f_k(z) \tag{3.13}$$

has partial sums

$$S_n(z) = \sum_{k=1}^{n} f_k(z) \tag{3.14}$$

which are, for all fixed values of n, analytic functions, since the sum of two analytic functions is analytic. The infinite series (3.13) is said to *converge** when there exists a function $S(z)$ such that

$$|S(z) - S_n(z)| < \varepsilon \qquad \text{for} \quad n > N(\varepsilon,z) \tag{3.15}$$

Then the remainder of the series (3.13) after n terms can be made arbitrarily small by deleting a number of terms dependent upon ε. The function $S(z)$ is called the *sum* of the infinite series (3.13).

In the special case where the minimal number N of terms required to make $|S(z) - S_n(z)| < \varepsilon$ is *independent* of z, for all z in a set $\mathscr{S}$, the infinite series (3.13) is said to *converge uniformly* to $S(z)$ in $\mathscr{S}$.

The series for which

$$f_k(z) = \frac{z}{(1 + kz)(1 + kz - z)} \tag{3.16}$$

and

$$S(z) - S_n(z) = \frac{1}{1 + nz} \tag{3.17}$$

converges (trivially for $z = 0$), since

$$|S(z) - S_n(z)| = \frac{1}{|1 + nz|} \leq \frac{1}{n\,|z| - 1} < \varepsilon \tag{3.18}$$

for

$$n > N(\varepsilon,z) = \frac{1}{|z|}\left(1 + \frac{1}{\varepsilon}\right)$$

but since N is *not* independent of z, the infinite series corresponding to (3.16) is *not uniformly convergent* in any finite closed region containing the origin $z = 0$. On the other hand, it *is uniformly convergent* in any closed region not containing any points within a circle of radius $a > 0$ about $z = 0$, since then (3.18) will be satisfied for

$$n > N(\varepsilon) = \frac{1}{a}\left(1 + \frac{1}{\varepsilon}\right)$$

independently of the value of z.

The notion of *uniform convergence* goes beyond that of ordinary convergence at a point (i.e., for a particular value of z) being concerned with convergence in a set independently of the location of its individual points.

* An often useful necessary and sufficient condition for convergence is given by the Cauchy convergence criterion: $|S_m(z) - S_n(z)| < \varepsilon$ for $m > n \geq N(\varepsilon,z)$. See G. H. Hardy, *Pure Mathematics*. Cambridge: Cambridge University Press, 1938, p. 162.

If the infinite series

$$\sum_{k=1}^{\infty} |f_k(z)| \tag{3.19}$$

is convergent, (3.13) is said to be *absolutely convergent*. Clearly, absolute convergence implies convergence, since

$$|S(z) - S_n(z)| = \left| \sum_{k=n+1}^{\infty} f_k(z) \right| \leq \sum_{k=n+1}^{\infty} |f_k(z)| < \varepsilon \tag{3.20}$$

but convergence does not imply absolute convergence. All points z for which $\lim_{n\to\infty} |f_n(z)|^{1/n} < 1$ are in the region of convergence of (3.19).

If the series (3.19) has the property that, from some value of k on,

$$|f_k(z)| \leq M_k \tag{3.21}$$

with the M_k independent of z and $\sum_k M_k$ convergent, then the uniform convergence of (3.13) follows. This is called the *Weierstrass M* test.

A uniformly convergent series of analytic functions may be integrated term by term to the same result as the integration of the sum function. That is, if the series

$$S(z) = \sum_{n=1}^{\infty} u_n(z) \tag{3.22}$$

is uniformly convergent, then

$$\int_L S(z)\, dz = \sum_{n=1}^{\infty} \int_L u_n(z)\, dz = \int_L \sum_{n=1}^{\infty} u_n(z)\, dz \tag{3.23}$$

where L is any path inside the region of uniform convergence.

Clearly, if $S_N(z) = \sum_{n=1}^{N} u_n(z)$,

$$\int_L S_N(z)\, dz = \sum_{n=1}^{N} \int_L u_n(z)\, dz$$

and

$$\left| \int_L [S(z) - S_N(z)]\, dz \right| \leq \int_L \varepsilon\, |dz| = \varepsilon L \tag{3.24}$$

where ε is independent of z (uniform convergence). Hence,

$$\int_L S(z)\, dz = \lim_{N\to\infty} \int S_N(z)\, dz = \sum_{n=1}^{\infty} \int u_n(z)\, dz$$

The sum of a uniformly convergent series of continuous functions is also continuous, since if

$$S(z) = \sum_{n=1}^{\infty} u_n(z)$$

$$|S(z+\varepsilon) - S(z)| \leq |S(z+\varepsilon) - S_N(z+\varepsilon)| + |S_N(z+\varepsilon) - S_N(z)| + |S_N(z) - S(z)| \tag{3.25}$$

The first and third terms on the right of (3.25) can be made arbitrarily small by choosing N sufficiently large independently of z, while the second term on the right can be made arbitrarily small because the partial sum $S_N(z)$ is continuous. Hence $S(z)$ is continuous.

A power series of the form

$$f(z) = \sum_{k=0}^{\infty} a_k z^k$$

with a positive radius of convergence, represents an analytic function inside its circle of convergence, which must thus pass through the singularity of the function nearest the origin. The argument follows:

$$\begin{aligned}\phi_\varepsilon(z) &= \frac{f(z+\varepsilon) - f(z)}{\varepsilon} - \sum_{k=1}^{\infty} k a_k z^{k-1} \\ &= \sum_{k=1}^{\infty} a_k \left[\frac{(z+\varepsilon)^k - z^k}{\varepsilon} - k z^{k-1}\right]\end{aligned}$$

$$(z+\varepsilon)^k - z^k - k z^{k-1}\varepsilon = k(k-1)\xi^{k-2}\varepsilon^2$$

by the mean value theorem, and

$$|\phi_\varepsilon(z)| \le |\varepsilon| \sum_{k=1}^{\infty} k(k-1)\,|a_k|\,|\xi|^{k-2}$$

Since $\lim_{k\to\infty} k^{1/k}(k-1)^{1/k}\,|a_k|^{1/k}\,|\xi|^{1-2/k} < 1$ for $|\xi| < R$, the series bounding $|\phi_\varepsilon(z)|$ converges. Hence, $\lim_{|\varepsilon|\to 0} |\phi_\varepsilon(z)| = 0$. Therefore a unique derivative of $f(z)$ exists and is equal to $\sum_{k=1}^{\infty} k a_k z^{k-1}$.

From the (Cauchy) multiplication of the series

$$\sum_{n=0}^{\infty} c_n z^n = \left(\sum_{n=0}^{\infty} a_n z^n\right)\left(\sum_{n=0}^{\infty} b_n z^n\right)$$

it follows by equating coefficients that

$$c_n = \sum_{k=0}^{n} a_{n-k} b_k \tag{3.26}$$

In particular, this (Cauchy convolution formula) becomes

$$\begin{aligned}c_0 &= a_0 b_0 \\ c_1 &= a_0 b_1 + a_1 b_0 \\ c_2 &= a_0 b_2 + a_1 b_1 + a_2 b_0 \\ c_3 &= a_0 b_3 + a_1 b_2 + a_2 b_1 + a_3 b_0\end{aligned}$$

etc.

These provide formulas for the coefficients in the (Cauchy) product of two power series. Here it is supposed that $\{a_n\}$ and $\{b_n\}$ are known and $\{c_n\}$ is calculated. If it is supposed instead that $\{c_n\}$ and $\{a_n\}$ are known, one obtains formulae for the coefficients $\{b_n\}$ of the ratio of two power series:

$$b_0 = \frac{c_0}{a_0} \qquad a_0 \neq 0$$

$$b_1 = \frac{c_1}{a_0} - \frac{c_0 a_1}{a_0^2}$$

$$b_2 = \frac{c_2}{a_0} - \frac{a_1}{a_0^2} c_1 + \frac{c_0 a_1^2}{a_0^3} - \frac{c_0 a_2}{a_0^2}$$

etc.

● EXAMPLES 3

1. Is the series $\sum_{n=1}^{\infty} e^{(1+in)^2}$ convergent?
   Ans. Yes. It is absolutely convergent:

$$a_n = e^{(1+in)^2} = e^{1-n^2+2in} = ee^{-n^2}e^{2in}$$

$$|a_n|^{1/n} = e^{1/n}e^{-n}$$

so

$$\lim_{n\to\infty} |a_n|^{1/n} = 0 < 1$$

2. Is the series $\sum_{n=1}^{\infty} (1/\sqrt{n})[(3 + 4i)/(5 + 12i)]^n$ divergent?
   Ans. No. It converges by the (D'Alembert) ratio test:

$$|a_n| = \frac{1}{\sqrt{n}} \left| \frac{3 + 4i}{5 + 12i} \right|^n = \left(\frac{5}{13}\right)^n \frac{1}{\sqrt{n}}$$

$$\left| \frac{a_{n+1}}{a_n} \right| = \left(\frac{5}{13}\right) \sqrt{\frac{n}{n + 1}}$$

so

$$\lim_{n\to\infty} \left| \frac{a_{n+1}}{a_n} \right| = \frac{5}{13} < 1$$

3. Find the radius of convergence of $\sum_{n=1}^{\infty} z^{2n+1}/(2n + 1)!$.
   Ans. $R = \infty$ since

$$\left| \frac{z^{2n+1}}{(2n + 1)!} \frac{(2n - 1)!}{z^{2n-1}} \right| = \frac{|z|^2}{(2n + 1)2n} \to 0 < 1$$

for any finite $|z|$.

4. Find the radius of convergence of $\sum_{n=2}^{\infty} [(n-1)/n]^{n^2} z^n$.
Ans. $R = e$ since $R = \lim_{n\to\infty} [n/(n-1)]^n$.

5. Find the radius of convergence of $\sum_{n=1}^{\infty} (nz)^n$.
Ans. $R = 0$ since $R = \lim_{n\to\infty} 1/n$.

6. Find the region of convergence of $\sum_{n=-\infty}^{\infty} e^{-nz}/(2^n + 3^n)$.
Ans. $-\log 3 < \operatorname{Re} z < -\log 2$

for $n > 0$

$$\lim_{n\to\infty} \left| \frac{e^{-nz}}{2^n + 3^n} \right|^{1/n} = \frac{e^{-x}}{3} < 1$$

For $n < 0$, let

$$m = -n \lim_{m\to\infty} \left| \frac{e^{mz}}{2^{-m} + 3^{-m}} \right|^{1/m} = 2e^x < 1$$

Hence for convergence of both portions ($n > 0$ and $m > 0$) of the series one must have

$$2 < e^{-x} < 3 \qquad \text{or} \qquad -\log 3 < \operatorname{Re} z < -\log 2$$

7. Is the series $\sum_{n=1}^{\infty} 1/n^2$ divergent?
Ans. No.

$$\sum_{n=1}^{\infty} \frac{1}{n^2} = \sum_{m=0}^{\infty} \frac{1}{(m+1)^2} \le 1 + \sum_{m=1}^{\infty} \frac{1}{m(m+1)}$$

$$= 1 + \sum_{m=1}^{\infty} \left(\frac{1}{m} - \frac{1}{m+1} \right) = 2$$

PROBLEMS 3

Discuss the convergence properties of the following series:

1. $\displaystyle\sum_{n=1}^{\infty} \frac{\sin n}{(n+1)\sqrt{n}}$

2. $\displaystyle\sum_{n=1}^{\infty} \sec(in)$

3. $\displaystyle\sum_{n=1}^{\infty} \left(\frac{5+12i}{3+4i} \right)^n$

4. $\displaystyle\sum_{n=1}^{\infty} \frac{(2n)!\, z^n}{(n!)^2}$

5. $\displaystyle\sum_{n=1}^{\infty} \frac{2^n z^{2n}}{(n!)^2}$

6. $\displaystyle\sum_{n=1}^{\infty} \frac{z^n}{n^n}$

7. $\displaystyle\sum_{n=1}^{\infty} \left(1 - \frac{\pi}{n} \right)^{n^2} z^n$

8. $\displaystyle\sum_{n=-\infty}^{\infty} \frac{z^n}{e^n + \pi^\pi}$

9. $\sum_{n=-\infty}^{\infty} \dfrac{e^{nz^2}}{e^n + \pi^n}$

10. $\sum_{n=1}^{\infty} z^n \left[\arctan\left(\dfrac{n+1}{n}\right)\right]^n$

11. $\sum_{n=2}^{\infty} \dfrac{n!\, z^n}{n^n}$

12. $\sum_{n=3}^{\infty} n!\, z^{n^2}$

13. $\sum_{n=3}^{\infty} \dfrac{z^{2^n}}{n^2}$

14. $\sum_{n=1}^{\infty} n\left(\dfrac{z-a}{z-b}\right)^n$

15. $\sum_{n=3}^{\infty} \dfrac{z^n}{n^2(1+z^n)}$

16. $\sum_{n=5}^{\infty} \dfrac{(-1)^n z^n}{n(1+z^n)}$

17. Establish the *Dirichlet test* for uniform convergence of $\sum_n u_n(z)v_n(z)$ if $|v_n(z)| < \varepsilon$ for $n > N(\varepsilon)$, $\sum_n [v_n(z) - v_{n+1}(z)]$ uniformly and absolutely convergent, and $|\sum_{k=1}^{s} u_k(z)| < K(s)$ [$N(\varepsilon)$ and $K(s)$ independent of z].
18. Show that if there exists $\lim_{k\to\infty} |a_k|^{1/k}$, then $f(z) = \sum_{k=0}^{\infty} a_k/z^k$ is an analytic function for $|z| > \lim_{k\to\infty} |a_k|^{1/k}$ including the point $z = \infty$.
19. Show that the convergence conclusions of this chapter for $\sum_{k=0}^{\infty} a_k z^k$ can be extended to Taylor series $\sum_{k=0}^{\infty} a_k(z-c)^k$.
20. Show that a Taylor series cannot be zero at an infinite number of points in a finite region without being identically zero (having all zero coefficients) throughout the region. Thus, show that there is only one Taylor series (unique) for any given function about any given point.
21. Show that in $\sum_{k=0}^{\infty} a_k(z-c)^k$ the coefficients $a_k = f^{(k)}(c)/k!$.
22. By comparing $\sum_{k=1}^{\infty} (1/k)$ to $\int_1^{\infty} dx/x$, show that the series diverges.

CHAPTER 4:

Cauchy's Theorem

4.1 COMPLEX INTEGRATION

A curve in the complex z plane may be specified by a pair of real parametric equations expressing the dependence of x and y upon a parameter t as in

$$x = x(t)$$
$$y = y(t) \qquad \text{for} \quad t_0 \leq t \leq t_1 \tag{4.1}$$

or by means of a single complex function $z(t)$,

$$z = z(t) = x(t) + iy(t) \qquad \text{for} \quad t_0 \leq t \leq t_1 \tag{4.2}$$

Thus $z = t + it^2$ for $0 \leq t \leq 1$ represents the portion of a parabola within a closed unit square with corners at 0, 1, $1 + i$, and i, as shown in Figure 4.1.

If we consider a particular path L in the z plane, the integral of a *complex* function $f(z)$ along L may be defined as a limit (assumed to exist)

$$\int_L f(z)\, dz = \lim_{\substack{n\to\infty \\ \Delta z_k \to 0}} \sum_{k=1}^{n} f(z_k)(\Delta z_k) \tag{4.3}$$

where it is supposed that n points z_k have been selected on the path L in such a way that the maximum modulus of the difference between two successive points $\Delta z_k = z_{k+1} - z_k$ approaches zero as the number n of points increases without limit.

It is readily observed that the modulus of such an integral cannot exceed the product of the length of the path L and the maximum modulus of $f(z)$ on L since

$$\left| \sum_k f(z_k)(\Delta z_k) \right| \leq |f(z_k)|_{\max} \sum_k |\Delta z_k| \tag{4.4}$$

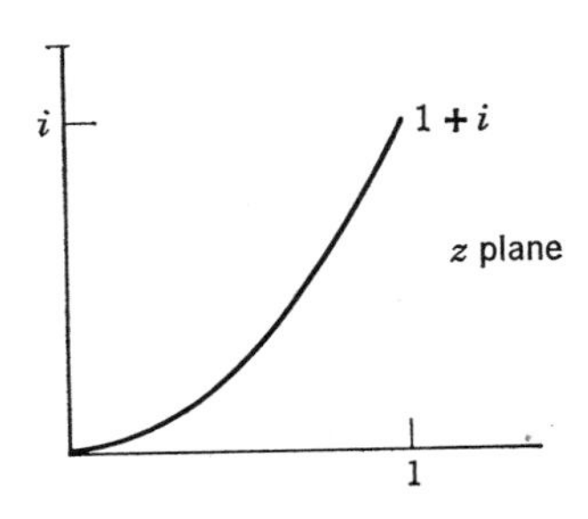

Figure 4.1. Locus of $z(t) = t + it^2$ for $0 \leq t \leq 1$.

The latter sum gives the length of a broken line with segments as chords approximating the arcs into which L is divided by the points z_k. Thus we may write after $n \to \infty$,

$$\left| \int f(z)\, dz \right| \leq ML \tag{4.5}$$

where M denotes the maximum modulus of $f(z)$ on L.

The value of such an integral of a complex function generally depends not only upon the values of z at the end points of L but also upon the shape of L. A simple example of this dependence upon the path of integration is afforded by $f(z) = 3\,|z|^2$ so that the integral is

$$3\int_L |z|^2\, dz$$

which for L extending along the real axis from $z = 0$ to $z = 1$ and thence parallel to the imaginary axis to $z = 1 + i$ yields ($dz = dx + i\, dy$)

$$3\int_L |z|^2\, dz = 3\int_0^1 x^2\, dx + 3i\int_0^1 (1 + y^2)\, dy = 1 + 4i \tag{4.6}$$

while the same integrand for L extending along the imaginary axis from $z = 0$ to $z = i$ and thence parallel to the real axis to $z = 1 + i$ yields

$$3\int_L |z|^2\, dz = 3i\int_0^1 y^2\, dy + 3\int_0^1 (x^2 + 1)\, dx = 4 + i \tag{4.7}$$

which clearly *differs* from the preceding even though the end points $z = 0$ and $z = 1 + i$ are the *same* for the two integrals.

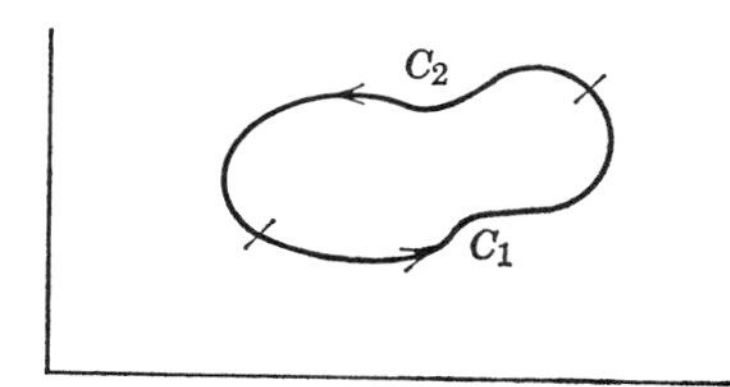

Figure 4.2. Decomposition of contour integral. Here, $\int_{C_1} + \int_{C_2} = 0$ and $\int_{C_1} = -\int_{C_2} = \int_{-C_2}$.

While it is thus generally true that the integral of a complex function depends upon the path connecting two given end points, it turns out that for the special class of analytic functions it is *independent* of the path, depending only on the end points and the choice of function $f(z)$. The property of the integral having the same value for any two paths between given end points is equivalent to the property of its having a zero value about any *closed* path (contour) passing through the two (former) end points. This follows from the fact that the integral about the closed contour can be expressed as the difference between two integrals extended from the initial point to the final point along the two portions into which these points divide the closed contour.

That the integral about any closed path interior to a closed simply connected region (any closed path within such a region encloses only points of the region) in which $f(z)$ is analytic must vanish is the conclusion of Cauchy's theorem:

$$\oint f(z)\, dz = 0 \tag{4.8}$$

It follows most easily from the application of Green's theorem (see Appendix B) in the x, y plane to the line integral obtained by replacing $f(z)$ by $u + iv$ and dz by $dx + i\, dy$. Thus,

$$\oint (u + iv)(dx + i\, dy) = \oint (u\, dx - v\, dy) + i \oint (v\, dx + u\, dy)$$

$$\oint f(z)\, dz = -\iint (\partial_x v + \partial_y u)\, dx\, dy + i \iint (\partial_x u - \partial_y v)\, dx\, dy$$

which both vanish by the Cauchy–Riemann equations (2.14). Notice that in Cauchy's theorem the closed contour may be deformed in any way provided that it does not pass over a singularity of $f(z)$.

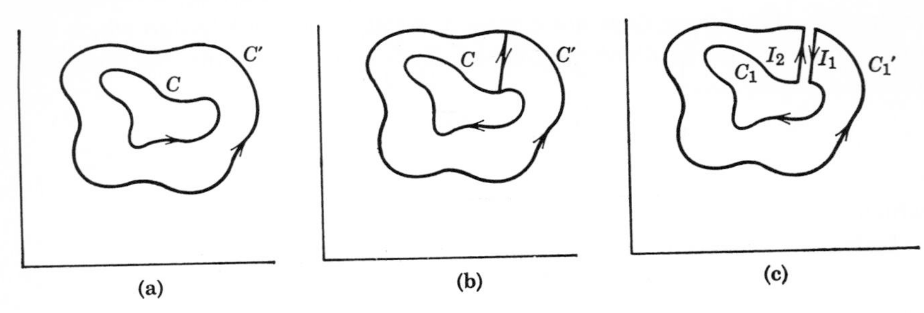

Figure 4.3. Equivalence of contours. In (a) $\int_C = \int_{C'}$; in (b) $\int_{-C} + \int_{C'} = 0$; in (c) $\int_{C_1'} + \int_{I_1} + \int_{I_2} + \int_{C_1} = 0$.

4.2 MORERA'S THEOREM

There is a converse to Cauchy's theorem due to Morera. In the form presented here it is assumed that $f(z) = u(x,y) + iv(x,y)$ has continuous derivatives with respect to x and y at all points of a closed simply connected region S and that

$$\oint f(z)\, dz = 0$$

for any closed path in this region. Then it follows that $f(z)$ is analytic and free of singularities in S. The proof proceeds from Green's theorem in the plane by replacing the line integrals

$$\oint (u + iv)(dx + i\, dy) = \oint u\, dx - v\, dy + i \oint v\, dx + u\, dy$$

by surface integrals

$$i \iint_S (\partial_x u - \partial_y v)\, dx\, dy - \iint (\partial_x v + \partial_y u)\, dx\, dy$$

which must then be separately zero. It then follows that the integrands

$$\partial_x u - \partial_y v = 0$$

$$\partial_x v + \partial_y u = 0$$

must be zero, since if there existed a small region ε of S throughout which either integrand were *single signed* and *not zero*, it would be possible to choose the original closed contour entirely within ε, and the corresponding surface integrals could not possibly be zero, in contradiction to the preceding argument. Thus there does not exist any small region of S within which the integrands fail to be identically zero. But these are simply the Cauchy–Riemann equations:

$$\partial_x u = \partial_y v$$
$$\partial_y u = -\partial_x v$$

which together with the continuity assumption imply that $f(z)$ is analytic.

Instead of supposing

$$\oint f(z)\, dz = 0$$

it can be equivalently assumed that for any a and z inside region S

$$\int_a^z f(z)\, dz$$

is independent of the integration path within S connecting a and z.

It follows from Morera's theorem that the sum of a uniformly convergent series of analytic functions is an analytic function. Thus,

$$\oint S(z)\, dz = \sum_{n=1}^{\infty} \oint u_n(z)\, dz = 0 \tag{4.9}$$

4.3 CURVATURE

Along any curve with equation $z = z(s)$ (s = arc length so that $ds = |dz|$) the complex number of unit modulus which has the direction of the curve at any point on it is

$$t = d_s z = \frac{dz}{ds} = \frac{dz}{|dz|} \tag{4.10}$$

called the *unit tangent*† to the curve. Since $t^*t = 1$,

$$t^* d_s t + t d t_s^* = 0$$

and $\text{Re}(t^* d_s t) = 0$, so t is *orthogonal* to $d_s t$. A complex number of unit modulus perpendicular to the curve is the (unit) *normal*

$$n = -it = -i\frac{dz}{|dz|} \qquad \text{with} \quad n^*n = 1 \tag{4.11}$$

and since $d_s t$ is orthogonal ($\perp$) to t, it must be parallel (or antiparallel) to n.

† Do not confuse this t with the real parameter t in the parametric equation $z = z(t)$ of a curve.

Hence there exists a real number κ, called the *curvature* of the curve, for which at each point of the latter the following (Frenet) formulae hold:

$$d_s t = -\kappa n = i\kappa t = i d_s n \tag{4.12}$$

Now $d_s t = d_s^2 z = d^2 z/|dz|^2$, so that the magnitude of the curvature is

$$|\kappa| = \frac{|d^2 z|}{|dz|^2} \tag{4.13}$$

The reciprocal of this is the *radius of curvature*. Generally the curvature is

$$\kappa = -it^* d_s t = -i\frac{dz^*\, d^2 z}{|dz|^3} = -i\frac{d^2 z}{|dz|\, dz}$$

and the *center of curvature* is

$$c = z - \frac{n}{\kappa} = z - \frac{dz^2}{d^2 z} = z - \frac{(d_s z)^2}{d_s^2 z} \tag{4.14}$$

so that the *circle of curvature* is given by

$$|\kappa|\,|z - c| = 1$$

at any particular point (where c and κ will have fixed values).

Integration of the expression for $d_s t$ in (4.12) yields

$$t = t_0 \exp\left(i\int_{s_0}^{s} \kappa\, ds\right) = in$$

When s has increased by the total perimeter of a *closed* curve, t must return to its original value t_0 (assuming the curve to have a continuous tangent). This is ensured by

$$\oint \kappa\, ds = 2\pi \tag{4.15}$$

a plane form of the *Gauss–Bonnet theorem* which follows from

$$\kappa\, ds = -i\frac{d^2 z}{dz} = -i\frac{d(dz)}{dz} = -i\, d(\log dz)$$

so that

$$\oint \kappa\, ds = -i\oint d(\log dz) = \oint d(\arg dz) = 2\pi = \frac{\oint \kappa\, ds}{\oint ds}\oint ds$$

The plane Gauss–Bonnet theorem for (closed) curves with continuous tangents may be stated as follows: The average curvature (with respect to arc length) of such a curve is 2π divided by its perimeter.

4.4 GAUSS–STOKES THEOREM

From Green's theorem in the plane (see Appendix B) one may derive the plane forms of *Stokes theorem* and the *Gauss divergence theorem* simultaneously.

Let $f(z)$ be a not necessarily analytic function of z, and define the complex *del operator* by

$$\partial = \partial_x + i\partial_y \tag{4.16}$$

(the corresponding vectorial notation would be ∇). Then ($f = \phi + i\psi$)

$$\partial f^* = (\partial_x + i\partial_y)(\phi - i\psi) = (\partial_x\phi + \partial_y\psi) - i(\partial_x\psi - \partial_y\phi)$$

Here, Re $\partial^* f$ is called the (two-dimensional) *divergence* of $f(z)$, and Im $\partial^* f$ is called the (two-dimensional) *curl* of $f(z)$.

Now consider a closed curve bounding a simply connected region. Then along this contour one has

$$\oint f^*\, dz = \oint (\phi\, dx + \psi\, dy) + i \oint (\phi\, dy - \psi\, dx)$$

According to Green's theorem in the plane this becomes (with $dS = dx\, dy$)

$$\oint f^*\, dz = \int (\partial_x\psi - \partial_y\phi)\, dS + i \int (\partial_x\phi + \partial_y\psi)\, dS$$

or

$$\oint f^*\, dz = -\text{Im} \int \partial f^*\, dS + i\, \text{Re} \int \partial f^*\, dS$$

so that

$$\oint f^*\, dz = i \int \partial f^*\, dS$$

which expresses simultaneously the plane forms of Stokes theorem and the Gauss divergence theorem, since the conjugate yields the *Gauss–Stokes theorem*:

$$\int \partial^* f\, dS = i \oint f\, dz^* = \oint n^* f\, ds \tag{4.17}$$

recalling that the unit normal to the curve is $n = -id_s z$. The real part of this

$$\int \text{Re}(\partial^* f)\, dS = \oint \text{Re}(n^* f)\, ds = -\text{Im} \oint f\, dz^* \tag{4.18}$$

is the Gauss divergence theorem, and the imaginary part

$$\int \operatorname{Im}(\partial^* f)\, dS = \operatorname{Re} \oint f\, dz^* = \oint \operatorname{Re}(t^* f)\, ds$$

is Stokes theorem.

A function $f(z)$ (not necessarily analytic) for which $\partial^* f$ is imaginary is called *divergenceless*, and one for which $\partial^* f$ is real is called *irrotational*. Clearly if $f(z)$ were the conjugate of an analytic function,

$$\oint f^*\, dz = 0$$

by Cauchy's theorem, and the function would be both divergenceless and irrotational. If $f(z)$ were a nonconstant analytic function, it could not be divergenceless and irrotational throughout a region because

$$\partial^* f = (\partial_x \phi + \partial_y \psi) + i(\partial_x \psi - \partial_y \phi) = 0$$

implies, together with the Cauchy–Riemann equations, that

$$\partial_x \phi = \partial_y \psi = -\partial_x \phi = 0 \qquad \partial_y \phi = -\partial_x \psi = \partial_x \psi = 0$$

and hence $f = \phi + i\psi =$ constant, contrary to assumption.

In applications of complex variables to irrotational and divergenceless vectorial fields it is therefore quite common to identify two-dimensional vectorial quantities with the *conjugate* of an analytic function (e.g., fluid velocity in hydrodynamics).

● EXAMPLES 4

1. Show that $\oint z\, dz = 0$ from the definition of the integral.

   Ans. The integral is the limit of

   $$\tfrac{1}{2}\sum_{k=1}^{n} (z_{k+1} + z_k)(z_{k+1} - z_k) = \tfrac{1}{2}\sum_{k=1}^{n} (z_{k+1}^2 - z_k^2) = \frac{z_{n+1}^2 - z_1^2}{2}$$

   For $z_{n+1} = z_1$ this is zero and remains so in the limit as $n \to \infty$ with z_k on any closed curve.

2. Show that

   $$\left| \int_L \sinh z\, dz \right| \le 13\sqrt{1 + \sinh^2(12)}$$

   where L is the straight line segment from $z = 0$ to $z = 5 + 12i$.

   Ans. The length of the segment is 13. The maximum modulus of

   $$|\sinh z| \le \sqrt{\sinh^2 y + \sin^2 x} \le \sqrt{1 + \sinh^2 y}$$

3. Would the bound for the integral in Example 2 have to be changed if the integration path were altered between the same two points $z = 0$ and $z = 5 + 12i$?
   Ans. No. The integral is independent of L.
4. Find an upper bound M for the maximum modulus of the function

$$f(z) = \frac{z^{1/3}}{z^2 + 4z + 3}$$

on a circle $|z| = R$ $(R \neq 1, R \neq 3)$ and the limit of $|f(z)|$ as $R \to \infty$ and as $R \to 0$.

Ans.
$$|f(z)| = \frac{|z|^{1/3}}{|z^2 + 4z + 3|} \leq \frac{|z|^{1/3}}{\big||z| - 3\big|\,\big||z| - 1\big|}$$
$$|f(z)| \leq \frac{R^{1/3}}{|R - 3|\,|R - 1|}$$

since

$$|z + 3| \geq \big||z| - 3\big|$$
$$|z + 1| \geq \big||z| - 1\big|$$
$$\lim_{R\to 0} |f(z)| \leq \lim_{R\to 0} \frac{R^{1/3}}{|R - 3|\,|R - 1|} = \lim_{R\to 0} \frac{R^{1/3}}{3} = 0$$
$$\lim_{R\to \infty} |f(z)| \leq \lim_{R\to \infty} \frac{R^{1/3}}{|R - 3|\,|R - 1|} = \lim_{R\to \infty} \frac{R^{1/3}}{R^2} = 0$$

5. Find an upper bound M for the maximum modulus of the polynomial

$$P(z) = 3z^5 + 7z^3 - 5z^2 + 2$$

on the circle of radius R sufficiently large to contain all roots of $P(z) = 0$ in its interior.
Ans. $P(z) = 3(z - z_1)(z - z_2)(z - z_3)(z - z_4)(z - z_5)$ where the z_k are the roots of $P(z) = 0$. Then

$$|P(z)| \leq 3\,|z - z_1|\,|z - z_2|\,|z - z_3|\,|z - z_4|\,|z - z_5|$$
$$\leq 3(2R)(2R)(2R)(2R)(2R) \leq 3{\cdot}2^5R^5 = 96R^5$$

● PROBLEMS 4

1. Show that $\oint dz = 0$ from the definition of the integral.
2. Why is the indefinite integral, with $z = x + iy$ and $f(z)$ analytic,

$$\int f(z)\, dz = F(z) + \text{constant}$$

if

$$\int f(x)\,dx = F(x) + \text{constant}$$

is the corresponding real integral?

3. Why do the two integrals

$$\oint \frac{dz}{z^2 - 3} \qquad \text{and} \qquad \oint \left[\frac{1}{(z^2 - 3)} + \sin z\right] dz$$

have the same value around any closed contour in the z plane, avoiding points for which $z^2 = 3$?

4. Evaluate around $|z| = r$:

$$\oint \arg z \, dz$$

5. Evaluate around $|z| = r$:

$$\oint |z| \, dz$$

6. Evaluate around $|z| = r$ for $0 \leq \arg z \leq \pi$ and $-r \leq \operatorname{Re} z \leq r$:

$$\int |z| \, dz$$

7. Evaluate around $|z| = r$:

$$\oint z^n \, dz$$

8. Evaluate on $|z| = \varepsilon$, $0 \leq \arg z \leq \alpha$ by substituting $z = \varepsilon e^{i\theta}$ and also by the indefinite integral:

$$\int \frac{dz}{z}$$

Evaluate from $z = 0$ to $z = 2(1 + i)$ along L:

$$\int_L \frac{dz}{z - (1 + i)}$$

9. For L: $\operatorname{Re} z = 0$ and $\operatorname{Im} z = 2$.
10. For L: $\operatorname{Im} z = 0$ and $\operatorname{Re} z = 2$.
11. For L: $|z - (1 + i)| = \sqrt{2}$ and $\arg z \geq \pi/4$.
12. For L: $|z - (1 + i)| = \sqrt{2}$ and $\arg z \leq \pi/4$.
13. Along which path L_1 or L_2 between $z = 0$ and $z = 8$ is

$$\int_L \tan z \, dz$$

largest in magnitude? L_1: $\operatorname{Im} z = 0$ and L_2: $|z - 4| = 4$.

14. If $f(z)$ is analytic for $|z| \leq 1$, show that for $r < 1$

$$\pi r f'(0) = \int_0^{2\pi} \mathrm{Re}[f(re^{i\theta})]e^{-i\theta}\, d\theta$$

15. If $f(z)$ and $g(z)$ are analytic and free of singularities in a region of the z plane, show that

$$\oint f[g(z)]\, dz = 0$$

around any closed contour inside the region.

16. If $f(z)$ and $g(z)$ are analytic and free of singularities in a region of the z plane, show that

$$\oint g(z)f(z)\, dz = 0$$

around any closed contour inside the region.

17. If $f(z)$ and $g(z)$ are analytic and free of singularities in a region of the z plane, under what conditions will

$$\oint \frac{f(z)}{g(z)}\, dz = 0$$

around any closed contour inside the region?

18. Show that $\partial z = \partial^* z^* = 0$ and that $\partial z^* = \partial^* z = 2$.
19. Show that $\partial f(z) = f'(z)\partial z$ and $\partial g(z^*) = g'(z^*)\partial z^* = 2g'(z^*)$.
20. By expressing the Laplacian operator $\Delta = \partial_x^2 + \partial_y^2$ as $\Delta = \partial^*\partial = \partial\partial^*$, show that the general solution of Laplace's equation $\partial^*\partial\phi = 0$ is given by $\phi = F(z) + G(z^*)$ where F and G are arbitrary (differentiable) functions.
21. By taking $f = \psi\partial\phi$ in the Gauss–Stokes theorem, show that

$$\int [\psi\partial^*\partial\phi + \partial^*\psi\partial\phi]\, dS = \oint \psi n^*\partial\phi\, ds$$

22. Prove that if u is real, then

$$i\oint u\partial u\, dz^* - \int u\partial^*\partial u\, dS$$

is real and positive unless u is constant.

23. For ϕ and ψ real show that

$$\int (\phi\partial^*\partial\psi - \psi\partial^*\partial\phi)\, dS = \oint [\phi n^*\partial\psi - \psi n\partial^*\phi]\, ds$$

24. For real ϕ show that $\oint \partial^*\partial\phi\, dS = \oint n^*\partial\phi\, ds$.

25. Show that if u is harmonic, $\oint \partial u \, dz^* = 0$.
26. Is the derivative of a harmonic function at a point independent of the direction from which the point is approached during the limiting process leading to the derivative?
27. From (4.17) deduce by choosing $f = z$ that the area enclosed is given by

$$S = \frac{i}{2} \oint z \, dz^* = \frac{-i}{2} \oint z^* \, dz$$

28. Show that $z = \int_0^s e^{i\phi} \, ds$, where $\phi = \arg t$.
29. Find t, n, and κ for the tractrix with $dz = (e^{-s} + \sqrt{1 - e^{-2s}}) \, ds$.
30. For the cycloid $z = 1 + i\theta - e^{i\theta}$, verify that $\kappa(4^2 - s^2) = 1$. Take $s = 0$ at $\theta = \pi$.
31. Find the average curvature of a closed curve formed of two parallel line segments of length l and two semicircular arcs of radius ε.

CHAPTER 5:

Cauchy Integral Theorem

5.1 INTERIOR AND EXTERIOR CAUCHY INTEGRAL FORMULAE

The Cauchy theorem makes it possible to solve the so-called Dirichlet problem in which the values of a function (analytic in a simply connected region R) are sought inside a region from a knowledge of its values on the contour (curve) bounding the region. Let us consider $f(z)$ to be free of singularities in the region R bounded by the contour C. Then if w specifies the location of any particular fixed point inside R,

$$g(z) = \frac{f(z) - f(w)}{z - w} \tag{5.1}$$

considered as a function of z, has no singularities inside R because as z approaches w, $g(z)$ approaches the derivative of $f(z)$ evaluated at $z = w$, that is,

$$g(w) = f'(w)$$

which exists since $f(z)$ was supposed to be analytic. Since $g(z)$ is thus analytic inside R, we may write

$$0 = \oint_C g(z)\, dz = \oint \left[\frac{f(z) - f(w)}{z - w}\right] dz \tag{5.2}$$

or

$$\oint_C \frac{f(z)\, dz}{z - w} = \oint_C \frac{f(w)\, dz}{z - w} = f(w) \oint_C \frac{dz}{z - w}$$

$$= f(w) \oint_C d[\log(z - w)] = 2\pi i f(w)$$

since only* $\arg(z - w)$ changes by 2π in a circuit of z completely about w. The result,

$$\frac{1}{2\pi i} \oint_C \frac{f(z)\,dz}{z - w} = f(w) \tag{5.3}$$

for w inside R is called the *Cauchy integral formula*, and in fact it may be used to calculate the value of $f(w)$ at any interior point w from the knowledge of the values of $f(z)$ at all points z along the contour C. If the point w were located outside the contour C, $f(z)/(z - w)$ would have no singularities for any value z inside the contour, and so by the Cauchy theorem (or by noting that $\arg(z - w)$ has no net change),

$$\frac{1}{2\pi i} \oint_C \frac{f(z)\,dz}{z - w} = 0$$

Thus we see that if $f(z)$ is analytic inside the contour, there may still be *two* possible outcomes to the integration along C:

$$\frac{1}{2\pi i} \oint_C \frac{f(z)\,dz}{(z - w)} = \begin{cases} f(w) & \text{for} \quad w \text{ inside } C \\ 0 & \text{for} \quad w \text{ outside } C \end{cases} \tag{5.4}$$

These formulae are collectively known as the Cauchy integral formula for functions analytic *inside* a contour.

If $f(z)$ is analytic *outside* C and w is a point inside C, then† in the neighborhood of $z = \infty$ (i.e., for sufficiently large $|z|$)

$$f(z) = f(\infty) + \frac{a_1}{z} + \frac{a_2}{z^2} + \cdots \tag{5.5}$$

and the contour may be deformed into a circle K chosen so large that

$$\oint_C \frac{f(z)}{z - w}\,dz \approx f(\infty) \oint_K \frac{dz}{z - w} = 2\pi i f(\infty) \tag{5.6}$$

If w is outside C, we can surround it by a small circular contour D, as shown in Figure 5.1. Then

$$\oint_C + \oint_D = \oint_K$$

But in the neighborhood of $z = w$

$$f(z) \cong f(w) + (z - w) f'(w)$$

* And *not* $\log|z - w|$.

† See Problem 18, Chapter 3.

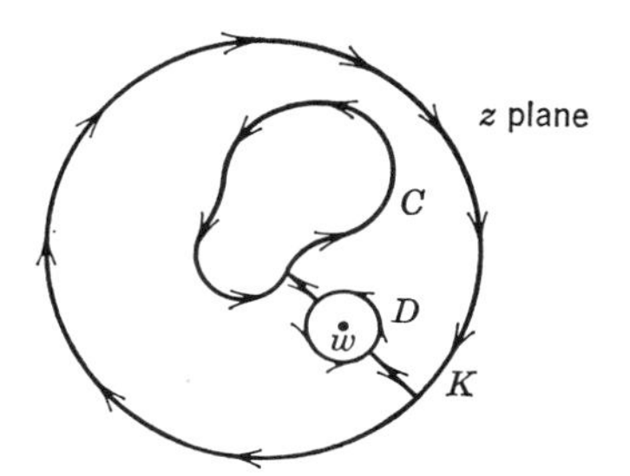

Figure 5.1. Contours in Cauchy integral theorem for functions analytic *outside* a contour C.

by the definition of the derivative, so

$$\oint_D \frac{f(z)}{z-w}\,dz \cong f(w)\oint_D \frac{dz}{z-w} + f'(w)\oint_D dz$$

$$\cong 2\pi i f(w) \tag{5.7}$$

Hence,

$$\oint_C \frac{f(z)}{z-w}\,dz = 2\pi i[f(\infty) - f(w)] \tag{5.8}$$

for w outside C. Together we may write

$$\frac{1}{2\pi i}\oint_C \frac{f(z)\,dz}{z-w} = \begin{cases} f(\infty) & \text{for} \quad w \text{ inside } C \\ f(\infty) - f(w) & \text{for} \quad w \text{ outside } C \end{cases} \tag{5.9}$$

as the Cauchy integral formulae for functions analytic *outside* a contour.

5.2 LIOUVILLE'S THEOREM

For functions analytic inside a contour and a point w inside the contour, the Cauchy integral formula may be differentiated to yield

$$d_w^n f = f^{(n)}(w) = \frac{n!}{2\pi i}\oint_C \frac{f(z)\,dz}{(z-w)^{n+1}} \tag{5.10}$$

If the contour is a circle of radius R about the point $z = w$ and the maximum value of $|f(z)|$ on the circle is M, we conclude by (4.5) that

$$|f^{(n)}(w)| \le \frac{Mn!}{R^n} \tag{5.11}$$

which is known as the *Cauchy inequality*. If the number M is independent of R,

$$|f'(w)| \leq \frac{M}{R} \to 0 \tag{5.12}$$

as $R \to \infty$, and since w could have been any point, we conclude that the derivative of an analytic function which is *everywhere bounded* (including $z = \infty$) is zero. The function itself must thus be a *constant*. This is *Liouville's theorem*.

A consequence of Liouville's theorem is that every algebraic equation* of positive degree $P(z) = 0$ possesses at least one root, since if there did not exist such a value z, we would have $P(z) \neq 0$ for all z. Then $1/P(z)$ would be bounded for all z, and since it is analytic with unique derivative $-P'(z)/[P(z)]^2$, it must be a constant so that $P(z)$ is a constant contrary to the assumption that it was of positive degree. This result is called the *fundamental theorem of algebra*.

5.3 LAGRANGIAN INTERPOLATION

We have seen how the Cauchy integral formula solves the Dirichlet problem in which functional values are sought inside a region from a knowledge of the corresponding values given on the boundary. A related problem is to find the values of a (complex) polynomial, given its values P_k at a finite set of distinct points z_k on the boundary of a region. This problem would be solved if we could find a set of polynomials $g_k(z)$ such that

$$g_k(z_m) = \delta_{km} = \begin{cases} 1 & \text{for} \quad k = m \\ 0 & \text{for} \quad k \neq m \end{cases} \tag{5.13}$$

since then

$$P(z) = \sum_k P_k g_k(z) \tag{5.14}$$

and

$$P(z_m) = \sum_k P_k g_k(z_m) = \sum_k P_k \delta_{km} = P_m$$

as required.

Let $Q(z) = \prod_k (z - z_k)$. Then $Q(z_k) = 0$ and

$$\frac{Q(z) - Q(z_k)}{z - z_k} \to Q'(z_k)$$

* Here $P(z)$ is a polynomial.

as $z \to z_k$, so that

$$\frac{Q(z)}{(z - z_k)Q'(z_k)} \to 1$$

as $z \to z_k$. Furthermore, $Q(z_m) = 0$, so

$$g_k(z) = \frac{Q(z)}{(z - z_k)Q'(z_k)}$$

satisfies the conditions and

$$P(z) = \sum_k P_k \frac{Q(z)}{(z - z_k)Q'(z_k)} \tag{5.15}$$

This is the *Lagrangian interpolation formula* which is naturally also valid for *real* polynomials.

Finally it should be noticed that the Cauchy integral formula (5.3) may be written as a formal average of functional values of $f(w)$ over the closed contour C with respect to (the weighting function) $\log(w - z)$:

$$f(z) = \langle f(w) \rangle_C = \frac{\oint f(w)\, d \log(w - z)}{\oint d \log(w - z)}$$

● EXAMPLES 5

1. Show that if $f(z)$ satisfies the conditions for the Cauchy integral theorem and is a constant everywhere on the contour C, then $f(z)$ has the same constant value at all points interior to the region bounded by C.
   Ans. If $f(z) = k$ on C, (5.3) implies that at interior points

   $$f(w) = \frac{1}{2\pi i} \oint \frac{k\, dz}{(z - w)} = \frac{k}{2\pi i} \oint d \log(z - w)$$

   $f(w) = k$, since the change in $\log(z - w)$ is $2\pi i$ when z circumscribes C.
2. Construct a function which is unity inside a contour C and zero outside it.
   Ans. Take $f(z) = 1$ in (5.4).
3. If $g(z)$ is analytic inside C and $f(z)$ is analytic outside C with $f(\infty) = 0$, evaluate

   $$\frac{1}{2\pi i} \oint \frac{g(z) - f(z)}{z - w}\, dz$$

   along C for w not on C.
   Ans. The result is $g(w)$ for w inside C and $f(w)$ for w outside C, which follows from (5.4) and (5.9).

4. Find a cubic polynomial $P(z)$ taking the values $0, 1, \omega, \omega^2$ at $z = 1, i, -1, -i$, respectively (ω and ω^2 are cube roots of unity).

Ans.
$$Q(z) = (z-1)(z-i)(z+1)(z+i) = (z^2-1)(z^2+1) = z^4 - 1$$
$$Q'(z) = 4z^3$$
$$P(z) = \frac{(z^4-1)}{4}\left\{\frac{1}{(z-i)i^3} - \frac{\omega}{(z+1)} + \frac{\omega^2}{(z+i)(-i)^3}\right\}$$
$$= \frac{(z^4-1)}{4}\left\{\frac{i}{(z-i)} - \frac{\omega}{(z+1)} - \frac{i\omega^2}{(z+i)}\right\}$$
$$4P = [i(z^2-1)(z+i) - \omega(z^2+1)(z-1) - i\omega^2(z-i)(z^2-1)]$$
or
$$4P(z) = (2i + i\omega - \omega)z^3 + 2\omega z^2 - (2i + i\omega + \omega)z$$

5. Evaluate
$$\oint \frac{z \sin(\pi/z)}{z-1}\, dz$$
about the circle $|z| = 2$.

Ans. $z \sin(\pi/z)$ is analytic except at $z = 0$; hence (5.9) yields $2\pi^2 i$.

● PROBLEMS 5

1. Prove that if $f(z)$ is analytic with no singularities inside or on a closed contour and if $f(z)$ has only simple zeros inside the contour, the average location of its zeros inside the contour is
$$\int z\, d[\log f(z)] \Big/ \int d[\log f(z)]$$
2. Evaluate around the unit circle $|z| = 1$:
$$\oint \frac{e^{1/z}\, dz}{(z-w)}$$
where $|w| \neq 1$.
3. Find a polynomial of degree four which takes the values $-i, -1, 0, 1, i$ at the points $z = 1, 2, 3, 4, 5$.
4. Find a polynomial of degree one which takes the values a and a^* for $z = 1$ and $z = -1$.

5. Derive Taylor's series for $f(z)$ from (5.10).
6. Let $f(z)$ be free of singularities inside and on a closed contour containing the points $z_1, \ldots, z_n$ inside it. Show that

$$P(z) = \frac{1}{2\pi i} \oint \frac{f(w)}{Q(w)} \left(\frac{Q(w) - Q(z)}{w - z} \right) dw$$

is a polynomial with $P(z_k) = f(z_k)$ if $Q(z) = \prod_{k=1}^{n} (z - z_k)$ and $Q'(z_k) \neq 0$.
7. If $w = f(z)$ only once within a region of the z plane bounded by a closed contour C, show that the inverse function to $f(z)$ is

$$F(w) = \frac{1}{2\pi i} \oint \frac{zf'(z)\, dz}{f(z) - w}$$

8. Verify that $F'(w)f'(z) = 1$ from the above.
9. Verify that $d_w^{n-1} F(w) = (n-1)!/2\pi i \oint dz/[f(z) - w]^n$.
10. Use the preceding to deduce that the Taylor series for $F(w)$ and $w = b = f(a)$ is

$$z = a + \sum_{n=1}^{\infty} \frac{(w-b)^n}{n!} d_z^{n-1}\{[\phi(z)]^n\}_{z=a}$$

with $\phi(z) = (z - a)/[f(z) - b]$. This is the *Lagrange reversion of series formula.*

CHAPTER 6:

Laurent Series and Residue Theorem

6.1 LAURENT'S THEOREM

The property of unique differentiability which characterizes the class of analytic functions implies, via the Cauchy integral theorem, that such functions may be expanded in (Laurent) series of positive and negative powers of the form

$$f(z) = \sum_{-\infty}^{\infty} a_n(z-a)^n \tag{6.1}$$

uniformly convergent in any closed annulus free of singularities of $f(z)$. In particular, let us suppose that such an annulus is

$$0 < c \leq |z-a| \leq b \tag{6.2}$$

Then if s is any interior point, as shown in Figure 6.1, the annulus may be approximated by the contour shown in Figure 6.2, on which, according to the Cauchy integral theorem,

$$2\pi i f(s) = \oint_{|z-a|=b} \frac{f(z)\,dz}{z-s} - \oint_{|z-a|=c} \frac{f(z)\,dz}{z-s} \tag{6.3}$$

with both integrals taken counterclockwise (the integrals along the straight portions cancel when they are brought together).

The integrands in (6.3) can also be written as

$$\frac{f(z)}{z-s} = \frac{f(z)}{(z-a)-(s-a)} \tag{6.4}$$

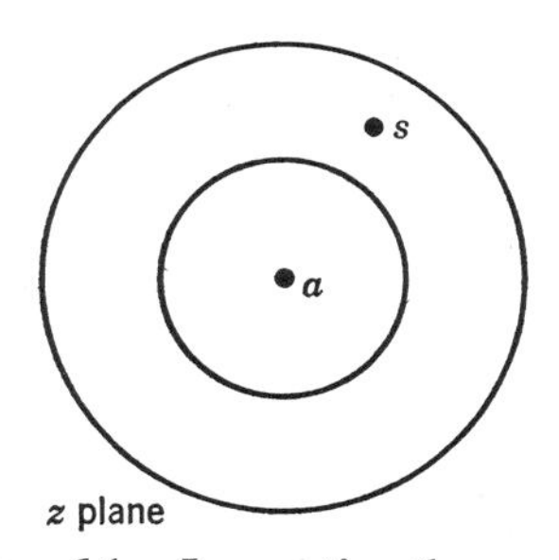

Figure 6.1. Laurent's theorem for functions analytic in annulus.

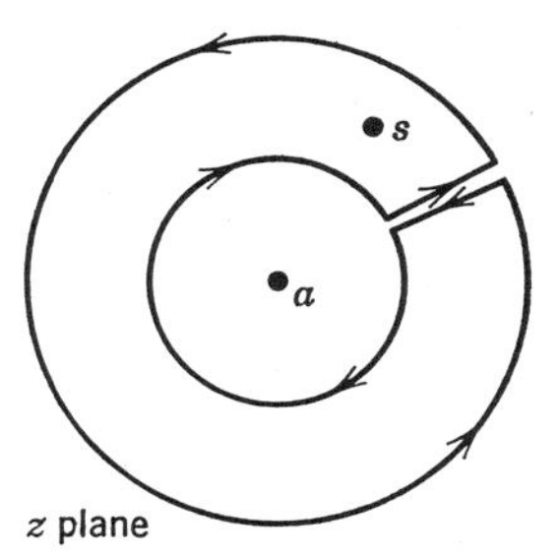

Figure 6.2. Approximate equivalent contour for Laurent theorem.

and (6.4) may be expanded as geometrical series for the case $b = |z - a| > |s - a|$:

$$\frac{1}{(z-a)-(s-a)} = \frac{1}{z-a}\left\{\frac{1}{1-(s-a)/(z-a)}\right\}$$
$$= \frac{1}{z-a}\sum_{n=0}^{\infty}\left(\frac{s-a}{z-a}\right)^n \tag{6.5}$$

and for the case $c = |z - a| < |s - a|$:

$$\frac{1}{(z-a)-(s-a)} = \frac{-1}{s-a}\sum_{n=0}^{\infty}\left(\frac{z-a}{s-a}\right)^n \tag{6.6}$$

Equation (6.5) is inserted into the first integral of (6.3) and (6.6) is inserted into the second integral of (6.3) to yield

$$f(s) = \sum_{n=0}^{\infty} a_n(s-a)^n + \sum_{n=0}^{\infty}\frac{b_n}{(s-a)^{n+1}}$$

where

$$2\pi i a_n = \oint \frac{f(z)\,dz}{(z-a)^{n+1}}$$

$$2\pi i b_n = \oint f(z)(z-a)^n\,dz$$

or, redefining coefficients,

$$c_n = \begin{cases} a_n & \text{for } n \geq 0 \\ b_{n-1} & \text{for } n < 0 \end{cases}$$

$$f(s) = \sum_{-\infty}^{\infty} c_n(s-a)^n \tag{6.7}$$

where

$$2\pi i c_n = \oint \frac{f(z)\,dz}{(z-a)^{n+1}} \tag{6.8}$$

The termwise integration of the series (6.5) and (6.6) is justified by uniform convergence. The fact that (6.7) and (6.8) are valid for any function regular in such an annulus is known as *Laurent's theorem*, and (6.7) is the *Laurent series* of $f(z)$. If $c_n = 0$ for $n < 0$, it is called a *Taylor series* (of *positive* powers), and if in addition $a = 0$, it is called a *MacLaurin series*. If $f(z)$ is free of singularities for

$$0 \leq |z - a| \leq b$$

the formula (6.8) may be replaced by the Taylor coefficient formula

$$c_n = \frac{f^{(n)}(a)}{n!} \tag{6.9}$$

which is obtained by repeated differentiation of the Cauchy integral formula (5.3). In this case only nonnegative powers can occur in (6.7).

If in (6.7) $c_n = 0$ for $n < m$ and $c_m \neq 0$ $(m > 0)$, $f(z)$ is said to have a *zero of order m* at $z = a$. Then $f^n(a) = 0$ for $0 \leq n < m$. If in (6.7) $c_n = 0$ for $n < m < 0$ and $c_m \neq 0$, $f(z)$ is said to have a *pole of order* $-m$ at $z = a$. Zeros and poles of order one are called simple zeros and simple poles. If there are an infinite number of negative powers, $f(z)$ is said to have an essential singularity at $z = a$.

6.2 RESIDUE THEOREM

The coefficient c_{-1} in (6.7) is called the *residue at* $z = a$. For $n = -1$, (6.8) becomes

$$\oint f(z)\, dz = 2\pi i c_{-1} \tag{6.10}$$

which shows that *only the residue* among all coefficients in (6.7) *survives* the integration of $f(z)$ about such a contour. Of course, if $f(z)$ were analytic inside the contour, c_{-1} would have to be zero. The shape of the contour may be deformed so long as it does not pass over a singularity of $f(z)$.

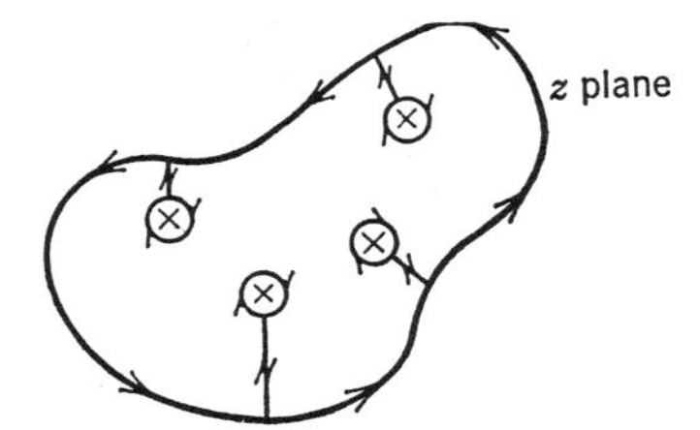

Figure 6.3. Reduction of contour in residue theorem to equivalent set of infinitesimal circles about poles.

If $f(z)$ possessed negative powers in its Laurent (series) expansions about several points inside the contour and the corresponding residues were ${}_kc_{-1}$, then a superposition of results like (6.10) would yield the *residue theorem*:

$$\oint f(z)\,dz = 2\pi i \sum_k {}_kc_{-1} \tag{6.11}$$

6.3 CALCULATION OF RESIDUES

The residue theorem makes it important to be able to calculate the residues of analytic functions. For those with poles of finite order this may be done as follows, even if the Laurent expansion is not explicitly known: If

$$f(z) = \frac{c_{-m}}{(z-a)^m} + \cdots + \frac{c_{-1}}{(z-a)} + c_0 + c_1(z-a) + \cdots$$

then

$$(z-a)^m f(z) = \sum_{k=-m}^{\infty} c_k(z-a)^{k+m}$$

is free of singularities in its circle of convergence and differentiation $(m-1)$ times yields

$$d_z^{m-1}[(z-a)^m f(z)] = \sum_{k=-1}^{\infty} \frac{(m+k)!}{(k+1)!} c_k(z-a)^{k+1}$$

so that at $z = a$ the derivative has the value (only the $k = -1$ term survives)

$$d_z^{m-1}[(z-a)^m f(z)]\big|_{z=a} = (m-1)!\, c_{-1} \tag{6.12}$$

which yields a formula for the residue of $f(z)$ with an mth-order pole at $z = a$. An example of this is

$$f(z) = \frac{\sin z}{z^3 - 3z^2 + 3z - 1} = \frac{\sin z}{(z-1)^3}$$

which has a triple pole at $z = 1$. The residue c_{-1} at $z = 1$ is thus given by

$$2c_{-1} = d_z^2(\sin z)\big|_{z=1} = -\sin z\big|_{z=1} = -\sin 1$$

For $m = 1$ the residue becomes

$$c_{-1} = \lim_{z\to a}[(z-a)f(z)] \tag{6.13}$$

An example of this is $f(z) = \sin z/z^2$, which has a simple pole $(m = 1)$ at the origin. Hence the residue c_{-1} at $z = 0$ is

$$c_{-1} = \lim_{z\to 0}\left[\frac{\sin z}{z}\right] = 1$$

If $f(z) = \psi(z)/\phi(z)$, with $\phi(a) = 0 \neq \phi'(a)$ so that the denominator has a simple zero at $z = a$, and if $\psi(z)$ has no singularity at $z = a$, then one has, for the residue of $f(z)$ at $z = a$,

$$c_{-1} = \lim_{z \to a} \left[\frac{(z - a)\psi(z)}{\phi(z) - \phi(a)} \right] = \frac{\psi(a)}{\phi'(a)} \tag{6.14}$$

since by definition of the derivative,

$$\phi'(a) = \lim_{z \to a} \left[\frac{\phi(z) - \phi(a)}{z - a} \right]$$

An example of this situation is

$$f(z) = \frac{\cos(\sin z)}{\sin z}$$

which has simple poles at $z = n\pi$ for integer n. Thus the residue at $z = n\pi$ is

$$c_{-1} = \frac{\cos(\sin n\pi)}{\cos n\pi} = \frac{1}{\cos n\pi} = (-1)^n$$

One should note that in the methods of this section explicit knowledge of the Laurent expansions has not been used.

6.4 CAUCHY PRINCIPAL VALUES

It often happens that one is involved with the evaluation of the limit (as $\varepsilon \to 0$) of an integral along an arc from which all points within a small distance ε of a given point $z = a$ of the arc have been excluded from the range of integration. Such a limit (which may exist even though the integral including all points of the arc is divergent) is called a *Cauchy principal value.* Some authors place a P in front of the integral sign to indicate the additional limiting process (beyond that of integration) required, but this practice will *not* generally be followed here. It is also not usually done for other types of improper integrals (so why here?). Instead, if the integral is improper in such a way that the Cauchy principal value exists, the latter will be understood to be meant by the notation for the integral. A simple example is

$$\int_{-2}^{3} \csc x \, dx$$

The integrand is undefined at $x = 0$ and the integral is thus improper. However,

$$\lim_{\varepsilon\to 0} \int_{-2}^{-\varepsilon} + \int_{\varepsilon}^{3} \csc x \, dx$$

exists and is given by

$$\int_{2}^{3} \csc x \, dx$$

since

$$\lim_{\varepsilon\to 0} \left[\int_{\varepsilon}^{2} \csc x \, dx + \int_{2}^{3} \csc x \, dx - \int_{\varepsilon}^{2} \csc y \, dy \right]$$

where $y = -x$. Hence the limit as $\varepsilon \to 0$ is

$$\log \left| \frac{\tan(\frac{3}{2})}{\tan 1} \right|$$

and it is this limit (the Cauchy principal value) that we understand by the symbol

$$\int_{-2}^{3} \csc x \, dx$$

for which some authors would use

$$P \int_{-2}^{3} \csc x \, dx$$

One should clearly note the important role played by the *symmetrical* range of integration excluded. Thus, if one tried to define a value for the given (improper) integral by

$$\lim_{\varepsilon\to 0} \int_{-2}^{-\varepsilon} + \int_{\varepsilon^2}^{3} \csc x \, dx$$

the result

$$\lim_{\varepsilon\to 0} \int_{\varepsilon}^{2} \csc x \, dx + \int_{2}^{3} \csc x \, dx + \int_{\varepsilon^2}^{\varepsilon} \csc x \, dx - \int_{\varepsilon}^{2} \csc y \, dy$$

which is equivalent to

$$\lim_{\varepsilon\to 0} \left\{ \log \left| \frac{\tan(\frac{3}{2})}{\tan 1} \right| + \log \left| \frac{\tan(\varepsilon/2)}{\tan(\varepsilon^2/2)} \right| \right\}$$

would fail to exist.

If one tried to define a value for the given integral by

$$\lim_{\varepsilon\to 0} \int_{-2}^{-\varepsilon} + \int_{2\varepsilon}^{3} \csc x \, dx$$

the result

$$\lim_{\varepsilon\to 0}\int_{\varepsilon}^{2}\csc x\,dx+\int_{2}^{3}\csc x\,dx-\int_{\varepsilon}^{2\varepsilon}\csc x\,dx-\int_{\varepsilon}^{2}\csc y\,dy$$

would yield

$$\log\left|\frac{\tan(\frac{3}{2})}{\tan 1}\right|-\lim_{\varepsilon\to 0}\log\left|\frac{\tan\varepsilon}{\tan(\varepsilon/2)}\right|$$

a limit which exists,

$$\log\left|\frac{\tan(\frac{3}{2})}{\tan 1}\right|-\log 2$$

but which is *different* from the Cauchy principal value. This illustrates the fact that the Cauchy principal value is not the only way to associate a value with an improper integral. Nevertheless, it is the one that is most commonly used and, as stated above, the convention will be adopted in this book that it is the Cauchy principal value (and *not* some other way of defining a value for an improper integral) which is to be understood whenever it exists.

Often the Cauchy principal value will arise from the limit of a contour integral with a semicircular excursion to avoid a singularity. Thus, for $f(z)$ free of singularities on the real axis and a real,

$$\int_{-\infty}^{\infty}\frac{f(x)\,dx}{x-a}$$

can be assigned a value [if $f(x)$ approaches zero sufficiently rapidly at $x=\pm\infty$] by

$$\lim_{\varepsilon\to 0}\int_{-\infty}^{a-\varepsilon}+\int_{a+\varepsilon}^{\infty}\frac{f(x)\,dx}{x-a}$$

which can be made part of an integral along a composite contour (before $\varepsilon\to 0$)

$$\int\frac{f(z)\,dz}{z-a}$$

$$I_1:\quad -\infty<z=z^*<a-\varepsilon$$

$$J:\quad |z-a|=\varepsilon\qquad \operatorname{Im} z>0$$

$$I_2:\quad a+\varepsilon<z=z^*<\infty$$

Since the integral along J approaches

$$-i\pi f(a)$$

(see Example 11), one can write formally

$$-i\pi f(a) = -i\pi \int f(z)\delta(z - a)\, dz$$

Hence (using the P notation exceptionally),

$$\int \frac{f(z)\, dz}{z - a} = P \int \frac{f(x)\, dx}{x - a} - i\pi \int f(z)\delta(z - a)\, dz \tag{6.15}$$

so that in a rather crudely formal way one has an "operator relation"

$$\frac{1}{z - a} = P \frac{1}{x - a} - i\pi\delta(z - a) \tag{6.16}$$

If the semicircular excursion were below $z = a$ instead of above it, the integration would be counterclockwise and (by Example 11) this result would be $i\pi f(a)$, so that the corresponding "operator relation" would be

$$\frac{1}{z - a} = P \frac{1}{x - a} + i\pi\delta(z - a) \tag{6.17}$$

The delta function then represents in both cases a rather extravagant way of remembering the result of Example 11.

The subject of Cauchy principal values will be taken up again in Chapter 11, but it will occur quite frequently without comment in the interim.

● EXAMPLES 6

1. Find the Laurent expansions of $1/(1 - z) + 2/(2 - z)$ about $z = 0$.

Ans. The function to be expanded in Laurent series has singularities at $z = 1$ and $z = 2$ (simple poles). Therefore there are three regions of convergence to consider:

(a) $|z| < 1$ for which

$$\frac{1}{1 - z} = 1 + z + z^2 + \cdots \qquad \text{valid for} \quad |z| < 1$$

$$\frac{2}{2 - z} = 1 + \frac{z}{2} + \frac{z^2}{2^2} + \cdots \qquad \text{valid for} \quad |z| < 2$$

so that

$$\frac{1}{1 - z} + \frac{2}{2 - z} = \sum_{n=0}^{\infty} \left(1 + \frac{1}{2^n}\right) z^n \qquad \text{for} \quad |z| < 1$$

(b) $|z| > 2$ for which

$$\frac{1}{1-z} = -\frac{1}{z[1-(1/z)]}$$

$$= -\left(\frac{1}{z} + \frac{1}{z^2} + \frac{1}{z^3} + \cdots\right) \qquad \text{valid for} \quad |z| > 1$$

$$\frac{2}{2-z} = -\frac{2}{z[1-(2/z)]}$$

$$= -\left(\frac{2}{z} + \frac{2^2}{z^2} + \frac{2^3}{z^3} + \cdots\right) \qquad \text{valid for} \quad |z| > 2$$

so that

$$\frac{1}{1-z} + \frac{2}{2-z} = -\sum_{n=1}^{\infty} \frac{(1+2^n)}{z^n} \qquad \text{for} \quad |z| > 2$$

(c) $1 < |z| < 2$ for which

$$\frac{1}{1-z} = -\left(\frac{1}{z} + \frac{1}{z^2} + \frac{1}{z^3} + \cdots\right) \qquad \text{valid for} \quad |z| > 1$$

$$\frac{2}{2-z} = 1 + \frac{z}{2} + \frac{z^2}{2^2} + \cdots \qquad \text{valid for} \quad |z| < 2$$

so that

$$\frac{1}{1-z} + \frac{2}{2-z} = 1 + \sum_{n=1}^{\infty}\left[\left(\frac{z}{2}\right)^n - \frac{1}{z^n}\right] \qquad \text{for} \quad 1 < |z| < 2$$

Notice that the explicit formula for the coefficients of the Laurent series has not been used here. It often turns out that (6.8) is of more theoretical than practical interest.

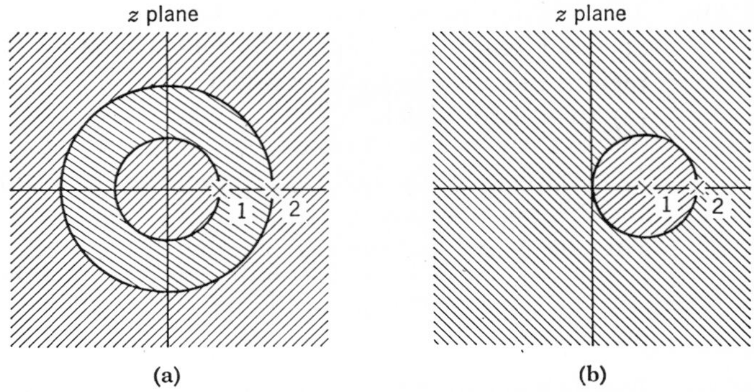

Figure 6.4. Annular regions for Laurent expansions (a) about origin; (b) about $z = 1$.

2. Find the Laurent series for the same function as in Example 1 about the point $z = 1$.

Ans. In Example 1 the annuli of convergence were as shown in Figure 6.4a, whereas here they are as shown in Figure 6.4b. The series in this case are found to be

$$\frac{1}{1-z} + \frac{2}{2-z} = \frac{1}{1-z} + 2\{1 - (1-z) + (1-z)^2 - (1-z)^3 + \cdots\}$$

$$\text{for} \quad 0 < |z-1| < 1$$

$$\frac{1}{1-z} + \frac{2}{2-z} = \frac{3}{1-z} - 2\left\{\frac{1}{(1-z)^2} - \frac{1}{(1-z)^3} + \frac{1}{(1-z)^4} + \cdots\right\}$$

$$\text{for} \quad 1 < |z-1|$$

3. Find the Laurent series for $e^{1/z}$ about $z = 0$.

Ans. From the MacLaurin series for $e^z = \sum_{n=0}^{\infty} z^n/n!$ one has $e^{1/z} = \sum_{n=0}^{\infty} 1/n!\, z^n$; hence $e^{1/z}$ has an essential singularity at $z = 0$.

4. Find an integral expression for the Laurent coefficients about $z = 0$ for the function

$$e^{(s/2)[z-(1/z)]}$$

treating s as a fixed complex number.

Ans. The given function has an essential singularity at $z = 0$ (also at $z = \infty$), so (6.8) yields

$$c_n = \frac{1}{2\pi i} \oint \frac{e^{(s/2)[z-(1/z)]}\, dz}{z^{n+1}} = J_n(s)$$

where the contour surrounds the origin $z = 0$. The notation $J_n(s)$ emphasizes the fact that the coefficients are functions of s, and $J_n(s)$ is called the *Bessel function of the first kind of order n.* Thus the Laurent expansion in terms of these Bessel functions is

$$e^{(s/2)[z-(1/z)]} = \sum_{n=-\infty}^{\infty} J_n(s) z^n$$

The function so expanded is also called the *generating function* for the $J_n(s)$.

5. By using the Laurent series of Example 4 establish the recurrence relations for the following Bessel functions $J_n(s)$:

   (a) $s[J_{n-1}(s) + J_{n+1}(s)] = 2nJ_n(s)$
   (b) $nJ_n(s) = s[J_{n-1}(s) - J'_n(s)]$
   (c) $nJ_n(s) = s[J_{n+1}(s) + J'_n(s)]$

   Ans. $e^{(s/2)[z-(1/z)]} = \sum J_n(s)z^n = \sum J_{n+1}(s)z^{n+1} = z\sum J_{n+1}(s)z^n$ so

$$\sum [J_{n+1}(s) + J_{n-1}(s)]z^n = \left(z + \frac{1}{z}\right)e^{(s/2)[z-(1/z)]}$$

and differentiating the generating function with respect to z and multiplying by z, one has

$$\sum_n nJ_n(s)z^n = \frac{s}{2}\left(z + \frac{1}{z}\right)e^{(s/2)[z-(1/z)]}$$

Hence (a) follows by equating coefficients. For (b) one differentiates the generating function with respect to s, getting

$$\frac{1}{2}\left(z - \frac{1}{z}\right)e^{(s/2)[z-(1/z)]} = \sum J'_n(s)z^n$$

and since

$$ze^{(s/2)[z-(1/z)]} = \sum J_{n-1}(s)z^n$$

then

$$\sum [J_{n-1}(s) - J'_n(s)]z^n = \frac{1}{2}\left(z + \frac{1}{z}\right)e^{(s/2)[z-(1/z)]}$$

so that (b) follows from (a). Similarly,

$$\sum [J_{n+1}(s) + J'_n(s)]z^n = \frac{1}{2}\left(z + \frac{1}{z}\right)e^{(s/2)[z-(1/z)]}$$

so that (c) follows.

6. By differentiating the generating function with respect to s, show that the Bessel functions $J_n(s)$ satisfy the differential equations

$$s^2J''_n + sJ'_n + (s^2 - n^2)J_n = 0$$

   Ans.

$$s^2\sum J''_n(s)z^n = \left[\frac{s}{2}\left(z - \frac{1}{z}\right)\right]^2 e^{(s/2)[z-(1/z)]}$$

$$s\sum J'_n(s)z^n = \left[\frac{s}{2}\left(z - \frac{1}{z}\right]\right)e^{(s/2)[z-(1/z)]}$$

$$\sum n^2J_n(s)z^n = z\partial_z[z\partial_z e^{(s/2)[z-(1/z)]}]$$

which is equal to

$$\left\{\left[\frac{s}{2}\left(z+\frac{1}{z}\right)\right]^2+\frac{s}{2}\left(z-\frac{1}{z}\right)\right\}e^{(s/2)[z-(1/z)]}$$

or

$$\left\{\left[\frac{s}{2}\left(z-\frac{1}{z}\right)\right]^2+\frac{s}{2}\left(z-\frac{1}{z}\right)+s^2\right\}e^{(s/2)[z-(1/z)]}$$

so that the *Bessel differential equation* follows.

7. Evaluate by the residue theorem:

$$I=\int_0^\infty e^{\cos x}\sin(\sin x)\frac{dx}{x}$$

Ans. In order to use the residue theorem, the given integral must be made part of a complex integral about a closed contour; I is closely related to

$$\oint e^{e^{iz}}\frac{dz}{z}$$

and if the closed contour is chosen to consist of four portions ($R \gg \varepsilon$),

$$I_1:\quad \varepsilon \le z = z^* \le R \qquad I_2:\quad -R \le z = z^* \le -\varepsilon$$

$$J_1:\quad |z| = \varepsilon \qquad 0 \le \arg z \le \pi$$

$$J_2:\quad |z| = R \qquad 0 \le \arg z \le \pi$$

Since there are no singularities inside the contour, the total integral about it is zero. The two integrals I_1 and I_2 together yield

$$\int_{-R}^{-\varepsilon} e^{\cos x+i\sin x}\frac{dx}{x}+\int_{\varepsilon}^{R} e^{\cos x+i\sin x}\frac{dx}{x}$$

$$=2i\int_{\varepsilon}^{R} e^{\cos x}\sin(\sin x)\frac{dx}{x}\to 2iI \qquad \text{as}\quad \varepsilon\to 0 \quad\text{and}\quad R\to\infty$$

Also the integral about J_1,

$$i\int_{\pi}^{0} e^{e^{i\varepsilon e^{i\theta}}}\,d\theta \to ie\int_{\pi}^{0} d\theta = -ie\pi \qquad \text{as}\quad \varepsilon\to 0$$

and the integral about J_2,

$$i\int_0^{\pi} e^{e^{iRe^{i\theta}}}\,d\theta = i\int_0^{\pi} e^{e^{-R\sin\theta}\,e^{iR\cos\theta}}\,d\theta \to i\int_0^{\pi} d\theta = i\pi$$

as $R \to \infty$. Thus, since the residues are zero inside the contour,

$$2iI - ie\pi + i\pi = 0$$

so that $I = (\pi/2)(e - 1)$.

8. Evaluate by the residue theorem:

$$\int_0^{\pi} \tan(\theta + c)\, d\theta \qquad \text{with} \quad \operatorname{Im} c \neq 0$$

Ans. A convenient contour is $|z| = e^{-\operatorname{Im} c}$ on which $z = e^{i(\theta+c)}$. Then

$$\int_0^{2\pi} \tan(\theta + c)\, d\theta = \oint \frac{[z - (1/z)]}{i[z + (1/z)]} \frac{dz}{iz} = \oint \left(\frac{1 - z^2}{1 + z^2}\right) \frac{dz}{z}$$

and the integrand has three simple poles at $z = 0$, $z = \pm i$. The residues are

$$\begin{array}{rll} 1 & \text{at} & z = 0 \\ -1 & \text{at} & z = i \\ -1 & \text{at} & z = -i \end{array}$$

(the last two) which are encircled by the contour only if $\operatorname{Im} c < 0$

so that

$$\oint \left(\frac{1 - z^2}{1 + z^2}\right) \frac{dz}{z} = 2\pi i \begin{cases} 1 & \text{for} \quad \operatorname{Im} c > 0 \\ -1 & \text{for} \quad \operatorname{Im} c < 0 \end{cases}$$

Also,

$$\int_0^{\pi} \tan(\theta + c)\, d\theta = \tfrac{1}{2} \int_0^{2\pi} \tan(\theta + c)\, d\theta$$

since $\tan(\theta + \pi) = \tan\theta$. Hence,

$$\int_0^{\pi} \tan(\theta + c)\, d\theta = \begin{cases} \pi i & \text{for} \quad \operatorname{Im} c > 0 \\ -\pi i & \text{for} \quad \operatorname{Im} c < 0 \end{cases}$$

9. Find the residues of $[z/(z^2 + c^2)]^4$.

Ans. The function has two fourth-order poles at $z = \pm ic$. At $z = ic$ the residue is

$$\frac{1}{3!} d_z^3 \left[\frac{z^4}{(z + ic)^4}\right]\Bigg|_{z=ic} = \frac{-i}{2^5 c^3}$$

while at $z = -ic$ it is $i/2^5 c^3$.

10. Find the residues of $e^{z^2} \csc z$ at singularities in the finite z plane.

Ans. The function has simple poles at $z = n\pi$; hence the residues are $(-1)^n e^{(n\pi)^2}$.

11. If $g(z)$ has only a simple pole of residue c at $z = a$, show that the limit of

$$\int_J g(z)\, dz$$

along the arc J counterclockwise,

$$J: \quad |z - a| = \varepsilon \qquad \alpha \leq \arg(z - a) \leq \beta$$

as $\varepsilon \to 0$, is given by $i(\beta - \alpha)c$.

Ans. Since $g(z)$ has only a simple pole at $z = a$,

$$g(z) - \frac{c}{z - a} = f(z)$$

is free of singularities within a sufficiently small circle about $z = a$. Hence,

$$\left| \int_J f(z)\, dz \right| \leq 2f(a)(\beta - \alpha)\varepsilon$$

which approaches zero as $\varepsilon \to 0$. Hence,

$$\lim_{\varepsilon \to 0} \int_J g(z)\, dz = \lim_{\varepsilon \to 0} \int_J \frac{c\, dz}{z - a} = ic \int_\alpha^\beta d\theta = ic(\beta - \alpha)$$

since $z - a = \varepsilon e^{i\theta}$ on J.

12. Show that if $|c| < \pi$,

$$\int_{-\infty}^{\infty} \frac{\sinh(cx)}{\sinh(\pi x)}\, dx = \tan\left(\frac{c}{2}\right)$$

Ans. The given integral is equivalent to the contour integral

$$I(c) = \oint \frac{\sin(icz)}{\sin(i\pi z)}\, dz$$

taken about the limit of the succession of contours

$$I_2: \quad \varepsilon < z = z^* < R$$

$$I_1: \quad -R < z = z^* < -\varepsilon$$

$$J_2: \quad |z| = R \qquad \text{with} \quad i(z - z^*) < 0$$

$$J_1: \quad |z| = \varepsilon \qquad \text{with} \quad i(z - z^*) < 0$$

as $\varepsilon \to 0$ and $R \to \infty$. On I_1 and I_2 the contour integral becomes

$$\int_{-R}^{-\varepsilon} + \int_\varepsilon^R \frac{\sinh(cx)}{\sinh(\pi x)}\, dx$$

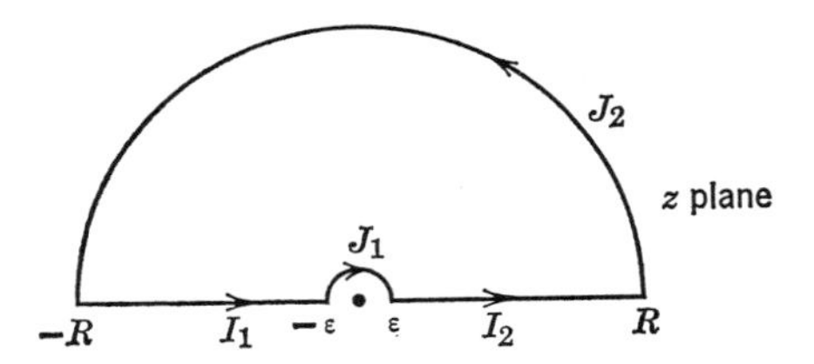

Figure 6.5. Contour for Example 12.

The limit of the sum of these portions of the contour integral is a Cauchy principal value and it can be identified with the original integral to be evaluated. The portion of $I(c)$ associated with J_2 is subject to

$$\left| \int \frac{\sin(icz)}{\sin(i\pi z)} \, dz \right| \le \pi R M(R)$$

where $M(R)$ is larger than $|\sin icz/\sin i\pi z|$ on J_2. On J_2,

$$\left| \frac{\sin icz}{\sin i\pi z} \right| = \left| \frac{e^{-cz} - e^{cz}}{e^{-\pi z} - e^{\pi z}} \right| \le \frac{2e^{|c|R}}{e^{\pi R}} = 2e^{-(\pi - |c|)R} = M(R)$$

for sufficiently large R. Hence, for $\pi > |c|$,

$$\lim_{R\to\infty} \pi R M(R) = \lim_{R\to\infty} 2\pi R e^{-(\pi-|c|)R} = 0$$

Thus the portion of $I(c)$ associated with J_2 goes to zero as $R \to \infty$. On J_1 one has

$$\left| \int \frac{\sin(icz)}{\sin(i\pi z)} \, dz \right| \le \pi \varepsilon M$$

with

$$\left| \frac{\sin icz}{\sin i\pi z} \right| = \left| \frac{\sin icz}{icz} \right| \left| \frac{i\pi z}{\sin icz} \right| \left| \frac{c}{\pi} \right| < \frac{2\,|c|}{\pi}$$

for sufficiently small ε. Hence $M = 2\,|c|/\pi$ here is independent of ε and

$$\lim_{\varepsilon\to 0} (\pi \varepsilon M) = \lim_{\varepsilon\to 0} \left(\pi\varepsilon \frac{2\,|c|}{\pi} \right) = 0$$

Thus the portion of $I(c)$ associated with J_1 goes to zero as $\varepsilon \to 0$. One might have anticipated this, since the integrand clearly has a limit as $z \to 0$. Thus, $(\varepsilon \to 0, R \to \infty)$ $I(c)$ is seen to be equivalent to the original integral, and since $I(c)$ may be evaluated by the residue theorem, one can obtain the result stated. The integrand has simple poles at $z = in$ ($n > 0$) as only singularities inside the contour and the corresponding residues are given by (see Section 6.3)

$$\left.\frac{\sin(icz)}{d_z \sin(i\pi z)}\right|_{z=in} = \frac{\sin(-cn)}{i\pi \cos(-n\pi)} = (-1)^n \frac{i \sin(nc)}{\pi}$$

and the sum of the residues is

$$\frac{i}{\pi} \sum_{n=1}^{\infty} (-1)^n \sin(nc) = \frac{i}{\pi} \operatorname{Im} \sum_{n=1}^{\infty} (-e^{ic})^n$$

which is

$$\frac{-i}{\pi} \operatorname{Im}\left(\frac{e^{ic}}{1 + e^{ic}}\right) = -\frac{i}{\pi} \operatorname{Im}\left(\frac{e^{ic/2}}{e^{ic/2} + e^{-ic/2}}\right) = \frac{\tan(c/2)}{2\pi i}$$

whence the stated result follows.

13. Show that if $|c| < \pi$,

$$\int_{-\infty}^{\infty} \frac{\cosh(cx)}{\cosh(\pi x)}\, dx = \sec\left(\frac{c}{2}\right)$$

Ans. Consider

$$I(c) = \oint \frac{\cosh(cz)}{\cosh(\pi z)}\, dz$$

about the limit of a contour consisting of four portions:

J_1:	$\lvert\operatorname{Re} z\rvert < R$	$\operatorname{Im} z = 0$
J_2:	$\operatorname{Re} z = R$	$\lvert\operatorname{Im} z\rvert \leq 1$
J_3:	$\lvert\operatorname{Re} z\rvert < R$	$\operatorname{Im} z = 1$
J_4:	$\operatorname{Re} z = -R$	$\lvert\operatorname{Im} z\rvert \leq 1$

as $R \to \infty$. The portion of the contour integral along J_1 becomes

$$\int_{-\infty}^{\infty} \frac{\cosh(cx)}{\cosh(\pi x)}\, dx$$

while the portion along J_3 becomes

$$-\int_{-\infty}^{\infty} \frac{\cosh(cx + ic)}{\cosh(\pi x + i\pi)}\, dx = \int_{-\infty}^{\infty} \frac{\cosh(cx)\cos c}{\cosh(\pi x)}\, dx$$

Since $\cosh(cx + ic) = \cosh cx \cos c + i \sinh cx \sin c$, then $\cosh(\pi x + i\pi) = -\cosh(\pi x)$, so that the portions along J_1

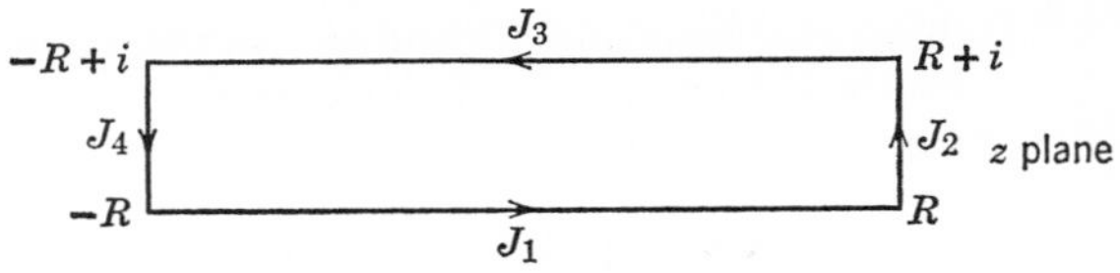

Figure 6.6. Contour for Example 13.

and J_3 together become

$$(1 + \cos c)\int_{-\infty}^{\infty} \frac{\cosh(cx)}{\cosh(\pi x)}\, dx$$

The portion along J_2 ($z = R + iy$) becomes

$$\left| i\int_0^1 \frac{\cosh(cR + icy)}{\cosh(\pi R + i\pi y)}\, dy \right| \leq 2e^{-(\pi-|c|)R}$$

since

$$\left| \frac{\cosh(cR + icy)}{\cosh(\pi R + i\pi y)} \right| \leq \frac{2e^{|c|R}}{e^{\pi R}}$$

for sufficiently large R. Thus J_2 and, similarly, J_4 yield zero integrals in the limit as $R \to \infty$. Finally, $I(c)$ may be evaluated by finding the residue of the simple pole of the integrand at $z = i/2$. This is

$$\left.\frac{\cosh(cz)}{d_z \cosh(\pi z)}\right|_{z=i/2} = \frac{\cosh(ic/2)}{\pi \sinh(i\pi/2)} = \frac{\cos(c/2)}{i\pi}$$

Hence, by the residue theorem,

$$I(c) = (1 + \cos c)\int_{-\infty}^{\infty} \frac{\cosh(cx)}{\cosh(\pi x)}\, dx = 2\cos\left(\frac{c}{2}\right)$$

so that the result follows. Naturally it could also have been obtained with a contour and technique similar to that of Example 12.

14. Show that for $a \geq b > 0$

$$\int_{-\infty}^{\infty} \frac{\sin[a(x - q)]\sin[b(x - p)]}{(x - p)(x - q)}\, dx = \frac{\pi \sin[b(p - q)]}{(p - q)}$$

Ans. Consider (supposing $p < q$ and $k > 0$)

$$\oint \frac{e^{ikz}\, dz}{(z - p)(z - q)}$$

about the limit of the contour consisting of six portions:

$$I_1: \quad -\infty < \operatorname{Re} z < p - \varepsilon \qquad \operatorname{Im} z = 0$$

$$J_1: \quad |z - p| = \varepsilon \qquad \operatorname{Im} z > 0$$

$$I_2: \quad p + \varepsilon < \operatorname{Re} z < q - \varepsilon \qquad \operatorname{Im} z = 0$$

$$J_2: \quad |z - q| = \varepsilon \qquad \operatorname{Im} z > 0$$

$$I_3: \quad q + \varepsilon < \operatorname{Re} z < \infty \qquad \operatorname{Im} z = 0$$

$$J_3: \quad |z| = R \qquad \operatorname{Im} z > 0$$

as $R \to \infty$ and $\varepsilon \to 0$. Along J_3 one has

$$\left| \int \frac{e^{ikz}\, dz}{(z - p)(z - q)} \right| \leq \pi R \frac{e^{-kR \sin\theta}}{(R - |p|)(R - |q|)}$$

since $z = Re^{i\theta} = R(\cos\theta + i\sin\theta)$ on J_3. Also, since [the chords between $(0,0)(\pi/2,1)$ and $(\pi/2,1)(\pi,0)$ lie under the sine curve]

$$\sin\theta \geq \frac{2}{\pi}\theta \qquad \text{for} \quad 0 \leq \theta \leq \frac{\pi}{2}$$

and

$$\sin\theta \geq \frac{2}{\pi}(\pi - \theta) \qquad \text{for} \quad \frac{\pi}{2} \leq \theta \leq \pi$$

$$e^{-kR\sin\theta} \leq e^{-(2/\pi)kR\theta} \qquad \text{for} \quad 0 \leq \theta \leq \frac{\pi}{2}$$

and

$$e^{-kR\sin\theta} \leq e^{-(2/\pi)kR(\pi-\theta)} \qquad \text{for} \quad \frac{\pi}{2} \leq \theta \leq \pi$$

so that the exponential approaches zero ($k > 0$) as $R \to \infty$ for $0 < \theta < \pi$ and is bounded for $\theta(\pi - \theta) = 0$ (at the end points). Hence, clearly the integral along J_3 approaches zero as $R \to \infty$,

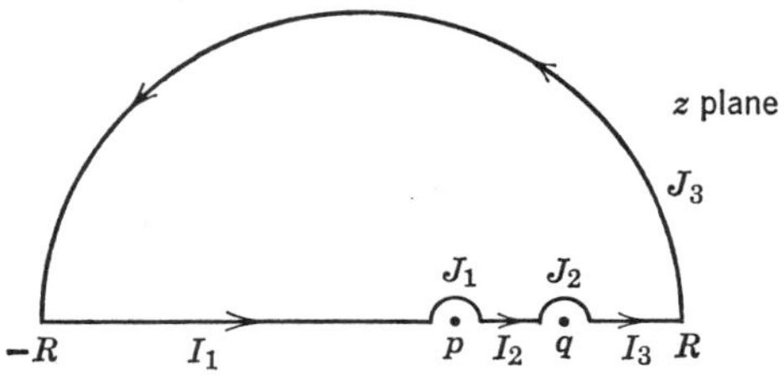

Figure 6.7. Contour for Example 14.

the exponential being bounded independently of θ. The argument here for J_3 is a special case of *Jordan's lemma*, which states that if $g(z) \to 0$ independently of $\theta = \arg z$ for $0 \leq \theta \leq \pi$ as $|z| \to \infty$, then

$$\lim_{|z| \to \infty} \int_{J_3} e^{ikz} g(z)\, dz = 0$$

The integrals along J_1 and J_2 become, with the method of Example 11,

$$\frac{-i\pi e^{ikp}}{p - q} \quad \text{and} \quad \frac{-i\pi e^{ikq}}{q - p}$$

while the sum of integrals along I_1, I_2, and I_3 becomes a Cauchy principal value, and since the integrand of the contour integral has no singularities *inside* the contour, one has

$$\int_{-\infty}^{\infty} \frac{e^{ikx}\, dx}{(x - p)(x - q)} - \frac{i\pi(e^{ikp} - e^{ikq})}{p - q} = 0$$

or

$$\int_{-\infty}^{\infty} \frac{e^{ikx}\, dx}{(x - p)(x - q)} = \frac{i\pi(e^{ikp} - e^{ikq})}{p - q}$$

If $k < 0$ rather than $k > 0$ as above, the mirror image (across the real axis) of the above contour would be used to ensure the integral along the large arc approaching zero. In this case $(k < 0)$,

$$\int_{-\infty}^{\infty} \frac{e^{ikx}\, dx}{(x - p)(x - q)} = -\frac{i\pi(e^{ikp} - e^{ikp})}{p - q}$$

since the images of J_1 and J_2 would now be traversed counterclockwise rather than clockwise as above. Now,

$$\begin{aligned} 4 \sin[a(x - q)]\sin[b(x - p)] & \\ &= [e^{ia(x-q)} - e^{-ia(x-q)}][e^{-ib(x-p)} - e^{ib(x-p)}] \\ &= e^{i(a-b)x} e^{i(bp-aq)} + e^{i(b-a)x} e^{i(aq-bp)} \\ &\quad - e^{-i(a+b)x} e^{i(aq+bp)} - e^{i(a+b)x} e^{-i(aq+bp)} \end{aligned}$$

whence by choosing k successively equal to $(a - b)$, $-(a - b)$, $-(a + b)$, $(a + b)$ and using the Cauchy principal values found above, the result follows.

15. Show that the one-dimensional (in space) wave equation

$$\partial_x^2 \phi = \partial_{ct}^2 \phi$$

reduces to the Helmholtz equation if $\phi(x,t) = e^{-i\omega t}\psi(x)$ (i.e., the wave is monochromatic of angular frequency ω).

Ans. Substitution yields the (one-dimensional) Helmholtz equation

$$d_x^2 \psi + k^2\psi \equiv 0$$

with $kc = \omega$ and general solution

$$\psi(x) = Ae^{ikx} + Be^{-ikx} \qquad k > 0$$

with the first term representing a wave *traveling to the right* and the second term representing a wave *traveling to the left* as is seen by reintroducing the factor $e^{-i\omega t}$ ($\omega t - kx =$ constant implies x increases as t does, while $\omega t + kx =$ constant implies x decreases as t increases).

16. If L represents a path of integration between $-\infty$ and $+\infty$ along the real axis in the complex w plane except for (vanishingly small) semicircular excursions avoiding the points $w = \pm k$ ($k > 0$), then find the values of

$$I(x) = \frac{1}{2\pi i}\int_L \frac{we^{iwx}\,dw}{w^2 - k^2}$$

for the four possible path choices.

Ans. The four possible paths are shown in Figure 6.8. If the corresponding four results for $I(x)$ are labeled $I_1(x)$, $I_2(x)$, $I_3(x)$, $I_4(x)$, one finds upon completing the path of integration by a large (infinitely) semicircular arc in the upper half w plane that the integration results are given in each case by the residues of singularities (poles) enclosed, since the integrals along the large semicircular arc will approach zero as its radius becomes infinite (see Jordan's lemma in Example 14). Thus for $x > 0$ (indicated by a $^+$)

$$I_1^+(x) = 0$$
$$I_2^+(x) = \cos kx$$
$$2I_3^+(x) = e^{ikx}$$
$$2I_4^+(x) = e^{-ikx}$$

since the residue at $x = k$ is $e^{ikx}/2$ and the residue at $x = -k$ is $e^{-ikx}/2$. For $x < 0$ it is necessary to complete the contour by a large semicircular arc in the lower half w plane, so that the integral along the latter (semicircular arc) will approach zero as its radius grows infinite. For $x < 0$ and Im $w < 0$, the exponential $e^{iwx} \to 0$ as $|w| \to \infty$. Hence, via Jordan's lemma for $x < 0$

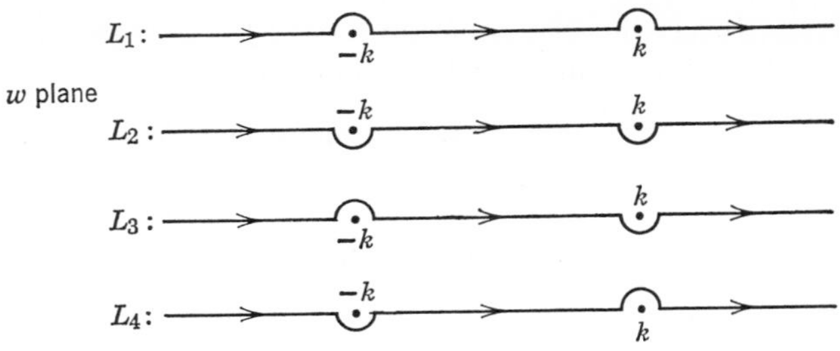

Figure 6.8. Four integration path choices for Example 16.

(indicated by a $^-$),

$$I_1^-(x) = -\cos kx$$
$$I_2^-(x) = 0$$
$$2I_3^-(x) = -e^{-ikx}$$
$$2I_4^-(x) = -e^{ikx}$$

where the contour is traversed *clockwise* rather than counter-clockwise as before (when x was positive). The results may be summarized together for x positive or negative by

$$I_1(x) = -u(-x)\cos kx$$
$$I_2(x) = u(x)\cos kx$$
$$2I_3(x) = [u(x) - u(-x)]e^{ik|x|}$$
$$2I_4(x) = [u(x) - u(-x)]e^{-ik|x|}$$

where

$$u(x) = \begin{cases} 1 & \text{for} \quad x \geq 0 \\ 0 & \text{for} \quad x < 0 \end{cases}$$

Each of these integrals is a solution of the Helmholtz equation

$$d_x^2 I_n + k^2 I_n = 0$$

for $x > 0$ and also for $x < 0$.

17. Show that for an appropriate choice for L (the integration path of Example 16) the solution (for $x > 0 < k$)

$$J(x) = \int_L \frac{w \sin wx \, dw}{\pi(w^2 - k^2)}$$

of the Helmholtz equation represents a wave traveling to the *right*.

Ans. Here one has simply

$$J(x) = \frac{1}{2\pi i}\int_L \frac{we^{iwx}\,dw}{w^2 - k^2} - \frac{1}{2\pi i}\int_L \frac{we^{-iwx}\,dw}{w^2 - k^2}$$

so

$$J(x) = I(x) - I(-x)$$

and thus, along the four choices of path that one has from Example 16,

$$J_1^+(x) = 0 - (-\cos kx) = \cos kx$$

$$J_2^+(x) = \cos kx - (0) = \cos kx$$

$$J_3^+(x) = \frac{e^{ikx}}{2} - \left(-\frac{e^{ikx}}{2}\right) = e^{ikx}$$

$$J_4^+(x) = \frac{e^{-ikx}}{2} - \left(-\frac{e^{-ikx}}{2}\right) = e^{-ikx}$$

where the second terms in each case can be calculated by applying the residue theorem to the second integral $I(-x)$ or by using the results for $I_n^-(x)$. Clearly *only* $J_3^+(x)$ represents a *wave traveling to the right* provided the time-dependence in the wave equation from which the Helmholtz equation was derived were of the form $e^{-i\omega t}$.

18. From the "spherical" wave equation

$$\frac{1}{r^2}\,\partial_r(r^2\partial_r\phi) = \partial_{ct}^2\phi$$

which is equivalent to

$$\partial_r^2(r\phi) = \partial_{ct}^2(r\phi)$$

derive the "spherical" Helmholtz equation for monochromatic waves.

Ans. Take $\phi(r,t) = e^{-i\omega t}\psi(r)$. Then

$$d_r^2(r\psi) + k^2(r\psi) = 0$$

with $kc = \omega$. The general solution of the latter is given by

$$r\psi(r) = Ae^{ikr} + Be^{-ikr}$$

with the first term representing an "explosion" wave and the second term representing an "implosion" wave; the explosion wave propagates *away* from the origin, while the implosion wave propagates *toward* the origin. In the case of an explosion wave alone ($B = 0$) it is readily seen that

$$d_r(r\psi) = ik(r\psi)$$

so that

$$r(d_r\psi - ik\psi) = \psi \to 0$$

as $r \to \infty$. The corresponding condition for ϕ,

$$r(\partial_r\phi - ik\phi) \to 0$$

as $r \to \infty$, is called an *explosion* or *radiation source* condition, while the condition (for $A = 0$)

$$r(\partial_r\phi + ik\phi) \to 0$$

as $r \to \infty$, is called an *implosion* or *radiation sink* condition. These are useful in the uniqueness theorems of wave propagation theory.

19. Show that

$$\frac{1}{\pi r}\int_L \frac{w\sin(wr)\,dw}{w^2 - k^2} = \psi(r)$$

represents a spherical implosion wave along an appropriate one of the four paths L_1, L_2, L_3, L_4 of Example 16.

Ans. Clearly L_4.

● PROBLEMS 6

1. Find the Laurent expansions of $1/(z + z^3)$ about $z = 0$
2. Find the Laurent expansions of $(z^2 - 1)/(z^3 + 3z^2 + 3z + 1)$ about $z = 1$ and about $z = 0$.
3. Find the Laurent series for $[(z + 1)/(z - 1)]^2$ valid for $|z - 1| > 0$.
4. Find the Laurent series for $1/(e^{2z} - e^z)$ convergent in the region $0 < |z| < 2\pi$.
5. Find the Laurent series for $1/(z^2 - 1)$ about $z = i$.
6. Find the Laurent expansions for $\sin(1/z)$ about $z = 0$ and about $z = 3$.
7. Find the Laurent series of $1/(1 + z^3)$ about $z = 0$ and about $z = -\frac{1}{2}$.
8. Find an integral expression for the Taylor coefficients about $z = 0$ for the function

$$\phi(s,z) = \frac{1}{\sqrt{1 - 2sz + z^2}}$$

treating s as a fixed complex number. These coefficients $P_n(s)$ are called Legendre polynomials (of degree n).

9. By showing that the generating function $\phi(s,z)$ for the Legendre polynomials satisfies the partial differential equation

$$\partial_z \phi = (s - z)\phi^3$$

establish the recurrence relation

$$nP_n(s) - (2n - 1)sP_{n-1}(s) + (n - 1)P_{n-2}(s) = 0$$

10. By showing that the generating function $\phi(s,z)$ of Problems 8 and 9 satisfies

$$z\partial_z \phi = (s - z)\partial_s \phi$$

establish that

$$sd_s P_n(s) - d_s P_{n-1}(s) = nP_n(s)$$

11. From the results of Problems 9 and 10 establish that

$$\begin{aligned} d_s P_n(s) - sd_s P_{n-1}(s) &= nP_{n-1}(s) \\ d_s P_{n+1}(s) - d_s P_{n-1} &= (2n + 1)P_n(s) \\ (s^2 - 1)\, d_s P_n(s) &= nsP_n(s) - nP_{n-1}(s) \\ (s^2 - 1)\, d_s P_n(s) &= (n + 1)P_{n+1}(s) - (n + 1)sP_n(s) \end{aligned}$$

12. Show that the $P_n(s)$ satisfy Legendre's differential equation

$$(1 - s^2)d_s^2 P_n - 2sd_s P_n + n(n + 1)P_n = 0$$

13. Find an integral expression for the Taylor coefficients about $z = 0$ for the function (Hermite generating function)

$$e^{2zs - z^2}$$

treating s as a fixed complex number. These coefficients are Hermite polynomials $\mathscr{H}_n(s)$ divided by $n!$

14. Establish the recurrence relations

$$d_s \mathscr{H}_n(s) = 2n\mathscr{H}_{n-1}(s)$$

$$\mathscr{H}_{n+1}(s) - 2s\mathscr{H}_n(s) + 2n\mathscr{H}_{n-1}(s) = 0$$

15. Show that the Hermite polynomials satisfy Hermite's differential equation

$$d_s^2 \mathscr{H}_n - 2sd_s \mathscr{H}_n + 2n\mathscr{H}_n = 0$$

16. Find an integral expression for the Taylor coefficients about $z = 0$ for (Laguerre generating function)

$$e^{-sz/(1-z)}$$

treating s as a fixed complex number. These coefficients are the Laguerre polynomials $L_n(s)$.

17. Establish the recurrence relation

$$nL_n(s) = (2n - 1 - s)L_{n-1}(s) - (n - 1)L_{n-2}(s)$$

18. Show that the $L_n(s)$ satisfy the Laguerre differential equation

$$sd_s^2L_n + (1 - s)d_sL_n + nL_n = 0$$

19. Evaluate $\oint dz/(1 + z^2)^{n+1}$ about the limiting contour of $\text{Im}\, z = 0$, $|\text{Re}\, z| \le R$ and $|z| = R$, $|\arg z - (\pi/2)| \le \pi/2$ as $R \to \infty$.
20. By integrating $\log \sin z$ around the rectangle with vertices at $0, \pi, \pi + iY, iY$, indented at 0 and π, prove that

$$\int_0^\pi \log \sin x \, dx = -\pi \log 2$$

21. Find the residue of $\sec z$ at $z = [n + (\frac{1}{2})]\pi$.
22. Find the residue of $\cot \pi z - (1/\pi z)$ at $z = n$.
23. Find the residue of $\sin z/z(z - 3)$ at $z = 0$ and $z = 3$.
24. Find the residue of $\csc^4 z$ at $z = 0$.
25. Find the residue of $1/z^3\sqrt{1 - z}$ at $z = 0$.
26. What is wrong with asking for the residue of $1/z\sqrt{z - z^2}$ at $z = 0$? (See Chapter 7.)
27. An analytic function takes the same values at symmetrically located points on opposite sides of a parallelogram. Find the sum of its residues inside the parallelogram.
28. Find the residues of $e^{iz}/(z^2 + 9)$ at $z = \pm 3i$.
29. Find the residues of $z^4/(z^2 + 1)^2(z^2 + 4)$ at all poles.
30. Evaluate $\int_0^{2\pi} d\theta/(4 + 3\cos\theta)$ by contour integration (residue theorem) about the unit circle $z = e^{i\theta}$.
31. Evaluate $\int_{-\infty}^{\infty} x^4\, dx/(x^2 + 3)^2(x^2 + 5)$ by contour integration (residue theorem) about an infinite semicircle with the real axis as diameter.
32. Does the arc have to be circular in Example 11?
33. Evaluate $\int_{-\infty}^{\infty} \cos 2x\, dx/(x^2 + 4)$ by integrating $\int e^{2iz}\, dz/(z^2 + 4)$ about an infinite semicircular contour with real axis as diameter. Does it matter whether the semicircle is in the upper half-plane ($\text{Im}\, z \ge 0$) or in the lower half-plane?
34. Evaluate $\int_{-\infty}^{\infty} e^{sx}\, dx/(3 + e^x)$ $(0 < S < 1)$ by integrating the corresponding complex integrand about an infinite rectangular contour symmetrical to the imaginary axis with the real axis as a side.
35. Show that $(a > 0 < b)$

$$\int_{-\infty}^{\infty} \frac{\cos bx - \cos ax}{x^2}\, dx = \pi(a - b)$$

by integrating

$$\oint \frac{e^{ibz} - e^{iaz}}{z^2}\, dz$$

around an infinite semicircle indented to avoid $z = 0$ but otherwise containing the real axis.

36. Show that

$$\int_{-\infty}^{\infty} \frac{x^2\, dx}{(x^2 + c^2)^3} = \frac{\pi}{8c^3} \qquad \text{Re}\, c > 0$$

37. Show that

$$\int_{-\infty}^{\infty} \frac{\sin[p(x-a)]\sin[q(x-b)]\, dx}{\pi(x-a)(x-b)} = \frac{\sin[p(a-b)]}{a-b}$$

for $q \geq p > 0$ by integrating $\int e^{i\mu z}\, dz/(z-a)(z-b)$ around an infinite semicircular contour indented to avoid $z = a$ and $z = b$.

38. Show that $(0 < m < \pi)$

$$2\int_0^{\infty} e^{imx^2} \frac{\cosh(mx)}{\cosh(\pi x)}\, dx = e^{im/4}$$

by integrating $e^{imz^2} \cosh \pi z$ about an infinite rectangle centered at $z = 0$.

39. Show that $(a > 0;\ k = k^*)$

$$\frac{1}{2\pi i}\int_{a-i\infty}^{a+i\infty} \frac{e^{kz}}{z^2}\, dz = \begin{cases} k & \text{for} \quad k > 0 \\ 0 & \text{for} \quad k \leq 0 \end{cases}$$

40. Show that for $(\text{Re}\, k > 0;\ |\text{Im}\, k| < \pi;$ and $0 < a < 1)$

$$\int_{a-i\infty}^{a+i\infty} e^{kz} \csc \pi z\, dz = \frac{2i}{1 + e^{-k}}$$

by integrating about an infinite rectangle of unit width with $\text{Re}\, z = a$ and $\text{Re}\, z = a + 1$ as sides.

41. Show that

$$\int_0^{\infty} \frac{\sin ky\, dy}{e^{2y} - 1} = \frac{\pi}{4}\coth\left(\frac{\pi k}{2}\right) - \frac{1}{2k}$$

by integrating $e^{kz}/(e^{-2iz} - 1)$ about an infinite rectangle of width π indented to avoid $z = 0$ and $z = \pi$ and having the positive imaginary axis as a side.

42. Show that $\int_{-\infty}^{\infty} \sinh kx\, dx/\sinh \pi x = \tan(k/2)$ for $|k| < \pi$ by integrating the corresponding complex integrand about an infinite semicircular contour indented to avoid $z = 0$ and with the real axis as diameter.

43. Show that

$$\int_{-\infty}^{\infty} \frac{\cos sx\, dx}{1 + x^4} = \frac{\pi}{\sqrt{2}} e^{-|s|\sqrt{2}} \left(\cos \left| \frac{s}{\sqrt{2}} \right| + \sin \left| \frac{s}{\sqrt{2}} \right| \right)$$

44. Show that

$$\int_{-\infty}^{\infty} \frac{dx}{x^6 + 1} = \frac{2\pi}{3}$$

45. Show that

$$\int_{-\infty}^{\infty} \frac{x^2\, dx}{x^4 + 1} = \frac{\pi}{\sqrt{2}}$$

CHAPTER 7:

Singularities and Analytical Continuation

.1 BRANCH POINTS AND RIEMANN SURFACES

We have already defined multiple (mth-order) and simple (first-order) poles in Chapter 6. If the Laurent expansion of $f(z)$ about $z = a$ contains an infinite number of negative powers, $f(z)$ is said to have an *essential singularity* at $z = a$. In this case there does *not* exist a positive integer k such that $(z - a)^k f(z)$ remains finite as z approaches a.

The poles and essential singularities are associated with infinities of the function at certain points, but there is another kind of singularity in the neighborhood of which the function may not be *uniquely* defined. This is called a *branch point* and in its neighborhood such a function is multivalued rather than single-valued. The function need not become infinite at such a branch point. Thus the function $w = \sqrt{z}$ may also be written

$$w = \rho e^{i\phi} = (re^{i\theta})^{1/2} = \sqrt{r}e^{i\theta/2} \tag{7.1}$$

so that on a small circle of radius ε about $z = 0$ the value of w changes through the values shown in Table 7.1 as θ increases from $\theta = 0$ to $\theta = 2\pi$. The important fact in Table 7.1 is that the initial ($\theta = 0$) and final ($\theta = 2\pi$) values of w are *not* identical, and this would be the result even if any other closed curve about $z = 0$ rather than the circle $|z| = \varepsilon$ were traversed. The function w is discontinuous at $z = \varepsilon$, and since ε may have any positive value, it is also discontinuous along the positive real axis. Thus if we consider a

Table 7.1

θ	0	$\pi/2$	π	$3\pi/2$	2π
z/ε	1	i	-1	$-i$	1
$w/\sqrt{\varepsilon}$	1	$\dfrac{1+i}{\sqrt{2}}$	i	$\dfrac{-1+i}{\sqrt{2}}$	-1

small angle δ, the values of w at $z = \varepsilon e^{i\delta}$ and $z = \varepsilon e^{i(2\pi-\delta)}$ yield a difference of

$$\sqrt{\varepsilon}\{e^{i\delta/2} - e^{i[\pi-(\delta/2)]}\} = 2\sqrt{\varepsilon}\cos\left(\frac{\delta}{2}\right) \tag{7.2}$$

which approaches $2\sqrt{\varepsilon} > 0$ as $\delta \to 0$.

On the other hand, if we were to perform another circuit of the circle $|z| = \varepsilon$, the sequence of values would be those shown in Table 7.2, so that the function w returns to its initial ($\theta = 0$) value at $\theta = 4\pi$. The above conclusion that w was discontinuous along the positive real axis was predicated on the assumption that the r, θ values selected points in a plane. If we modify this assumption to one in which we suppose that the range $2\pi \leq \theta < 4\pi$ corresponds to a distinct superposed sheet of values with the two sheets joined at $\theta = 0$ and $\theta = 4\pi$, then w *will be continuous* in this double-sheeted surface, which is called a *Riemann surface.* On this surface the points

$$z = \varepsilon e^{i\delta} \qquad \text{and} \qquad z = \varepsilon e^{i(2\pi-\delta)}$$

are *not* adjacent, so that the demonstration of (7.2) no longer applies. By extending the notion of the surface on which values of z are defined from the complex (z) plane to the double-sheeted Riemann surface, we have arrived at a way of defining unique values for w so that no discontinuity in (w) values occurs for z on this Riemann surface. If we wish to restrict the z values to a plane, there will be an ambiguity in the values of w. An alternative way of avoiding this ambiguity is to *forbid transits* of the positive real axis

Table 7.2

θ	2π	$5\pi/2$	3π	$7\pi/2$	4π
z/ε	1	i	-1	$-i$	1
$w/\sqrt{\varepsilon}$	-1	$\dfrac{-1-i}{\sqrt{2}}$	$-i$	$\dfrac{1-i}{\sqrt{2}}$	2

so that the z plane is replaced by a *cut* z plane separated by a slit along the positive real axis. With such a procedure the points $z = \varepsilon e^{i\delta}$ and $z = \varepsilon e^{i(2\pi-\delta)}$ are again *not* adjacent, and the discontinuity conclusion (7.2) is thus *invalid*.

We have singled out the positive real axis as the locus of discontinuities or for a cut, but if we had measured our angular location θ from any other half-line from the origin to infinity, the result would have been the same. Furthermore, the restriction to a straight line is unnecessary, since the property is a discontinuity on a circuit of the origin. Thus we can take the cut to be any non-self-intersecting curved line joining the origin to infinity. However, we would often find *straight* cuts to be of convenience.

The case of the function $\sqrt{z}$ is not an isolated example. Branch points are encountered wherever roots occur and also with logarithmic functions such as $w = \log z$ for which, as before, w does not return to the same value as z traverses the circle $|z| = \varepsilon$. Here,

$$w = \log \varepsilon + i\theta \tag{7.3}$$

so that not only does w *not* return to the value $\log \varepsilon$ as θ increases from 0 to 2π, but it also does *not* return to that value after any integer number of circuits of $|z| = \varepsilon$. In this case the Riemann surface (for $w = \log z$) must contain an *infinite* number of superposed distinct sheets. However, a cut from $z = 0$ to $z = \infty$ will suffice to make $\log z$ single-valued in the cut z plane. The point $z = 0$ is called a logarithmic branch point of $\log z$.

Since functions with branch points are not uniquely defined in annuli about them, Laurent's theorem is inapplicable and they do *not* possess Laurent series about these points. Thus $\sqrt{z}$, $\log z$, $1/\sqrt{z}$ do not have such expansions about $z = 0$, although they may have such expansions about other points for which $z \neq 0$.

Finally we should define what is meant by a singularity of $f(z)$ at $z = \infty$. By this is meant a singularity of $f(1/z)$ at $z = 0$. Thus $f(z) = z^3$ has a third-order *pole at infinity*, $f(z) = 1/\sqrt{z}$ has a *branch point at infinity*, and $f(z) = \log z$ has a logarithmic branch point at $z = \infty$. Similarly, $f(z) = 1/z^5$ is said to have a fifth-order *zero at infinity*.

An analytic function totally free of singularities in the finite z plane is called an *entire function* (or an *integral function*), and according to Liouville's theorem it must have a singularity at $z = \infty$ unless it is a constant everywhere. Entire functions have *infinite* radii of convergence, since the circle of convergence of their power series must pass through the nearest singularity ($z = \infty$). The ratio of two entire functions is not necessarily entire, since the denominator may have zeros. The ratio of two entire functions is called a *meromorphic function* which may thus have poles but no essential singularities in the finite plane.

7.2 ANALYTICAL CONTINUATION BY POWER SERIES

In Example 1 of Chapter 6 it has been seen that a function such as

$$f(z) = \frac{1}{1-z} + \frac{2}{2-z} \tag{7.4}$$

possesses three Laurent series expansions about $z = 0$:

$$S_1(z) = \sum_{n=0}^{\infty} \left(1 + \frac{1}{2^n}\right) z^n \qquad \text{for} \quad |z| < 1$$

$$S_2(z) = 1 + \sum_{n=1}^{\infty} \left[\left(\frac{z}{2}\right)^n - \frac{1}{z^n}\right] \qquad \text{for} \quad 1 < |z| < 2$$

$$S_3(z) = -\sum_{n-1}^{\infty} \left(\frac{1 + 2^n}{z^n}\right) \qquad \text{for} \quad 2 < |z|$$

The value of $f(z)$ could be calculated from an appropriate one of these series provided z were in the corresponding annulus of convergence and would yield the same value as (7.4). Of course it would be ridiculous to calculate such a value from an infinite series when the simpler expression (7.4) is available. However, it often happens that one does not possess a simple general expression like (7.4) (valid at all nonsingular points) but only has a series (or integral or product) valid in a limited subregion (such as one of the annuli above). In this case it is desirable to obtain an analytic expression valid in a more extended region so that values may be calculated and behavior of the function deduced outside the original region even in ignorance of a "universal" expression such as (7.4). If such an analytical expression valid for the calculation of the values of the *same* function in an *extended* region can be obtained, it is called an *analytical continuation* of the function. The process whereby it is obtained is also called analytical continuation. The definition of the function from the original expression is said to have been continued by the new expression into the extended region.

There are several methods of analytical continuation of which the simplest (and most tedious) is by power (Taylor) series. This is done as follows: With every (nonsingular) point of the original region there will be associated a circle of convergence of the Taylor series about this point. This circle will pass through the nearest singularity of the function. By selecting a new point of expansion in this circle another Taylor series with a new circle of convergence can be constructed. Continuing in this way there are two possibilities.

If the union of sets of all such circles does not extend beyond the boundary of the original region, this is called a *natural boundary* and analytical continuation across it is impossible. More usually it will be possible to construct an extension of the original region by these circles. In fact whenever the original boundary has an interval *free* of singularities, it will be possible to continue across this interval. Only if the singularities are (everywhere dense) so distributed that there does not exist an interval of the boundary free of them, will a natural boundary occur. Functions such as (7.4) do not have a natural boundary, and one can always continue around their (isolated) singularities. At this stage it might be suspected that the functions with natural boundaries form a null set. To allay this suspicion an example of one is

$$f(z) = \sum_{n=0}^{\infty} z^{2^n} + 1 \tag{7.5}$$

This is convergent for $|z| < 1$. Furthermore, since

$$f(z^2) = f(z) - z$$
$$f(z^4) = f(z) - z - z^2$$

then

$$f(z^{2^n}) = f(z) - \sum_{k=0}^{n-1} z^{2^k}$$

This shows that if $z = e^{i\theta}$ is a singularity of $f(z)$ (e.g., $\theta = 0$), the roots of $z^{2^n} = e^{i\theta}$ are also singularities for any integer n. Hence, the existence of a single singularity on $|z| = 1$ (the circle of convergence) implies an infinite number of singularities in every θ interval. That there actually is at least one singularity follows from the fact that $|z| = 1$ and not some larger circle is the circle of convergence. Hence, $|z| = 1$ is seen to be a natural boundary of (7.5), which therefore cannot be analytically continued from $|z| < 1$ into $|z| > 1$.

If there is no natural boundary, the process of analytical continuation may be used to extend the original definition of the function to all regions where it is analytic provided only that analytic expressions may be found valid in the various regions. Naturally, singularities of the function can never be interior to any of the circles of convergence used in the process. The process of analytical continuation is unique and always leads to the same value for a function at a point, since the power series representation of the function about this point is unique. (See Problem 20, Chapter 3.)

7.3 UNIFORM CONVERGENCE OF INTEGRALS

Although the analytical continuation by power (Taylor) series is conceptually simple, it is often cumbersome and tedious in practice. It is therefore common to effect the continuation by an integral expression. In order to use these, one must have means of concluding that such integral expressions actually represent analytic functions.

As was seen by (4.9), the sum of a uniformly convergent series of analytic functions is analytic. There is an analogous result for integrals. If $G(z,t)$ is a continuous function of the real variable t and an analytic function of the complex variable z free of singularities within a closed contour C in the z plane, then

$$F(z) = \int_a^b G(z,t)\, dt \tag{7.6}$$

is also analytic and free of singularities inside the contour C.

This is proved by observing that since $G(z,t)$ is continuous in t:

$$F_n(z) = \sum_{k=1}^{n} G(z,t_k)(t_{k+1} - t_k) \qquad t_1 = a;\ t_n = b$$

approaches $F(z)$ as $n \to \infty$. If it can be shown that the approach to the limit is uniform (i.e., uniform convergence), then it will follow that $F(z)$ is analytic:

$$F(z) - F_n(z) = \sum_{k=1}^{n} \int_{t_k}^{t_{k+1}} [G(z,t) - G(z,t_k)]\, dt$$

so that

$$|F(z) - F_n(z)| \le \sum_{k=1}^{n} \int_{t_k}^{t_{k+1}} |G(z,t) - G(z,t_k)|\, dt$$

and since $G(z,t)$ is continuous, t_{k+1} and t_k can be chosen so close together that $|G(z,t) - G(z,t_k)| < \varepsilon$ for $t_k \le t \le t_{k+1}$. Then

$$|F(z) - F_n(z)| \le \varepsilon \sum_{k=1}^{n} \int_{t_k}^{t_{k+1}} dt = \varepsilon \int_a^b dt = \varepsilon(b - a)$$

for sufficiently large n independently of z. Hence (7.6) follows.

The limit of integration b in the preceding result may be allowed to approach infinity without changing the conclusion provided the integral converges uniformly in any closed region inside C:

$$\int_a^{\infty} G(z,t)\, dt$$

is said to converge uniformly if

$$\left| \int_T^{\infty} G(z,t)\, dt \right| < \varepsilon = \varepsilon(T)$$

can be made arbitrarily small independently of z by choosing T sufficiently large.

There is a Weierstrass M test for integrals analogous to that for series (3.21). Here, if $G(z,t)$ is continuous in $t \geq a$ when z lies in a closed finite domain throughout which

$$|G(z,t)| \leq M(t) \tag{7.7}$$

with $\int_a^\infty M(t)\,dt$ convergent, then $\int_a^\infty G(z,t)\,dt$ is uniformly and absolutely convergent in the domain. This follows immediately by applying (7.7) to the preceding definition.

It is also sometimes useful to have sufficient conditions for uniform convergence of integrals which do not necessarily converge absolutely. These are provided by the Dirichlet test and the Abel test. The former states that if $\phi(z,t)$ and $\partial_t\psi(z,t)$ are continuous in $t \geq a$ and if

(1) $\left|\int_a^T \phi(z,t)\,dt\right| \leq K$ with K independent of z and T

(2) $\int_a^\infty \partial_t\psi(z,t)\,dt$ uniformly and absolutely convergent

(3) $\psi(z,t) \to 0$ uniformly as $t \to \infty$

then

$$\int_a^\infty \phi(z,t)\psi(z,t)\,dt$$

is uniformly convergent in the closed finite domain of the z plane, where the conditions hold.

The proof is straightforward. If

$$\Phi(z,t) = \int_a^t \phi(z,t)\,dt$$

then by partial integration (by parts) using (3), one has

$$\int_T^\infty \phi\psi\,dt = -\Phi(z,T)\psi(z,T) - \int_T^\infty \Phi\partial_t\psi\,dt$$

and since $|\Phi(z,T)| \leq K$, one can choose T so large that [by (1), (2), (3)]

$$\left|\int_T^\infty \phi\psi\,dt\right| \leq K(\varepsilon_1 + \varepsilon_2)$$

becomes arbitrarily small independently of z, which proves the result.

The Abel test arrives at the same conclusion, but replaces (1) and (3) by

(1′) $\int_a^\infty \phi(z,t)\,dt$ uniformly convergent

(3′) $|\psi(z,t)| < K$

It should be understood that the application of these criteria for uniform convergence to analytical continuation are for the purpose of establishing that certain integral expressions are in fact analytic functions of z. Once two distinct analytical expressions have been shown to coincide at an infinite set of points in a finite region, they must uniquely represent the *same* function wherever they converge.

● EXAMPLES 7

1. Show that the function $z^{2\alpha} = e^{2\alpha \log z}$ takes an infinite number of values for each value of $z \neq 0$ in the (uncut) z plane, provided α is not a rational number and that the Riemann surface for $f(z)$ has a finite number of sheets if α is rational.

   Ans. Since $\log z$ has an infinite sheeted Riemann surface, every circuit of the origin changes the value of $\log z$ by $2\pi i$, since

   $$\Delta \log z = \Delta \log |z| + i\Delta \arg z = 2\pi i$$

   For α not rational, $2\alpha \log z$ will change by $4\pi\alpha i$, which cannot be an integer multiple of $2\pi i$, so $z^{2\alpha}$ cannot return to the same value after any integer number of circuits of $z = 0$. For $\alpha = p/q =$ rational, the change in $2\alpha \log z$ will be $4\pi p i/q$ for each circuit, so after q circuits $z^{2\alpha}$ must return to the same value.

2. Why is the function $z^{1/2}(1 - z)^{1/3}$ six-valued?

   Ans. Three circuits about $z = 1$ return the function to its original value, while two circuits about $z = 0$ do the same, and *six* circuits about both $z = 0$ and $z = 1$ are required to return the function to its original value at a point. The latter may be seen by taking a large circle $z = Re^{i\theta}$ about the origin. Then

   $$z^{1/2}(1 - z)^{1/3} \sim R^{5/6}e^{i5\theta/6}$$

3. Describe the singularities of

   $$\frac{1}{z}\log\left(\frac{1}{1 - 2z}\right)$$

   Ans. The function has logarithmic branch points at $z = \frac{1}{2}$ and $z = \infty$ and a simple pole at $z = 0$ for every branch except the principal one.

4. Find the radii of convergence of the Taylor series for $1/(1 \pm \sqrt{3 - z})$ about $z = 0$.

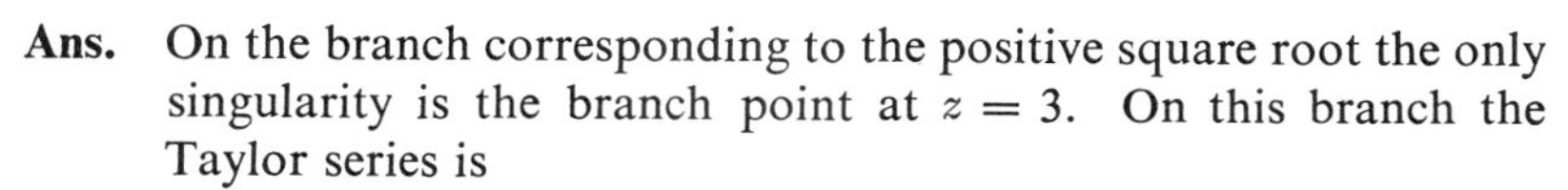

Ans. On the branch corresponding to the positive square root the only singularity is the branch point at $z = 3$. On this branch the Taylor series is

$$\frac{1}{1+\sqrt{3}}\left\{1 + \frac{z\sqrt{3}}{6(1+\sqrt{3})} + \cdots\right\}$$

and its radius of convergence is 3. On the branch corresponding to the negative square root there is the pole at $z = 2$. The Taylor series for this branch is

$$\frac{1}{1-\sqrt{3}}\left\{1 - \frac{z\sqrt{3}}{6(1-\sqrt{3})} + \cdots\right\}$$

and its radius of convergence is 2, since $z = 2$ is the nearest singularity to the origin on this branch.

5. Verify that

$$\int_0^\infty \frac{x^{-k}}{1+x}\,dx = \pi \csc(\pi k)$$

by contour integration $(0 < k < 1)$.

Ans. The contour must be chosen so as to avoid encircling the branch point at the origin. Thus, consider

$$\oint \frac{z^{-k}}{1+z}\,dz$$

around the contour shown in Figure 7.1, which consists of portions on a small circle C_ε of radius ε and on a large circle C_R of radius R, and a straight section I above the real axis and another straight section J below the real axis. On C_R

$$\left|\int \frac{z^{-k}\,dz}{1+z}\right| \leq \frac{2R^{-k}2\pi R}{R} \to 0$$

as $R \to \infty$. On C_ε

$$\left|\int \frac{z^{-k}\,dz}{1+z}\right| \leq 2\varepsilon^{-k}2\pi\varepsilon \to 0$$

as $\varepsilon \to 0$. On I it is simply the given integral as $\varepsilon \to 0$ and $R \to \infty$. On J there is $z = xe^{2\pi i}$, so the integral along J approaches $e^{-2\pi i k}\int_{-\infty}^{0} x^{-k}\,dx/(1+x)$ and hence the sum of the integrals along I and J is $(1 - e^{-2\pi i k})\int_0^\infty x^{-k}\,dx/(1+x) = 2\pi i e^{-i\pi k}$. Hence,

$$\int_0^\infty \frac{x^{-k}\,dx}{1+x} = \pi\left(\frac{2i}{e^{i\pi k} - e^{-i\pi k}}\right) = \pi \csc(\pi k)$$

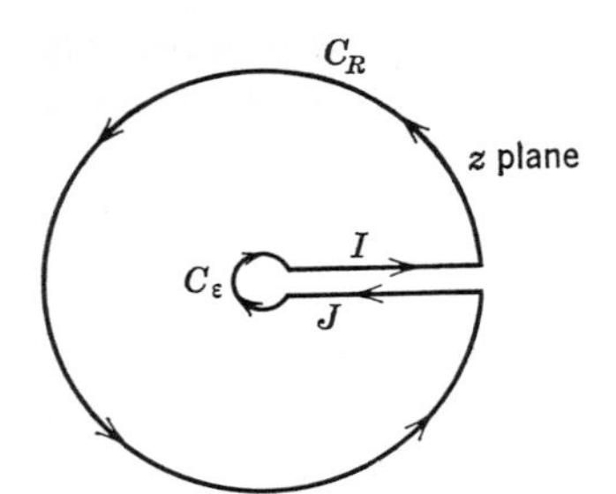

Figure 7.1. Contour for Example 5.

6. Describe the singularities of

$$J_0(z) = \sum_{k=0}^{\infty} \frac{(-1)^k z^{2k}}{2^{2k}(k!)^2}$$

Ans. Since the radius of convergence of the series is infinite, $J_0(z)$ is an entire function whose only singularity is at $z = \infty$. Since $J_0(1/z)$ has an infinite number of negative powers of z, $J_0(z)$ has an essential singularity at $z = \infty$.

7. Describe the singularities of

$$e^z\sqrt{z} = \sum_{k=0}^{\infty} \frac{z^{k+(1/2)}}{k!}$$

Ans. Although the series converges for all finite z, there is a branch point at $z = 0$ since fractional powers occur. Thus $e^z\sqrt{z}$ is *not* an entire function. There is also a branch point at $z = \infty$.

8. Which of the following are entire and which are meromorphic?

$$J_0(z)\csc z \qquad z^3 \cos z \qquad e^{\sin z} \qquad e^{1/z}$$

Ans. $J_0(z)\csc z$ is meromorphic, $z^3 \cos z$ and $e^{\sin z}$ are entire, while $e^{1/z}$ is neither.

9. Having established the relation

$$\int_0^{\infty} e^{-zu^3}\, du = \frac{\Gamma(\frac{1}{3})}{3} z^{-1/3} \qquad \text{for} \quad z = z^* > 0$$

prove that it is also valid for $\operatorname{Re} z \geq \delta > 0$. Regard $\Gamma(\frac{1}{3})$ as a known constant or refer to Section 9.3 for the definition of $\Gamma(z)$.

Ans. The integral is uniformly convergent in $\operatorname{Re} z \geq \delta > 0$, since

$$\left| \int_0^{\infty} e^{-zu^3}\, du \right| \leq \int_0^{\infty} e^{-\delta u^3}\, du$$

is convergent independently of z. Thus the integral is an *analytic* function of z in $\operatorname{Re} z \geq \delta > 0$. Also $z^{-1/3}$ is analytic in the same region. Since these two expressions coincide for $z = z^* \geq \delta > 0$, they must then coincide throughout $\operatorname{Re} z \geq \delta > 0$. (See Sections 7.3 and 7.2.)

10. Prove that

$$F(z) = \int_0^\infty e^{-t} \cosh zt \, dt$$

represents an analytic function in the region $|\operatorname{Re} z| \leq 1 - \delta < 1$ and that the region of definition may be extended to the whole z plane except $z = \pm 1$ by the expression

$$\frac{1}{1 - z^2}$$

Ans.

$$2F(z) = \int_0^\infty e^{(z-1)t} \, dt + \int_0^\infty e^{-(z+1)t} \, dt$$

$$2\,|F(z)| \leq \int_0^\infty e^{-\delta t} \, dt + \int_0^\infty e^{-\delta t} \, dt = \frac{2}{\delta}$$

so the integral is uniformly convergent for $|\operatorname{Re} z| \leq 1 - \delta < 1$ and represents an analytic function in this region. On the other hand, substitution of the power series for

$$\cosh zt = \sum_{n=0}^{\infty} \frac{(zt)^{2n}}{(2n)!}$$

upon termwise integration, yields

$$F(z) = \sum_{n=0}^{\infty} \frac{z^{2n}}{(2n)!} \int_0^\infty e^{-t} t^{2n} \, dt = \sum_{n=0}^{\infty} z^{2n}$$

which is equal to $1/(1 - z^2)$ for $|z| \leq 1 - \delta$. This function is analytic except for $z = \pm 1$, and since it coincides with the integral for $|z| \leq 1 - \delta$ they must coincide wherever both expressions have a meaning. Hence the value of the integral is $1/(1 - z^2)$ throughout $|\operatorname{Re} z| \leq 1 - \delta < 1$.

11. Show that if

$$\int_0^\infty e^{-zt} \cos(2zt) \, dt = \frac{1}{5z}$$

for real positive $z \geq \delta > 0$, then it is also true for complex z with $\operatorname{Re} z \geq \delta > 0$.

Ans. Use the Dirichlet test with

$$\phi(z,t) = \cos 2zt \qquad \text{and} \qquad \psi(z,t) = e^{-zt}$$

so that

$$(1) \quad \left| \int_0^T e^{-zt}\, dt \right| = \left| \frac{1 - e^{-zT}}{z} \right| \leq \frac{1 - e^{-\delta T}}{\delta} \leq \frac{1}{\delta}$$

$$(2) \quad \left| z \int_0^\infty e^{-zt}\, dt \right| \leq |z| \int_0^\infty e^{-\delta t}\, dt = \frac{|z|}{\delta}$$

$$(3) \quad |e^{-zt}| \leq e^{-\delta t} \to 0 \qquad \text{as} \quad t \to \infty$$

Hence the integral is uniformly convergent in $\operatorname{Re} z \geq \delta > 0$ and represents a function analytic in this region. Since $1/z$ is analytic except at $z = 0$ and coincides with the integral for real positive $z \geq \delta > 0$, it must coincide for all z in $\operatorname{Re} z \geq \delta > 0$.

12. Evaluate (for $\alpha = \alpha^* > 1$)

$$\int_0^\infty e^{-x^\alpha/\sqrt{2}} \sin(x^\alpha/\sqrt{2})\, dx$$

by considering a contour integral about the limit of a closed contour consisting of the following portions:

$$I_1: \quad \varepsilon < z = z^* < R$$

$$J_1: \quad |z| = R \qquad 0 < \arg z < \frac{\pi}{4\alpha}$$

$$I_2: \quad \varepsilon < |z| < R \qquad \arg z = \frac{\pi}{4\alpha}$$

$$J_2: \quad |z| = \varepsilon \qquad 0 < \arg z < \frac{\pi}{4\alpha}$$

as $R \to \infty$ and $\varepsilon \to 0$.

Ans. Consider

$$\oint e^{-z^\alpha}\, dz = 0$$

The integrand has *no* singularities inside the contour and is hence null. Along J_1

$$\left| \int e^{-z^\alpha}\, dz \right| \leq e^{-R^\alpha \cos(\pi/4)} \frac{\pi}{4\alpha} R \to 0$$

as $R \to \infty$. Along J_2

$$\left| \int e^{-z^\alpha}\, dz \right| \le e^{-\varepsilon^\alpha} \cos(\pi/4)\, \frac{\pi}{4\alpha}\, \varepsilon \to 0$$

as $\varepsilon \to 0$. Hence the integral along I_2 must be the negative of that along I_1, which is

$$\int_0^\infty e^{-x^\alpha}\, dx = \int_0^\infty e^{-t} t^{(1/2)-1}\, \frac{dt}{\alpha} = \frac{\Gamma(1/\alpha)}{\alpha}$$

[see the definition of $\Gamma(z)$, the gamma function, in Chapter 9]; I_2 yields, with $z = xe^{i\pi/4\alpha}$,

$$\begin{aligned} e^{i\pi/4\alpha} \int_\infty^0 e^{-x^\alpha e^{i\pi/4}}\, dx &= e^{i\pi/4\alpha} \int_\infty^0 e^{-x^\alpha(1+i)/\sqrt{2}}\, dx \\ &= -e^{i\pi/4\alpha} \int_0^\infty e^{-x^\alpha/\sqrt{2}} \left[\cos\left(\frac{x^\alpha}{\sqrt{2}}\right) - i \sin\left(\frac{x^\alpha}{\sqrt{2}}\right) \right] dx \end{aligned}$$

Hence,

$$\int_0^\infty e^{-x^\alpha/\sqrt{2}} \left[\cos\left(\frac{x^\alpha}{\sqrt{2}}\right) - i \sin\left(\frac{x^\alpha}{\sqrt{2}}\right) \right] dx = e^{-i\pi/4\alpha}\, \frac{\Gamma(1/\alpha)}{\alpha}$$

so

$$\int_0^\infty e^{-x^\alpha/\sqrt{2}} \sin\left(\frac{x^\alpha}{\sqrt{2}}\right) dx = \sin\left(\frac{\pi}{4\alpha}\right) \frac{\Gamma(1/\alpha)}{\alpha}$$

13. Evaluate

$$\oint \sqrt{c^2 - z^2}\, dz$$

about the limit of the closed elliptical contour

$$\left(\frac{x}{a}\right)^2 + \left(\frac{y}{\varepsilon}\right)^2 = 1$$

with foci at $\pm c$ and $a = \sqrt{c^2 + \varepsilon^2}$ as $\varepsilon \to 0$.

Ans. Without changing the value (since no singularities are traversed), the elliptical contour can be replaced by a dumbbell-shaped one consisting of portions of two circles J_1, J_2 of radius ε centered at $z = \pm c$ and connecting horizontal straight-line segments

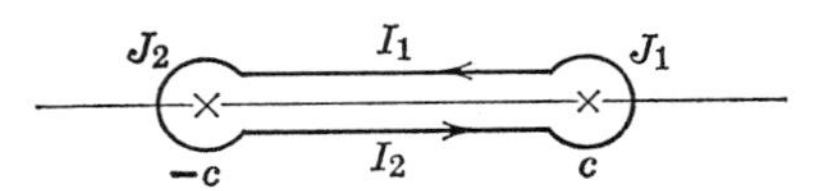

Figure 7.2. Contour for Example 13.

I_1, I_2 with $\operatorname{Im} z = \pm\varepsilon/3$, as shown in Figure 7.2. The limit of I_1 yields

$$-\int_{-c}^{c} \sqrt{c^2 - x^2}\, dx = -\frac{\pi c^2}{2}$$

On J_2, $z = -c + \varepsilon e^{i\theta}$ $(c > 0)$,

$$\left| \int \sqrt{c^2 - z^2}\, dz \right| \leq 2\sqrt{2c\varepsilon}\, 2\pi\varepsilon \to 0$$

as $\varepsilon \to 0$. On J_1, $z = c + \varepsilon e^{i\theta}$, and the integral there also tends to zero as $\varepsilon \to 0$. In passing around J_2 to reach I_2 the integrand changes sign because of the square root. Hence the integral on I_2 approaches

$$\int_{-c}^{c} (-\sqrt{c^2 - x^2})\, dx = -\frac{\pi c^2}{2}$$

as $\varepsilon \to 0$. Thus the contour integral must yield $-\pi c^2$ (for counterclockwise integration). Notice how the square root produces *doubling* of the value along I_1 instead of the cancellation that would occur if the square root were single-valued on both sides of the cut between the two branch points ($-c$ and c) of the integrand.

14. Evaluate $(a > b > 0)$

$$\int_0^1 \frac{dx}{(a - bx)\sqrt{x(1 - x)}}$$

by integrating a corresponding integral in the z plane about a closed contour which is the limit of one composed of the following portions:

$$\begin{array}{llr}
I_1: & \operatorname{Im} z = \varepsilon & -R \leq \operatorname{Re} z \leq 1 \\
J_1: & |z| = R & |\arg z| \leq \pi - \arctan(\varepsilon/R) \\
I_2: & \operatorname{Im} z = -\varepsilon & -R \leq \operatorname{Re} z \leq 1 \\
J_2: & |z - 1| = \varepsilon & |\arg(z - 1)| \leq \pi/2
\end{array}$$

as $\varepsilon \to 0$ and $R \to \infty$.

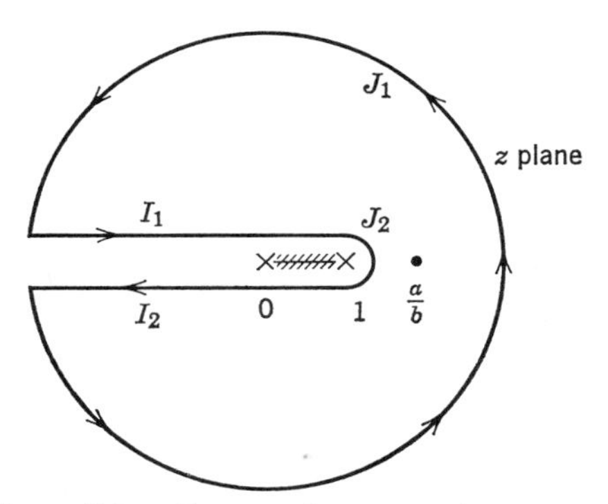

Figure 7.3. Contour for Example 14.

Ans. Consider

$$\oint \frac{dz}{(a - bz)\sqrt{z(1 - z)}}$$

The integrand has branch points at $z = 0$ and $z = 1$ and a simple pole at $z = a/b > 1$. As $\varepsilon \to 0$ the integral along J_2 yields

$$\left| \int \frac{dz}{(a - bz)\sqrt{z(1 - z)}} \right| \leq \frac{2\pi\varepsilon}{a\sqrt{\varepsilon}} \to 0$$

a zero limit. As $R \to \infty$ the integral along J_1 yields

$$\left| \int \frac{dz}{(a - bz)\sqrt{z(1 - z)}} \right| \leq \frac{2(2\pi R)}{bR^2} \to 0$$

also a zero limit. Furthermore, the limiting portions of I_1 and I_2 which are *not* adjacent to the cut between the branch points 0 and 1 will yield cancellation, since the integrand has a *unique* value as the negative real z axis is approached from above or below. However, the portions directly above and below the cut will yield *double* the desired integral because of the change in sign of the square root upon traversal about $z = 1$ (that traversal could quite as easily be effected by a small circular arc instead of a semicircular one). The limiting contour encloses only the simple pole at $z = a/b > 1$ of which the residue is

$$\frac{-1}{b\sqrt{(a/b)[1 - (a/b)]}} = \frac{-1}{-i\sqrt{a(a - b)}}$$

where the square root of -1 has here been chosen to be $-i$ so that the result of the integration will be positive, as examination of the original integral indicates it should be. Multiplication of the residue by $2\pi i$ then yields twice the original integral, so that finally

$$\int_0^1 \frac{dx}{(a - bx)\sqrt{x(1 - x)}} = \frac{\pi}{\sqrt{a(a - b)}}$$

● PROBLEMS 7

1. Show that

$$\int_0^\infty \frac{x^\alpha \, dx}{1 + 2x \cos \beta + x^2} = \frac{\pi \sin(\alpha\beta)}{\sin(\alpha\pi)\sin \beta} \qquad |\alpha| < 1 \qquad 0 < \beta < \pi$$

2. Show that

$$\int_0^\infty t^m e^{-t \cos \beta} \sin(t \sin \beta) \, dt = m! \sin[(m + 1)\beta]$$

by integrating about a contour along the positive real axis and $\arg z = \beta > 0$ avoiding the origin.

3. Evaluate by the residue theorem:

$$\int_0^\infty \frac{x^{1/5} \, dx}{x^2 + 3x + 2}$$

4. Describe the singularities of $\sec z$ and $\sqrt{(z - 3)(z - 5)}$ in the whole z plane.
5. Describe the singularities of e^{-1/z^2} and $\tanh z$ in the whole z plane.
6. Prove that e^z has no zeros in the finite z plane.
7. Describe the singularities of $(\sin \sqrt{z})/\sqrt{z}$. Is it entire?

Describe the Riemann surfaces for the following:

8. $(z - 2)^{1/5}$
9. $\left(\frac{z - 2}{z - 3}\right)^{1/3}$
10. $\log[(z - 2)(z - 3)]$
11. $\arctan z$

Describe the singularities of the following:

12. $e^{\log z}$
13. $J_0(z)\log(z - 3)$
14. $\log P(z)$ $[P(z) = \text{polynomial}]$
15. $\frac{\sqrt{z}}{z^4 + 1}$

16. Calculate by the residue theorem:

$$\int_0^\infty \frac{x^\alpha \, dx}{(1 + x)^m}$$

17. A function $f(z)$ is analytic for $\operatorname{Im} z \geq 0$ and real for $\operatorname{Im} z = 0$. An analytical continuation of $f(z)$ into $\operatorname{Im} z < 0$ is $f^*(z^*)$. Prove this.
18. A function $f(z)$ is analytic for $\operatorname{Re}(a^*z) \geq 0$ and is real for $\operatorname{Re}(a^*z) = 0$. Find an analytical continuation of $f(z)$ into $\operatorname{Re}(a^*z) < 0$ (Schwarz principle of reflection).
19. A function $f(z)$ is analytic for $|z| \leq 1$ and real for $|z| = 1$. Show that an analytical continuation of $f(z)$ to $|z| > 1$ is $f^*(a^2/z^*)$.
20. Establish by analytical continuation:

$$(1 - z^4)\int_0^\infty e^{-t}(\cosh zt + \cos zt)\,dt = 2$$

for

$$|\operatorname{Re} z| \leq 1 - \delta < 1 \qquad |\operatorname{Im} z| \leq 1 - \delta < 1$$

21. Show that

$$\int_0^\infty e^{-t}t^{z-1}\,dt$$

is an analytic function for $\operatorname{Re} z \geq \delta > 0$.
22. Show that

$$\int_0^{2\pi} \frac{d\theta}{1 + \cos z \sin \theta} = \frac{2\pi}{\sqrt{1 - \cos^2 z}}$$

23. Show that

$$\int_{-\infty}^\infty e^{-t^2+2zt}\,dt = \sqrt{\pi}e^{z^2}$$

24. Evaluate by contour integration:

$$\int_0^1 \frac{dx}{\sqrt{x(1 - x)}}$$

25. Show by contour integration that

$$\int_0^1 \frac{dx}{(1 + ax^2)\sqrt{1 - x^2}} = \frac{\pi}{2\sqrt{1 + a}}$$

26. Evaluate by contour integration:

$$\int_a^b \frac{dx}{(c - x)\sqrt{(b - x)(x - a)}}$$

for $c > b > a > 0$.
27. Evaluate by contour integration:

$$\int_0^1 \frac{dx}{(\cosh x)\sqrt{x(1 - x)}}$$

CHAPTER 8:
Conformal Mapping

8.1 DISTORTION AND INVARIANCE IN ANALYTIC MAPPINGS

It has already been mentioned that every analytic (uniquely differentiable) function $f(z)$ of a complex variable z can be regarded as defining a mapping between the z plane of complex values and the corresponding w plane of complex values where $w = f(z)$. Thus to any particular known point z_0 in the z plane there will correspond a particular point (at least one) $w_0 = f(z_0)$ which is also known as soon as the function $f(z)$ is known. Furthermore, the set of values of z corresponding to a segment of any curve passing through z_0 will correspond to a (image) set of values of w on a segment of an image curve in the w plane.

First let us inquire into the linear distortion or magnification of distances in passing from the z plane to the w plane. Since $w = f(z)$, the *change* in w along a small arc through z becomes

$$dw = f'(z)\,dz \tag{8.1}$$

so the length $|dz|$ in the z plane becomes $|dw| = |f'(z)|\,|dz|$ in the w plane. Hence the magnification factor is the modulus of the derivative of the mapping function, and since this does not depend on direction of the arc through z, we see that the magnification is *isotropic* but not necessarily homogeneous since $|f'(z)|$ can vary from point to point in the z plane.

Suppose now that there are two definite distinct curves passing through z and that z changes by dz_1 and dz_2 as points on the first and second curve slightly removed from z are chosen. There will be two image curves through the point $w = f(z)$ on the w plane, and close points on these curves will differ by dw_1 and dw_2 from w.

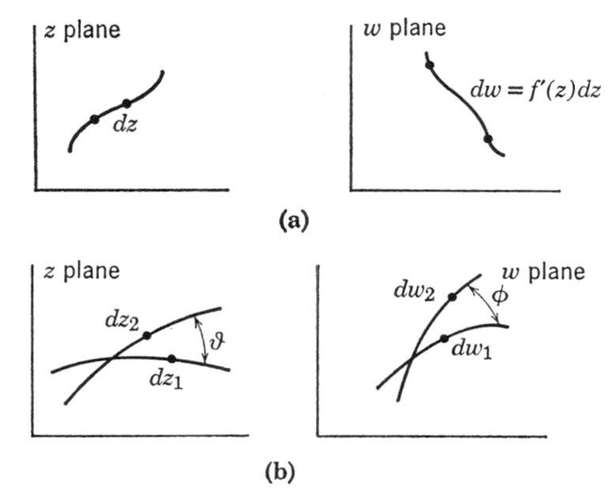

Figure 8.1. (a) Magnification of distance; (b) preservation of angle $\vartheta = \phi$.

The changes in z and w along the two curves and their w plane images are related by

$$dw_1 = f'(z)\, dz_1$$

$$dw_2 = f'(z)\, dz_2 \tag{8.2}$$

where the same value of the derivative can be used in the two equations (8.2) because $f(z)$ is analytic. It is easy to show that the analytic mapping $w = f(z)$ is *conformal*, i.e., preserves the angle between (the tangents of) two curves through a point. The angle between the two curves along which dz_1 and dz_2 are taken is called θ. The corresponding angle in the w plane is called ϕ. Then

$$\phi = \arg(dw_2) - \arg(dw_1) = \arg\left(\frac{dw_2}{dw_1}\right) = \arg\left(\frac{dz_2}{dz_1}\right) = \theta \tag{8.3}$$

since the ratio of the two equations in (8.2) yields

$$\frac{dw_2}{dw_1} = \frac{dz_2}{dz_1} \tag{8.4}$$

unless $f'(z)$ is zero or infinite, in which case dw_2/dw_1 is indeterminate. Thus we conclude that $w = f(z)$ is a conformal mapping except at points of the z plane for which $f'(z) = 0$ or ∞.

We might suspect that the distortion of a small area in an analytic mapping would be given by $|f'(z)|^2$, since linear distortion is given by $|f'(z)|$. In fact the area of a parallelogram of sides dz_1 and dz_2 is given by

$$A_z = \mathrm{Im}(dz_1^* \, dz_2) \tag{8.5}$$

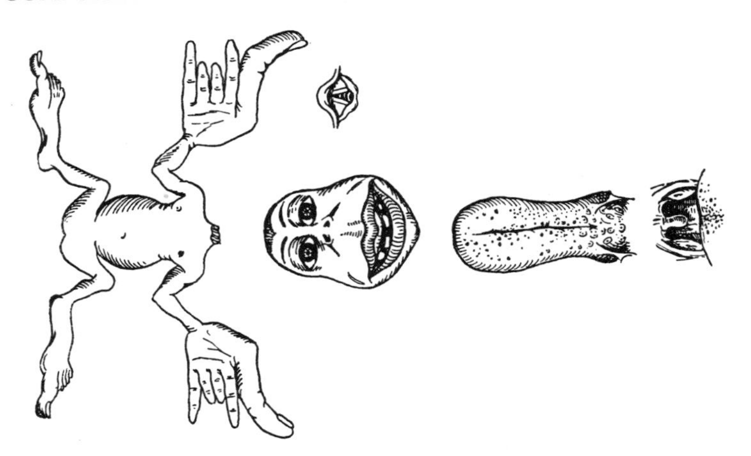

Figure 8.2. The homunculus R^3, an example of distortion in mapping.

while the area of the image "parallelogram" of sides dw_1 and dw_2 is given by (8.2),

$$A_w = \operatorname{Im}(dw_1^* \, dw_2) = \operatorname{Im}(|f'(z)|^2 \, dz_1^* \, dz_i) = |f'(z)|^2 \, A_z \qquad (8.6)$$

as suspected.

Having discussed the distortion of lengths and areas and the preservation of angles under analytic mappings, we turn now to some general properties of such mappings which are easily discerned if we view certain topological properties of these mappings as obvious.

8.2 MAXIMUM MODULUS PRINCIPLE

The *maximum modulus theorem* states that a function $f(z)$ cannot have a maximum value of its modulus $|f(z)|$ at an *interior* point of any region in which it is analytic. Suppose on the contrary that there exists such a point p interior to (not on the boundary of) the region R in which $f(z)$ is analytic. The point p could be isolated from the boundary C of R by constructing a small circle K of radius ε about p. The image point of p under $w = f(z)$ would be $q = f(p)$, some point *interior* to the image region R' of R and *isolated* from the boundary C' of R' by the closed image curve K' of the small circle K. But the maximum value of $|w| = |f(z)|$ must be *on the boundary* C' of R' at the point of C' which is farthest from the origin of the w plane. Thus we have the *contradiction* that q is on the boundary and isolated from it. This is impossible, so it must have been impossible for p to be chosen interior to R.

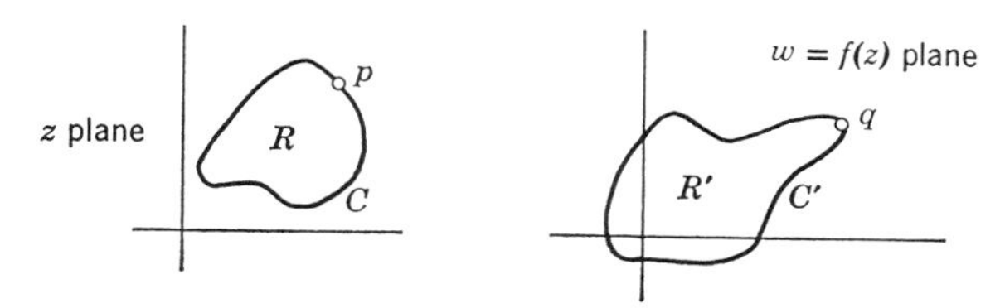

Figure 8.3. Maximum modulus principle.

Thus the p among all points in R for which $|f(p)| =$ maximum must be on C (see Fig. 8.3).

In any conformal mapping $w = f(z)$ of the interior and boundary of the region $|z| \leq R$ which preserves the origin [i.e., for which $z = 0$ is mapped into $w = 0 = f(0)$] the ratio of the modulus of an image point $|w|$ to the modulus of an original point $|z|$ cannot exceed the ratio of the maximum modulus M of $f(z)$ on $|z| = R$ to R. This result is called *Schwarz's lemma*, and it is proved by noting that if $f(z)$ is free of singularities for $|z| \leq R$ and if $f(0) = 0$, then

$$g(z) = \frac{f(z)}{z}$$

also has no singularities for $|z| \leq R$ since $g(0) = f'(0)$. Thus the maximum modulus of $g(z)$ for $|z| \leq R$ must occur on the boundary $|z| = R$, and if $|f(z)| \leq M$ on $|z| = R$, one has $|w/z| = |g(z)| = |f(z)|/|z| \leq M/R$, which was to be shown.

8.3 PRINCIPLE OF ARGUMENT

From the knowledge that any polynomial of the nth degree can be expressed as a product of linear factors corresponding to its roots,

$$w = P_n(z) = A \prod_{k=1}^{n} (z - z_k) \tag{8.7}$$

one can conclude that 2π times the number of roots of $P_n(z) = 0$ inside a *closed* contour C is given by the change in argument of w as z traverses the contour C completely. Thus,

$$\log w = \log A + \sum_{k=1}^{n} \log(z - z_k) \tag{8.8}$$

and the imaginary part of this is

$$\arg w = \arg A + \sum_{k=1}^{n} \arg(z - z_k) \tag{8.9}$$

Figure 8.4. Principle of argument.

As z goes around C completely there is *no* change in arg A, and the change in $\arg(z - z_k)$ depends only on whether z_k is interior to the region surrounded by C or exterior to C. If z_k is interior, the change in $\arg(z - z_k)$ when z travels completely around C will clearly be 2π radians. If z_k is exterior, one can construct two extreme tangents T_1 and T_2 to C, so that C falls entirely in the angle between the two. The two tangent points divide C into two portions, and as z travels from T_1 to T_2 along C the change in $\arg(z - z_k)$ will be equal in magnitude but *opposite in sign* to the change in $\arg(z - z_k)$ as z travels from T_2 to T_1 along C. Thus in this case the *total change* in arg $(z - z_k)$ will be zero (see Fig. 8.4).

For *each root* interior to the region bounded by C a circuit of C by z contributes 2π radians increment to arg w. Thus,

$$\Delta_C \arg w = 2\pi N \tag{8.10}$$

where N is the number of roots of w inside and Δ_C is the change incident upon a circuit of C by z. Multiple roots are multiply counted. This is called the *principle of argument*.

The result may be extended to the case where $w = f(z)$ is a single-valued analytic function free of zeros on C and free of singularities within and on C. Then $f(z)$ may be written in the form

$$w = f(z) = P_n(z)g(z)$$

where $g(z) \neq 0$ inside or on C, and $P_n(z)$ is a polynomial. Thus,

$$\Delta_C \arg w = \Delta_C \arg P_n(z) + \Delta_C \arg g(z)$$

and $\Delta_C \arg g(z) = 0$, since the (simply connected) image (in the g plane) of R (in the z plane) does *not* include $g = 0$. Thus one concludes that (8.10) applies to analytic functions in addition to polynomials.

If $f(z)$ possesses poles as well as zeros in R, one has

$$w = f(z) = \frac{P_n(z)g(z)}{Q_m(z)}$$

where $Q_m(z)$ is a polynomial with zeros in R. Then

$$\Delta_C \arg w = \Delta_C \arg P_n(z) - \Delta_C \arg Q_m(z)$$

so that $(1/2\pi)\Delta_C \arg w$ gives the *difference* between the number of zeros and the number of poles of w in R.

A corollary of the principle of argument in *Rouché's theorem*, which states that if $|g(z)| < |f(z)| \neq 0$ on C and $f(z)$ and $g(z)$ are uniquely defined analytic functions within R and on C, then $f(z)$ and $f(z) + g(z)$ have the same number of zeros in R. Since

$$\Delta_C \arg(f + g) = \Delta_C \arg f + \Delta_C \arg\left(1 + \frac{g}{f}\right)$$

the last term must be zero, because $1 + g/f$ traverses a closed curve *not* including the origin as z traverses C. Thus,

$$\Delta_C \arg(f + g) = \Delta_C \arg f$$

which proves Rouché's theorem.

8.4 BILINEAR (MÖBIUS) MAPPINGS

An important special class of mappings $w = f(z)$ is the class of *bilinear* (linear fractional, Möbius, homographic) mappings:

$$w = \frac{az + b}{cz + d} = \frac{a}{c} + \frac{bc - ad}{c(cz + d)} \tag{8.11}$$

for which $bc - ad \neq 0$. These mappings are *bi-unique*, with exactly one image point in the w plane for each given point in the z plane and vice versa. The inverse of (8.11) is

$$z = \frac{b - dw}{cw - a} \tag{8.12}$$

Instead of considering (8.11) as giving a mapping of the z plane onto the w plane, we can rename the w plane as a kind of transformed z plane. Then we refer to a mapping of the z plane *onto itself* by which we mean that each point of the z plane is moved to a location given by (8.11). If we so view (8.11), we might then ask if any points of the z plane *remain fixed* during the mapping (8.11). For such points

$$z = \frac{az + b}{cz + d} \tag{8.13}$$

and since this is a quadratic equation for z,

$$cz^2 + (d - a)z - b = 0 \tag{8.14}$$

it is clear that no more than *two* points may be fixed. The condition that the point at infinity be fixed is that $c = 0$, as may be seen from (8.11). To specify that (8.11) preserve the location of the points z_1 and z_2 we may set, in (8.11),

$$\begin{aligned} b &= -cz_1z_2 \\ d &= a - c(z_1 + z_2) \end{aligned} \tag{8.15}$$

since the points z_1 and z_2 are then the roots of (8.14). Another way of writing the particular transformation (mapping) which preserves z_1 and z_2 is

$$\frac{w - z_1}{w - z_2} = \left(\frac{a - cz_1}{a - cz_2}\right)\left(\frac{z - z_1}{z - z_2}\right) \tag{8.16}$$

In the general mapping (8.11) three constants are essential since the numerator and the denominator may always be divided by any nonvanishing fourth constant. Thus (8.11) can be solved for the three essential coefficients if three points z_1, z_2, z_3 in the z plane and their three image points w_1, w_2, w_3 in the w plane are specified. Another way of expressing this is to set equal to zero the determinant of the coefficients of a, b, c, d in the linear equations obtained from (8.11) by replacing z and w by z_1, w_1; z_2, w_2; z_3, w_3, successively. Then

$$\begin{vmatrix} 1 & z & w & zw \\ 1 & z_1 & w_1 & z_1w_1 \\ 1 & z_2 & w_2 & z_2w_2 \\ 1 & z_3 & w_3 & z_3w_3 \end{vmatrix} = 0 \tag{8.17}$$

An alternative way of writing the mapping (8.17) is

$$\left(\frac{w - w_1}{w - w_3}\right)\left(\frac{w_2 - w_3}{w_2 - w_1}\right) = \left(\frac{z - z_1}{z - z_3}\right)\left(\frac{z_2 - z_3}{z_2 - z_1}\right) \tag{8.18}$$

which exhibits the correspondence of z_1, z_2, z_3 to w_1, w_2, w_3.

The combination

$$\lambda = \frac{(w - w_1)/(w - w_3)}{(w_2 - w_1)/(w_2 - w_3)} \tag{8.19}$$

of the (complex) coordinates of four points w, w_2, w_1, w_3 is called the *cross ratio* of the four points, and it is of considerable importance in projective geometry where it is shown to be an *invariant* under projective transformations. Equation (8.18) shows that cross ratio to be invariant under (8.11). In other words, the cross ratio calculated for four points in the z plane will be identical with the cross ratio calculated for the four image points in the w plane.

The bilinear mappings (8.11) send circles and straight lines into circles and straight lines. That is, under (8.11) a circle can only become another circle or a straight line and a straight line can only become another straight line or a circle. This follows from the fact that the form of the equation for a circle (straight line for $A = 0$) is

$$Az^*z + Bz + B^*z^* + C = 0 \tag{8.20}$$

where A and C are real.

A simple case of the bilinear mappings is *conjugate inversion* with respect to a circle of radius r and center at c. This is an analytic mapping

$$w - c = \frac{r^2}{z - c} \tag{8.21}$$

Since $z - c = \rho e^{i\phi}$ has as image $w - c = (r^2/\rho)e^{-i\phi}$, it is clear that the circle $|z - c| = r$ of inversion is reflected in a line through $z = c$ parallel to the real axis, while any point outside the circle ($\rho > r$) is imaged inside the circle $|w - c| = r$ since for such a point $|w - c| = r^2/\rho < r$. Similarly, any point inside becomes a point outside. In the terminology of describing (8.21) as a mapping of the z plane onto itself we would say that the exterior and interior of the circle $|z - c| = r$ were interchanged by the mapping. The closely related *inversion*

$$w - c = \frac{r^2}{z^* - c^*} \tag{8.22}$$

is *not* an analytic function and it simply "moves" each point to a new radial location without changing its direction from the center of inversion c. Equation (8.21) moves each point to such a new location, and then "reflects" it across a line through c parallel to the real axis.

The effects of inversion (8.22) on a straight line diametral, secantal, tangential, or not intersecting the circle of inversion are illustrated in Figures 8.5–8.8. The corresponding effects of (8.22) on a circle are shown in Figures 8.9–8.12.

It may be noted in passing that the rigid body motions in the plane (rotations and translations) are a special case of the bilinear mappings, since a

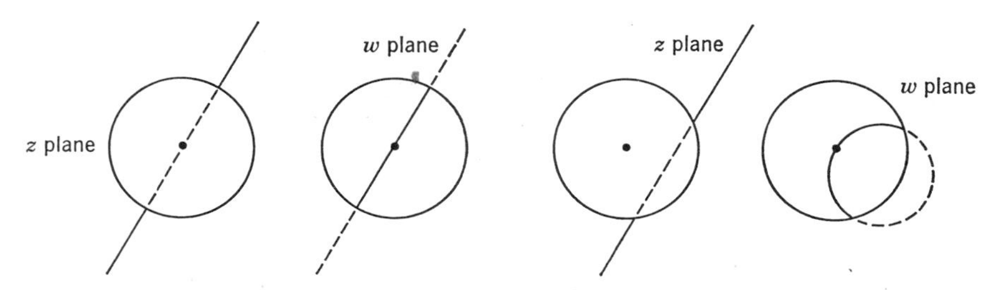

Figure 8.5. Inversion of diametral line. **Figure 8.6.** Inversion of secantal line.

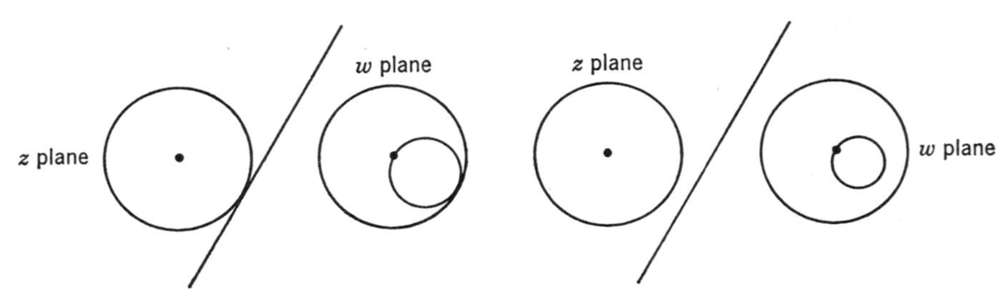

Figure 8.7. Inversion of tangential line. **Figure 8.8.** Inversion of external line.

rotation of the z plane through angle α about point $z = s$ followed by a translation specified by t is given by

$$w = e^{i\alpha}(z - s) + t \tag{8.23}$$

The distortion in a bilinear mapping is given by the modulus of

$$d_z w = w' = \frac{ad - bc}{(cz + d)^2} \tag{8.24}$$

It is readily verified that (8.23) yields a derivative of *unit* modulus which corresponds to the fact that dimensions of figures are preserved in euclidean geometry.

It is possible to map a half-plane onto the interior of a circle of radius r and center c by the following argument in which it is recalled that $\text{Re}(a^*b)$ is equivalent to the scalar product of $\vec{a}$ and $\vec{b}$ (Section 1.1): Suppose the half-plane consists of points z for which

$$x \cos \alpha + y \sin \alpha = \text{Re}(n^*z) \geq p \tag{8.25}$$

where $n = e^{i\alpha}$ and the real number p are known, and let the point a in this region be mapped into the origin of the w plane. The distance of a from the

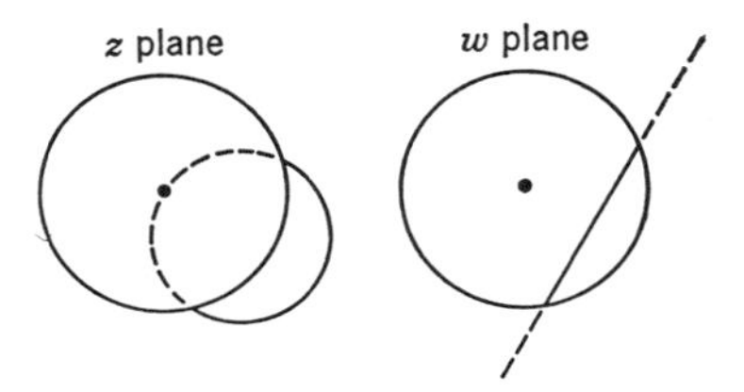

Figure 8.9. Inversion of circle through center cutting circumference twice.

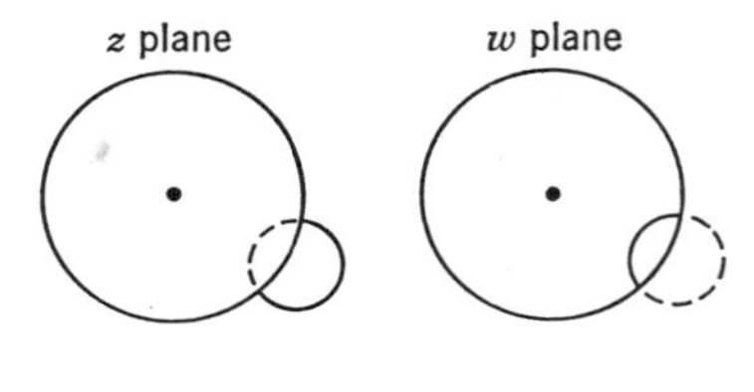

Figure 8.10. Inversion of circle not through center cutting circumference twice.

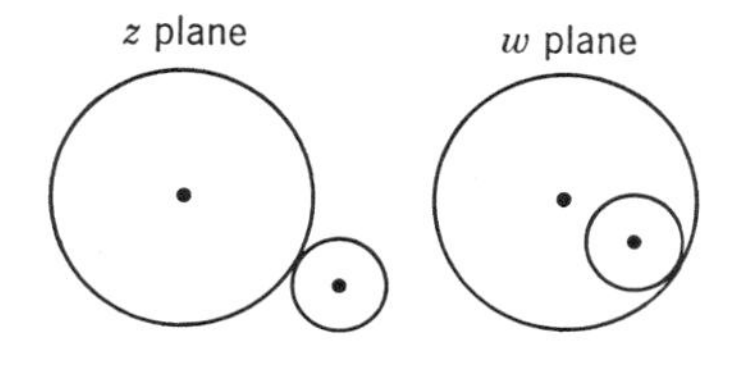

Figure 8.11. Inversion of externally tangent circle.

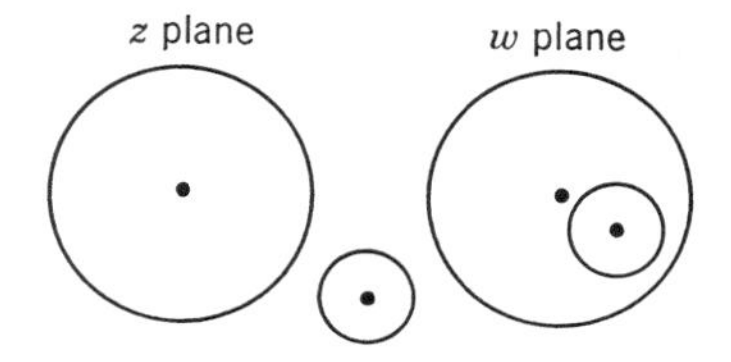

Figure 8.12. Inversion of external circle.

boundary line of the half-plane is $\text{Re}(n^*a - p)$, so the "reflection" of a in this line must be located at

$$b = a - 2e^{i\alpha}\,\text{Re}(n^*a - p) \tag{8.26}$$

Then a bilinear mapping which takes $z = a$ into $w = 0$ and makes $|w| = 1$ for $\text{Re}(n^*z) = p$ is clearly

$$w = \frac{z - a}{z - b} \tag{8.27}$$

To map the half-plane onto a circle of radius r with center at $w = c$ (8.27) is replaced by

$$w = c + r\left(\frac{z - a}{z - b}\right) \tag{8.28}$$

with b given by (8.26).

It is easily verified that the mapping

$$w = Rre^{i\phi}\left(\frac{z - a}{a^*z - r^2}\right) \tag{8.29}$$

maps the interior of the circle $|z| = r$ in the z plane into the interior of the circle $|w| = R$ in the w plane and the point $z = a$ into $w = 0$. Thus, for

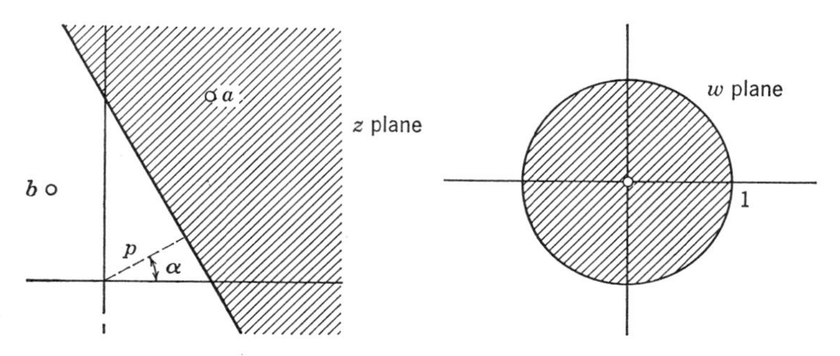

Figure 8.13. Conformal mapping of half-plane onto interior of unit circle.

$z = re^{i\theta}$, (8.29) becomes

$$|w| = Rr\frac{|re^{i\theta} - a|}{|a^*re^{i\theta} - r^2|} = R\frac{|re^{i\theta} - a|}{|r - a^*e^{i\theta}|} = R\frac{|re^{i\theta} - a|}{|re^{-i\theta} - a^*|} = R \quad (8.30)$$

since the modulus of a complex number and that of its conjugate are identical.

8.5 SCHWARZ–CHRISTOFFEL MAPPINGS

Another useful mapping is that of the wedge

$$0 \leq \arg z \leq \alpha \quad (8.31)$$

into the half-plane Im $w \geq 0$, which is achieved by

$$w = z^{\pi/\alpha} \quad (8.32)$$

since $z = re^{i\alpha}$ corresponds to $w = -r$ (the negative real axis as r increases from zero to infinity). Here the conformal property fails at $z = 0$.

A most important mapping is that of a polygon into a half-plane. This is called the *Schwarz–Christoffel mapping*. Suppose the polygon is in the z plane and has vertices at the points $\{z_k\}$ with interior angles α_k, and it is desired to map its interior onto Im $w > 0$ in the w plane. Then the boundary of the polygon must be mapped into the real axis of the w plane with the image points of the vertices at $\{w_k\}$. In order to achieve this as z "moves" along a side of the polygon *between* vertices, w must move along the real axis, and after z moves around a vertex, w must still be moving in the direction of the real axis. Thus, between vertices,

$$\arg dz = \text{constant} \quad (8.33)$$

or, since $\arg dw = 0$,

$$\arg\left(\frac{dz}{dw}\right) = \text{constant} \quad (8.34)$$

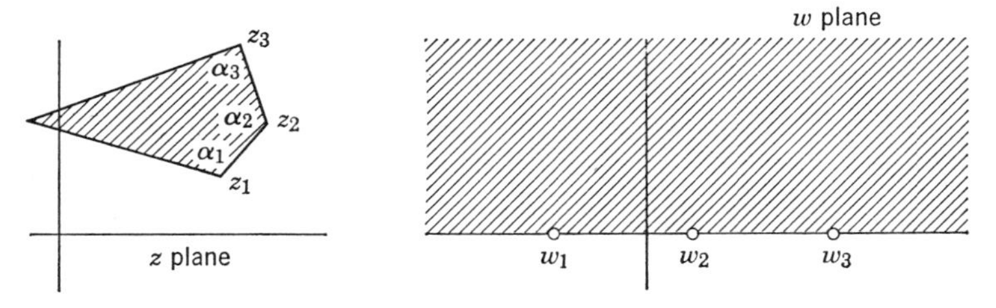

Figure 8.14. Schwarz–Christoffel mapping of polygon interior on half-plane.

while at z_k the change in arg dz must be $(\pi - \alpha_k)$ and the change in $\arg(w - w_k)$ must be $-\pi$. Thus, with Δ denoting change,

$$\Delta \arg dz = \Delta \arg\left(\frac{dz}{dw}\right) = (\pi - \alpha_k) \tag{8.35}$$

$$\Delta \arg(w - w_k) = -\pi \tag{8.36}$$

so that

$$\Delta \arg\left(\frac{dz}{dw}\right) = \left(\frac{\alpha_k - \pi}{\pi}\right)\Delta \arg(w - w_k) \tag{8.37}$$

This increase in argument corresponds to a relation of the form

$$\frac{dz}{dw} \sim (w - w_k)^{(\alpha_k - \pi)/\pi}$$

and a factor such as that on the right is needed for each vertex. Hence,

$$\frac{dz}{dw} = C \prod_{k=1}^{n} (w - w_k)^{(\alpha_k - \pi)/\pi} \tag{8.38}$$

which satisfies (8.37) at each vertex and (8.33) between each pair of adjacent vertices. This is the differential equation for the Schwarz–Christoffel mapping.

.6 RIEMANN MAPPING THEOREM AND GREEN'S FUNCTIONS

The polygons need not be closed or convex to effect the mapping. It appears plausible that one could increase indefinitely the number of sides of a polygon, with vertices equally dividing a closed curve bounding a simply connected region, and so conclude the existence of a mapping taking the interior of such a region into a half-plane. For any preassigned large number of vertices the Schwarz–Christoffel mapping would give an *approximate* mapping onto the half-plane (see Fig. 8.15). Since the mapping of a half-plane into the

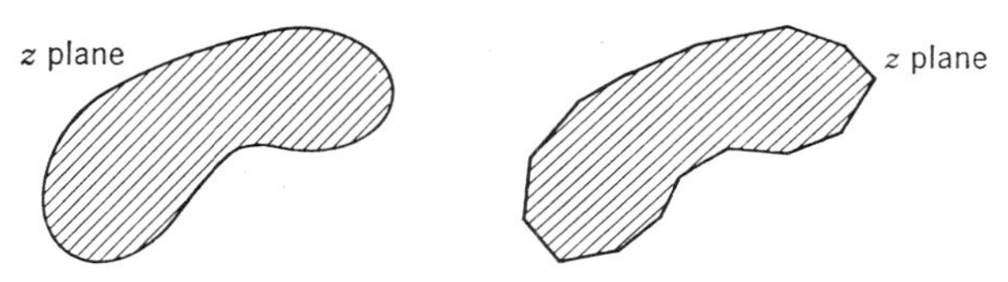

Figure 8.15. Schwarz–Christoffel mapping as an approximate Riemann mapping.

interior of a circle (8.28) has already been given, one might then expect the result of the *Riemann mapping theorem*: There exists a unique analytic function which conformally maps a (simply connected) region bounded by a continuous closed (non-self-intersecting) curve onto the interior of a unit circle $|w| < 1$ while transforming the point $z = a$ into $w = 0$ and a given direction at $z = a$ into the positive real direction at $w = 0$. For a proof the reader is referred to Appendix A.

The physical significance of this theorem is that it establishes the existence of a Green's function (for the Dirichlet problem with Laplace's equation) for the cylindrical region whose cross section is the region of the theorem. The Green's function is then the potential of a line charge (perpendicular to the z plane) inside the region and parallel to the generators of the cylindrical boundary C, which latter is supposed to be grounded (zero potential). If $w = f(z)$ is a function mapping the region onto the (interior of) unit circle $|w| \leq 1$ and taking $z = a$ into $w = 0$, then $g(z,a) = 2 \log |w| = 2 \log |f(z)|$ is the Green's function for this region since

(1) for z on C, $|w| = 1$ so that $g(z,a) = 0$;
(2) $g(z,a)$ is harmonic (solution of Laplace equation) except at $z = a$ since it is the real part of the analytic function $\log f(z)$;
(3) for z close to a, $g(z,a) \approx 2 \log |z - a|$ [i.e., $g(z,a)$ approximates the potential of a line charge at a].

Conclusion (3) follows from the Taylor expansion (series) for $f(z)$ near $z = a$:

$$f(z) \approx f(a) + (z - a)f'(a) = (z - a)f'(a) \tag{8.39}$$

and from the fact that the potential of a line charge carrying q units of charge per unit length is $2q \ln r$, where r is the distance away from the line. The latter result follows from the Gauss electrostatic theorem* (total electrostatic flux proportional to total charge enclosed):

$$\oint \bar{E} \cdot \bar{n}\, dS = -4\pi q l \tag{8.40}$$

applied to a cylinder of length l and radius r with axis on the line charge. Here dS is an area element on the surface of the cylinder. Then (8.40) implies, since the field $\bar{E}$ is radial and $E = |\bar{E}|$,

$$2\pi r l E = -4\pi q l \tag{8.41}$$

so that

$$E = -\frac{2q}{r} \tag{8.42}$$

* O. D. Kellog, *Potential Theory*. Berlin: Springer-Verlag, 1967, p. 43.

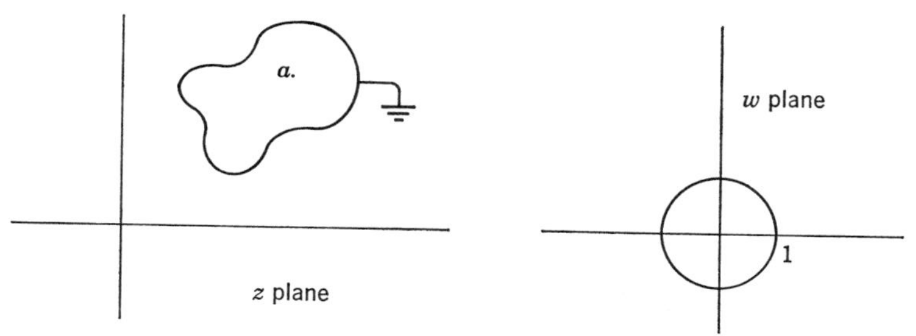

Figure 8.16. Mapping for Green's function for harmonic Dirichlet problem.

and the potential is

$$\phi = 2q \log r \tag{8.43}$$

Thus it is seen that the defining properties (1)–(3) of the Green's function are all fulfilled by $g = 2 \log |f(z)|$, with $w = f(z)$ the mapping function described in the Riemann mapping theorem.

The utility of the Green's function is a result of the fact that the potential at any point z inside the region may be expressed in terms of the potential on boundary C and the Green's function as follows:

$$\phi(z) = \frac{1}{4\pi} \oint \phi(s)\partial_n g(z,s)\, ds \tag{8.44}$$

where the integration is carried out around the closed contour C (any typical point of which is located at arc length s from some fixed initial point on C) and ds is the corresponding differential arc length along C. Note then that $\phi(s)$ and $\partial_n g(z,s)$ are evaluated only along C. This solves the Dirichlet problem for the region bounded by C.

The result (8.44) may be established from the form

$$\int (\phi\Delta g - g\Delta\phi)\, dS = \oint (\phi\partial_n g - g\partial_n\phi)\, ds \tag{8.45}$$

of Green's theorem (see Appendix B). Here dS is the area element in the z plane, while ds is as above. Because g vanishes on C and ϕ is harmonic ($\Delta\phi = 0$) inside C, this becomes

$$\int \phi\Delta g\, dS = \oint \phi\partial_n g\, ds \tag{8.46}$$

Now since $\Delta g = 0$ except within a circle of arbitrarily small radius ε about $z = a$, the left side may be reduced to an area integral over such a small circle that ϕ approximates $\phi(a)$ over this circle (the approximation improves for $\varepsilon \to 0$), and the left side thus becomes

$$\phi(a)\int \Delta g \, dS$$

Since

$$\int \Delta g \, dS = \oint \partial_n g \, ds \tag{8.47}$$

is a special case of (8.46) for $\phi = 1$, one notes that the left side of (8.47) may be evaluated from its right side, which, since $(r = |z - a|)$,

$$g \approx 2 \ln r$$

and

$$\partial_n g = \partial_r g \approx \frac{2}{r} = \frac{2}{\varepsilon}$$

on the circumference of the small circle, yields

$$\int \Delta g \, dS = 2\pi\varepsilon\left(\frac{2}{\varepsilon}\right) = 4\pi$$

so that

$$4\pi\phi(a) = \oint \phi \partial_n g \, ds$$

Since the point a could be located anywhere in the interior of the region, a slight change of notation yields (8.44).

The Green's function can also be used to determine the potential when charges are present in the region. In this case (8.40) applied to

$$\bar{E} = -\nabla\phi$$

yields (with $dS = l \, ds$)

$$\oint \bar{n} \cdot \nabla\phi \, ds = \oint \partial_n \phi \, ds = 4\pi q$$

whence (8.47) applied to ϕ yields

$$\int (\Delta\phi - 4\pi\rho) \, dS = 0$$

where dS is now an area element in the z plane and ρ is charge per unit volume. Since this is true even for arbitrarily small regions of integration,

$$\Delta\phi = 4\pi\rho$$

Poisson's equation, follows. In subregions where ($\rho = 0$) charge is absent this reduces to Laplace's equation:

$$\Delta\phi = 0$$

Under the assumption that ρ is not necessarily zero one can return to (8.45) to obtain

$$\int \phi \Delta g \, dS - 4\pi \int \rho g \, dS = \oint \phi \partial_n g \, ds \tag{8.48}$$

instead of (8.46), so that (8.44) is replaced by

$$\phi(z) = \int \rho g \, dS + \frac{1}{4\pi} \oint \phi \partial_n g \, ds \tag{8.49}$$

The Riemann mapping theorem, which establishes the existence of a Green's function, is important in all potential problems in which the fields are identical on every section perpendicular to a fixed direction. Physically it simply amounts to asserting that there exists a potential for a line charge inside a parallel grounded cylindrical metal container. It also shows that the solution of the problem of finding the potential inside an (nonmetallic) identically shaped surface in the absence of internal line charges is established by (8.44) as soon as the mapping of a normal cross section into the interior of the unit circle is known. While the theorem (see Appendix A) asserts the existence of such a mapping, it does not divulge the explicit function $f(z)$ which accomplishes it. Except for certain simple mappings (e.g., half-plane to circle, circle to circle, etc.) one must then usually rely upon a Schwarz–Christoffel approximate mapping, as explained previously.

8.7 HARMONIC MEASURE

Consider a simply connected region R of the z plane bounded by a closed contour C. If C is divided into two mutually exclusive portions (sets) C_0 and $\tilde{C}_0$, there exists a function (harmonic in R) $\phi_{C_0}(z)$ which is zero on $\tilde{C}_0$ and unity on C_0. This is called the *harmonic measure* of C_0 in R, and it clearly depends upon the region as well as on the set C_0. Since in the Dirichlet problem of potential theory one is interested in finding a function harmonic in R, taking prescribed values on C, this may be considered to be a limit of the superposition of a set of harmonic measures for subsets of C. Thus, if the parametric equation for C is

$$z = g(t) \qquad \text{with} \qquad g(0) = g(1) \qquad\qquad 0 \leq t \leq 1$$

and if $h[g(t)] = \psi(t)$ is the function prescribed on C, it is clear that one has approximately

$$\psi(t) \cong \sum_k \psi_k(t)$$

where

$$\psi_k(t) = \begin{cases} \psi\left(\dfrac{t_k + t_{k-1}}{2}\right) & \text{for} \quad t_{k-1} \leq t < t_k \\ 0 & \text{otherwise} \end{cases}$$

for the set $\{t_k\}$ of a large number of points equidistant on C. If now $\phi_k(z)$ is the harmonic measure of the portion of C with $t_{k-1} \leq t < t_k$ in R, the Dirichlet problem is clearly approximately solved by

$$f(z) = \sum_k \psi_k\left(\frac{t_k + t_{k-1}}{2}\right) \phi_k(z)$$

and as the number of points increases, the approximation improves. Notice the analogy of the method to that of Lagrangian interpolation, although here a preassigned value is taken on a whole interval of the bounding contour rather than at a point.*

8.8 APPLICATIONS OF SOURCES AND SINKS

If the z plane is imagined to possess a distribution of mass or electrical charge which does not change in totality with time but may redistribute itself (concentrating or diminishing in particular regions), it is of interest to ask how the density may change as a function of position and time. This will be governed by the so-called *equation of continuity* which might better be called the *equation of conservation*:

$$\partial_t \rho + \nabla \cdot (\rho \mathbf{v}) = \partial_t \rho + \operatorname{Re}\{\partial^*(\rho v)\} = 0 \tag{8.50}$$

where v is the (complex) velocity of the material (mass or charge) at a point and ρ is the real quantity (mass or charge) per unit area. The (mass or charge) current density ρv represents the quantity (mass or charge) crossing a unit length in a unit of time in the direction arg v. The proof is quite simple. Consider any fixed region of area S. If mass (or charge) cannot be created or destroyed, it can only enter or leave S through the boundary of the region. Thus,

$$d_t \int \rho \, dS = \int \partial_t \rho \, dS = -\oint \rho \operatorname{Re}(n^* v) \, ds$$

* For a general development of the Dirichlet problem in terms of harmonic measure the reader is referred to R. Nevanlinna, *Eindeutige Analytische Funktionen*, Berlin: Springer-Verlag, 1936.

where $\rho\ \mathrm{Re}(n^*v)$ is the mass (or charge) crossing a unit length of the boundary in a unit time orthogonally. The component of velocity parallel to the boundary $\mathrm{Im}(n^*v)$ does not transport mass (or charge) across the boundary:

$$\int \partial_t \rho \, dS = -\int \mathrm{Re}\{\partial^*(\rho v)\} \, dS$$

by the Gauss divergence theorem. Hence,

$$\int [\partial_t \rho + \mathrm{Re}\{\partial^*(\rho v)\}] \, dS = 0$$

and since this is also true for an *arbitrarily small portion* of S, it must follow that

$$\partial_t \rho + \mathrm{Re}\{\partial^*(\rho v)\} = 0$$

for if the latter expression could be positive throughout a small finite region, integration over a subregion of this could not lead to a zero result. Similarly the integrand could not be negative. Hence it can only be zero.

The equation can also be written in the form

$$\mathrm{Re}\{\partial_t \rho + v\partial^* \rho\} = -\rho \, \mathrm{Re}(\partial^* v)$$

where the left side gives the total time rate of change of the density ρ corresponding to

$$\partial_t \rho \, dt + \partial_x \rho \, dx + \partial_y \rho \, dy$$

the total differential of density. Thus if the total change in density due to change in time and change in position is zero, it follows that

$$\mathrm{Re}(\partial^* v) = 0 \tag{8.51}$$

the velocity of flow is divergenceless.

The del operator ∂ may also be expressed in terms of polar coordinates:

$$\partial = e^{i\theta}\left(\partial_r + \frac{i}{r}\,\partial_\theta\right) \tag{8.52}$$

by a simple transformation of $(\partial_x + i\partial_y)$. Thus if the velocity of a fluid is

$$v = \frac{c}{z^*} \qquad c = \text{real}$$

it is readily verified to be divergenceless at all points $z \neq 0$ except the origin. Thus,

$$v = \frac{c}{r}\, e^{i\theta}$$

and

$$\partial^* v = e^{-i\theta}\left(\partial_r - \frac{i}{r}\partial_\theta\right)\frac{c}{r}e^{i\theta}$$

$$\partial^* v = -\frac{c}{r^2} + \frac{c}{r^2} = 0 \qquad r \neq 0$$

For this reason the net mass transport rate (mass/time) out of any region not containing $z = 0$ must be zero. The mass transport out of a small circle about the origin is given by

$$\int \text{Re}\{\partial^*(\rho v)\}\, dS \approx \rho \oint \text{Re}(n^* v)\, ds$$

(the circle is supposed to be so small that ρ does not differ appreciably from its value at the center). Then

$$n^*\, ds = i\, dz^*$$

so $n^* v\, ds = ic(dz^*/z^*) = c\, d\theta$ since, on the circle, $dr = 0$. Thus the mass transport rate is $2\pi\rho c$, which gives c dimensions of area/time.

Hydrodynamical problems involving the irrotational flow of incompressible fluids in two dimensions can often be profitably studied by complex variable methods. The velocity of incompressible fluids is divergenceless ($\nabla \cdot \bar{v} = 0$), which follows from the equation of (continuity) mass conservation. For irrotational flow ($\nabla \times \bar{v} = 0$) there exists a *velocity potential* ϕ with $\bar{v} = \nabla\phi$ so that the divergencelessness implies that ϕ is harmonic ($\Delta\phi = 0$). The irrotationality implies via Stokes theorem that the complex velocity v satisfies

$$\text{Re}\left\{\oint v^*\, dz\right\} = 0 \tag{8.53}$$

which will be automatically fulfilled for $f(z)$ analytic:

$$v^* = f'(z) \tag{8.54}$$

where $f(z) = \phi + i\psi$, with ϕ the velocity potential of v. Thus, by the Cauchy–Riemann equations,

$$v = \partial_x\phi - i\partial_x\psi = \partial_x\phi + i\partial_y\phi$$

and $f(z)$ is called a *complex flow function*; Re $\oint v^*\, dz$ is called the *circulation.*

Although the flow has been assumed to be irrotational and divergenceless generally, there will be exceptional points at which singularities of the complex flow function occur. These are of great significance, and they essentially determine the resulting flow pattern. The simplest singularity in the finite z plane is a logarithmic one at $z = 0$ (c = source strength = real):

$$f(z) = c \log z = c[\log |z| + i \arg z]$$

so

$$\phi = c \log |z| \tag{8.55}$$

and the equipotential lines are circles on which $\phi =$ constant. The fluid velocity is orthogonal to the *equipotential* lines (the fluid flows along *streamlines* orthogonally to the equipotentials):

$$v = f'(z)^* = \frac{c}{z^*} = \frac{ce^{i\theta}}{r}$$

showing that the flow is radially directed outward (point source) if $c > 0$ and inward (point sink) if $c < 0$. Obviously, if fluid is "appearing" or "disappearing" at $z = 0$, the incompressibility assumption is violated there. If ($c =$ real)

$$f(z) = ic \log z = c[-\arg z + i \log |z|]$$

so

$$\phi = -c \arg z$$

and

$$v = f'(z)^* = \frac{-ic}{z^*} = \frac{-ice^{i\theta}}{r}$$

The flow is tangentially directed clockwise if $c > 0$ and counterclockwise if $c < 0$. Here the equipotentials are straight half-lines from the origin. Such a source is called a *vortex point source*, but it is not a "source" of fluid. Rather it is a source of rotation in the fluid. Notice that one passes from point source to vortex point source simply by multiplying the complex flow function by i.

8.9 MULTIPOLES

More complicated flows can be generated by superposing individual flows due to point sources and/or vortex point sources. In this way a dipole flow field can be generated as the limiting case of a flow due to a point source and point sink of equal strengths when the two are brought together in such a way that the product of source strength and separation is constant. Thus,

$$f(z) = c \log(z + a) - c \log(z - a)$$

or

$$f(z) = c \log\left(\frac{z + a}{z - a}\right) = c \log\left(1 + \frac{2a}{z - a}\right)$$

$$f(z) = c\left\{\left(\frac{2a}{z - a}\right) + \frac{1}{2}\left(\frac{2a}{z - a}\right)^2 + \cdots\right\} \approx \frac{2ac}{z - a}$$

$$g(z) = \frac{m}{z} = \lim_{\substack{a \to 0 \\ c \to \infty}}\left(\frac{2ac}{z - a}\right) \qquad \text{with} \qquad m = \lim (2ac) \tag{8.56}$$

and $g(z)$ is the complex flow function for a *dipole flow field*, with m the *dipole moment*.

The final result $g(z)$ for the complex flow function for a dipole field can also be written as

$$g(z) = md_z(\log z) = f_2(z)$$

If two dipoles of opposite orientations and equal strengths m are located at $z = a$ and $z = -a$, one has a limiting complex flow function $f_4(z)$ for a *quadrupole flow field* with *quadrupole moment* n as $a \to 0$ and $2ma \to n$:

$$\frac{m}{z+a} - \frac{m}{z-a} = \frac{-2ma}{z^2 - a^2} \to -\frac{n}{z^2} = f_4(z)$$

Also

$$f_4(z) = nd_z\left(\frac{1}{z}\right) = nd_z^2(\log z)$$

If two quadrupoles of opposite orientations and equal strengths n are located at $z = a$ and $z = -a$, one has

$$\frac{n}{(z-a)^2} - \frac{n}{(z+a)^2} = \frac{4azn}{(z^2-a^2)^2} \to \frac{2s}{z^3} = f_8(z)$$

an *octupole flow field* with *octupole moment* $s = \lim_{\substack{a\to 0\\ n\to\infty}} 2an$. Also

$$f_8(z) = sd_z^3(\log z)$$

and so on for higher *multipoles* whether they are hydrodynamical, electrostatic, or of other types.

On the other hand, a general distribution of charge of density $\rho(a)$ [i.e., $\rho(a)\,dS$ is the charge per unit length of the cylindrical volume element of cross section dS containing the point a] would generate a potential corresponding to a complex flow function

$$f(z) = 2\int \log(z-a)\rho(a)\,dS \tag{8.57}$$

Suppose that charges are confined to a finite region [i.e., $\rho(a) = 0$ for $|a|$ sufficiently large]. Then

$$f(z) = 2\log z\int \rho(a)\,dS + 2\int \log\left(1 - \frac{a}{z}\right)\rho(a)\,dS$$

and z can be chosen so large that

$$-\log\left(1 - \frac{a}{z}\right) = \sum_{k=1}^{\infty}\frac{1}{k}\left(\frac{a}{z}\right)^k$$

and

$$f(z) = 2q_0 \log z + 2\sum_{k=1}^{\infty} \frac{q_k}{z^k} \tag{8.58}$$

with

$$q_0 = \int \rho(a)\, dS = \text{total charge}$$

$$q_k = \frac{-1}{k} \int a^k \rho(a)\, dS = \text{multipole moment}$$

which exhibits the flow function as a superposition of a point charge and a collection of multipoles (multipole expansion).

Another way of describing this is to say that from a very great distance (so far that the reciprocal of $|z|$ is negligible) the charge distribution appears to be approximately a point (line) charge. From a somewhat closer distance (so that the reciprocal of $|z|^2$ is negligible) it appears to be a point charge plus a dipole. As one comes still closer, quadrupole and octupole and higher multipoles must be added to represent $f(z)$ accurately. In this way a multipole expansion (8.58) can be associated with the general distribution of charge specified by $\rho(a)$.

.10 COMPLEX FLOWS AND MAPPINGS

Numerous flows with singularities at $z = \infty$ are important. One of the simplest is the uniform flow

$$f(z) = v_\infty e^{-i\alpha} z$$

with $v = v_\infty e^{i\alpha}$ uniform velocity in direction α. The flow in a rectangular corner is given by

$$f(z) = cz^2 \qquad v = 2cz^*$$

for which v is imaginary for $z = iy$ and real for $z = x$. The flow around a plate along the positive real axis is given by

$$f(z) = c\sqrt{z} \qquad v = \frac{c}{2\sqrt{z^*}}$$

for which v is real if z is real and positive, and v is imaginary if z is real and negative.

If $f(z)$ is a complex flow function for which no singularities (sources) are within the circle $|z| \leq a$, one can find the corresponding (modified) flow when a solid cylinder is inserted into the flow at $|z| \leq a$. The *theorem of Milne–Thomson* yields

$$f^+(z) = f(z) + f^*\left(\frac{a^2}{z^*}\right) \tag{8.59}$$

the required modified flow function, since

(1) $f^+(z)$ has no singularities other than those of $f(z)$ outside the cylinder; a^2/z^* inverts the exterior to the interior of the cylinder and $f^*(a^2/z^*)$ is analytic, as may be seen from the power series for $f(z)$;

(2) on the cylinder, $f^+(z) = f(z) + f^*(z)$ is real so $\operatorname{Im} f^+(z) = 0$ and the cylinder must be a streamline (orthogonal trajectory of equipotentials); points of zero velocity on the cylinder are called stagnation points.

Since the real and imaginary parts of analytic functions are harmonic, there is an intimate connection between the theory of analytic functions and the two-dimensional problems of electrostatics. The existence of an electrostatic potential ϕ whose negative gradient is the electric field intensity is implied by the irrotationality of the latter. In this case all configurations of charges, conductors, and insulators are regarded as of infinite extent in the two directions of a straight line, and every section of such a configuration orthogonal to the line is supposed to be identical. It is in the context of a typical section that the terminology is given. Thus one refers to a line charge parallel to the given line as a "point charge," etc. One may also apply complex variable methods to the stationary flow of currents in a plane conducting sheet obeying Ohms law, in which case the current is proportional to the gradient of the potential. In the former case the irrotationality of the complex electric intensity E implies, via Stokes theorem (or from Faraday induction law),

$$\operatorname{Re} \oint E^* \, dz = 0$$

which is automatically fulfilled for

$$-E^* = f'(z)$$

with $f(z)$ analytic and

$$f(z) = \phi + i\psi$$

where ϕ is the electrostatic potential. Thus,

$$-E = \partial_x\phi - i\partial_x\psi = \partial_x\phi + i\partial_y\phi$$

and $f(z)$ may be called the *complex electropotential function*. At a metal surface, $4\pi\sigma = |\partial_n\phi| = |f'(z)|$ gives the surface charge density, since $f'(z) = \partial_n\phi + i\partial_n\psi = \partial_n\phi$ by the Cauchy–Riemann equations. The curves $\phi =$ constant are called *equipotentials*, and the orthogonal trajectories $\psi =$ constant are called *lines of force*.

As in the hydrodynamical case the singularities of $f(z)$ essentially determine the field. The simplest case is that of a line charge of strength q (charge per unit length) at $z = 0$ for which the complex electropotential function is

$$f(z) = -2q \log z$$

with corresponding electric intensity

$$E = \frac{2q}{z^*} = \frac{2qe^{i\theta}}{r}$$

For $q > 0$ the lines of force are directed *away* from $z = 0$, with E being understood to be the force (due to q) that *would* act on a *unit positive charge* if the latter were placed at a particular location $z = re^{i\theta}$ in the *field* of q. The charge q is said to *generate* the electric field. For $q < 0$ the lines of force are directed toward $z = 0$. The quantity $\operatorname{Im} \oint E^*\, dz$ is called the *total flux* through the contour about which it is extended.

Although multiplication of the complex function $f(z)$ by i interchanges the point sources and vortex point sources of a hydrodynamical flow, the requirement that the electric field lines of force originate on electrostatic charges excludes imaginary charges, so the multiplication of the complex electropotential function by i leads instead to a description of a magnetostatic field. The latter may be systematically introduced as follows: The irrotationality of the complex magnetic intensity H follows from the assumption of stationarity (no explicit dependence on time and therefore no displacement current) and the assumption of no convective current at nonsingular points of the field. It implies, via Stokes theorem (or from ampere circuital law), that

$$\operatorname{Re} \oint H^*\, dz = 0 \qquad (8.60)$$

which is automatically satisfied for

$$-H^* = f'(z) \qquad (8.61)$$

with $f(z)$ the *complex magnetostatic potential function*

$$f(z) = \phi + i\psi$$

and ϕ the magnetostatic potential. Thus,

$$-H = \partial_x\phi - i\partial_x\psi = \partial_x\phi + i\partial_y\phi$$

The simplest singularity is that of a linear current I at $z = 0$ for which

$$f(z) = 2Ii \ln z$$

with corresponding magnetic intensity

$$H = \frac{2Ii}{z^*} = \frac{2Iie^{i\theta}}{r} \tag{8.62}$$

(the Biot–Savart law). The z plane is understood to be a typical cross section of the configuration in which the current I is passing orthogonally through at $z = 0$. For $I > 0$ the current is piercing $z = 0$ from below and the direction of H is that of increasing θ (counterclockwise). For $I < 0$ the current is piercing $z = 0$ from above and the direction of H is clockwise.

For the case of stationary flow of electric currents in thin plane sheets of constant (unit) conductivity, Ohms' law and the Faraday induction law imply that the electric current density (current per unit area flowing across a normal section of the sheet) is irrotational. Thus Stokes theorem implies that the complex current density J satisfies

$$\text{Re} \oint J^* \, dz = 0 \tag{8.63}$$

which holds for

$$J^* = f'(z) \tag{8.64}$$

with $f(z)$ the *complex electrokinetic potential function*

$$f(z) = \phi + i\psi$$

and ϕ the electrokinetic potential. The simplest singularity is that of a "point" source of current I at $z = 0$ for which

$$f(z) = \frac{I}{2\pi h} \log z$$

where h is the sheet thickness. Thus the current density is

$$J = \frac{I}{2\pi h z^*} = \frac{Ie^{i\theta}}{2\pi h r}$$

with $I > 0$ corresponding to a current source (input) and $I < 0$ corresponding to a current sink (output).

Naturally, in all the above cases, superpositions of variously located singularities lead to functions $f(z)$ describing more complicated fields. Also, the various interpretations of the different fields for any particular choice of $f(z)$ lead to analogies such as that between magnetostatics and hydrodynamical vortex flows.

8.11 KOLOSOV POTENTIALS AND STRESS PROBLEMS

In the theory of elasticity the stresses σ_x, σ_y, τ_{xy} in a thin plate under forces acting in its plane may be expressed (in the absence of body forces) in terms of the *Airy function* $U(x,y)$, according to

$$\sigma_x = \partial_y^2 U \qquad \sigma_y = \partial_x^2 U \qquad \tau_{xy} = -\partial_x \partial_y U \tag{8.65}$$

The existence of the Airy function may be inferred from the static stress equilibrium equations

$$\partial_x \sigma_x = -\partial_y \tau_{xy} \qquad \partial_x \tau_{xy} = -\partial_y \sigma_y$$

which are satisfied by

$$\sigma_x = \partial_y V \qquad \tau_{xy} = -\partial_x V$$

and

$$\sigma_y = \partial_x W \qquad \tau_{xy} = -\partial_x W$$

respectively, so that

$$\partial_x V = \partial_y W$$

satisfied by

$$W = \partial_x U \qquad V = \partial_y U$$

By combining the compatibility condition,

$$\partial^* \partial(\sigma_x + \sigma_y) = 0 \tag{8.66}$$

with

$$\sigma_x + \sigma_y = \partial^* \partial U = \Delta U \tag{8.67}$$

it is shown that the Airy function is biharmonic, i.e., a solution of the biharmonic equation*

$$\Delta^2 U = (\partial^* \partial)^2 U = 0 \tag{8.68}$$

By transforming to $z = x + iy$ and $z^* = x - iy$ from x and y, (8.68) becomes (due to Goursat)

$$\partial_z^2 \partial_{z^*}^2 U = 0 \tag{8.69}$$

The general real solution of this (biharmonic) equation is seen to be

$$U(z,z^*) = \text{Re}[z^* \phi(z) + \chi(z)] \tag{8.70}$$

since $\partial_{z^*}(\partial_{z^*} \partial_z^2 U) = 0$ implies that

$$2\partial_{z^*} \partial_z^2 U = \phi''(z)$$

* N. I. Muskhelishvili, *Some Basic Problems of the Mathematical Theory of Elasticity*. Groningen: Noordhoff, 1953, p. 104.

is a function (say $\phi''(z)$) of z alone (i.e., not of z^*). Then

$$2\partial_z^2 U = z^*\phi''(z) + \chi''(z)$$

with χ'' independent of z^* because this is the most general expression yielding $\phi''(z)$ upon differentiation with respect to z^*. The next integration yields

$$2\partial_z U = z^*\phi'(z) + \chi'(z) + \psi(z^*)$$

whence

$$2U = z^*\phi(z) + \chi(z) + z\psi(z^*) + \xi(z^*)$$

However, since $U = U^*$ is supposed to be real, one chooses

$$\psi(z^*) = \phi^*(z)$$
$$\xi(z^*) = \chi^*(z)$$

so that

$$U = \mathrm{Re}[z^*\phi(z) + \chi(z)]$$

Also,

$$\begin{aligned}
\partial_x U &= \mathrm{Re}[z^*\phi'(z) + \phi(z) + \chi'(z)] \\
\partial_y U &= \mathrm{Re}[-i\phi(z) + iz^*\phi'(z) + i\chi'(z)] \\
\partial_y U &= -\mathrm{Im}[z^*\phi'(z) - \phi(z) + \chi'(z)]
\end{aligned} \tag{8.71}$$

so that

$$\partial^* U = (\partial_x - i\partial_y)U = z^*\phi'(z) + \chi'(z) + \phi^*(z)$$
$$\partial U = (\partial_x + i\partial_y)U = z[\phi'(z)]^* + [\chi'(z)]^* + \phi(z)$$

Furthermore,

$$\begin{aligned}
\partial_x^2 U &= \mathrm{Re}[z^*\phi''(z) + 2\phi'(z) + \chi''(z)] \\
\partial_y^2 U &= \mathrm{Re}[2\phi'(z) - z^*\phi''(z) - \chi''(z)]
\end{aligned} \tag{8.72}$$

so

$$\Delta U = \partial^*\partial U = 4\ \mathrm{Re}\ \phi'(z) = \sigma_x + \sigma_y$$

Also,

$$\partial_x\partial_y U = \mathrm{Re}[iz^*\phi''(z) + i\chi''(z)] = -\mathrm{Im}[z^*\phi''(z) + \chi''(z)] = -\tau_{xy}$$

and since

$$\sigma_y - \sigma_x = 2\ \mathrm{Re}[z^*\phi''(z) + \chi''(z)]$$

one has

$$\sigma_y - \sigma_x + 2i\tau_{xy} = 2[z^*\phi''(z) + \chi''(z)]$$

with the *Kolosov potentials*

$$\Phi(z) = \phi'(z) \qquad \Psi(z) = \chi''(z)$$

the stress expressions become

$$\sigma_x + \sigma_y = 4 \operatorname{Re} \Phi(z) = \sigma_r + \sigma_\theta$$

$$\sigma_y - \sigma_x + 2i\tau_{xy} = 2[z^*\Phi'(z) + \Psi(z)] \tag{8.73}$$

According to Hooke's law* (case of plane stress) the stresses and displacements are related by ($E =$ Young's modulus, $\nu =$ Poisson's ratio, $G =$ shear modulus)

$$E\partial_x u = \sigma_x - \nu\sigma_y$$

$$E\partial_y v = \sigma_y - \nu\sigma_x \tag{8.74}$$

$$G(\partial_y u + \partial_x v) = \tau_{xy}$$

Thus,

$$E\partial_x u = \partial_y^2 U - \nu\partial_x^2 U = \sigma_x - \nu\sigma_y$$

So by substitution,

$$E\partial_x u = \operatorname{Re}[2(1 - \nu)\phi' - (1 + \nu)z^*\phi'' - (1 + \nu)\chi''] \tag{8.75}$$

and

$$E\partial_y v = \partial_x^2 U - \nu\partial_y^2 U = \sigma_y - \nu\sigma_x$$

so that substituting one has

$$E\partial_y v = \operatorname{Re}[(1 + \nu)z^*\phi'' + 2(1 - \nu)\phi' + (1 + \nu)\chi''] \tag{8.76}$$

Hence,

$$Eu = \operatorname{Re}[\alpha z^*\phi' + \beta\phi + \gamma\chi']$$

yields

$$E\partial_x u = \operatorname{Re}[\alpha\phi' + \alpha z^*\phi'' + \beta\phi' + \gamma\chi'']$$

which is consistent with (8.75) for

$$\alpha = -(1 + \nu) = \gamma \qquad \beta = (3 - \nu)$$

so

$$Eu = \operatorname{Re}[-(1 + \nu)z^*\phi' + (3 - \nu)\phi - (1 + \nu)\chi'] \tag{8.77}$$

Similarly,

$$Ev = \operatorname{Re}[\alpha' z^*\phi' + \beta'\phi + \gamma'\chi']$$

so

$$E\partial_y v = \operatorname{Re}[-i\alpha'\phi' + i\alpha' z^*\phi'' + i\beta'\phi' + i\gamma'\chi'']$$

and this checks with (8.76) for

$$i\alpha' = 1 + \nu = i\gamma' \qquad i\beta' = 3 - \nu$$

* See S. Timoshenko and J. N. Goodier, *Theory of Elasticity*, 2nd ed., New York: McGraw-Hill, 1951.

so that

$$Ev = \text{Im}[(1 + \nu)z^*\phi' + (3 - \nu)\phi + (1 + \nu)\chi'] \tag{8.78}$$

and

$$Es = E(u + iv) = (3 - \nu)\phi - (1 + \nu)(z[\phi']^* + [\chi']^*) \tag{8.79}$$

This gives the complex displacement $s = u + iv$ for plane stress (i.e., no stress normal to the z plane). The corresponding formula for s in case of plane strain (i.e., no displacement normal to the z plane) is

$$Es = E(u + iv) = (1 + \nu)(3 - 4\nu)\phi - (1 + \nu)(z[\phi']^* + [\chi']^*) \tag{8.80}$$

In addition to these the *Muskhelishvili formulae* for complex force F and real moment M exerted on an arc

$$z = z(s) \qquad a \leq s \leq b$$

are useful. These are

$$F = \int_a^b F_n\, ds = -i\int_a^b d(\partial U) = -i(\partial U)\Big|_a^b \tag{8.81}$$

since F_n, the force per unit length (per unit area if the direction orthogonal to the z plane also is considered) in the direction of the normal to the curve, is given by

$$F_n = (\sigma_x \cos\alpha + \tau_{xy} \sin\alpha) + i(\tau_{xy} \cos\alpha + \sigma_y \sin\alpha)$$

for α the argument of the unit normal to the curve. Then

$$n = e^{i\alpha} = -id_s z = d_s y - id_s x$$

and

$$F_n = (\partial_y^2 U d_s y + \partial_x \partial_y U d_s x) + i(-\partial_x \partial_y U d_s y - \partial_x^2 U d_s x)$$

so that

$$F_n = d_s(\partial_y U - i\partial_x U) = -id_s(\partial U)$$

which establishes (8.81). This may also be written

$$F = -i(z[\phi']^* + [\chi']^* + \phi)|_a^b \tag{8.82}$$

The moment acting on the arc is

$$M = \text{Im}\int_a^b z^* F_n\, ds$$

or

$$-M = \text{Re}\int_a^b z^*\, d(\partial U) = \text{Re}\left[z^*(\partial U)\Big|_a^b - \int_a^b \partial U\, dz^*\right]$$

Since $\mathrm{Re}(\partial U\, dz^*) = dU$, one has

$$-M = \mathrm{Re}(z^*z[\phi']^* + z^*[\chi']^* + z^*\phi) - U \Big|_a^b$$

and $U = \mathrm{Re}[z^*\phi + \chi]$, so

$$-M = \mathrm{Re}[z^*z\phi' + z\chi' - \chi] \Big|_a^b$$

$$M = \mathrm{Re}(\chi - z\chi' - z^*z\phi') \Big|_a^b \tag{8.83}$$

The Muskhelishvili formulae are (8.82) and (8.83).

The (Kirsch) formulae for stresses in a plate with circular hole of radius R subjected to uniaxial tension $\sigma = \sigma_x$ at infinity may be derived as follows: The expressions for $\Phi(z)$ and $\Psi(z)$ must reduce to

$$\Phi(\infty) = \frac{\sigma}{4} \qquad \Psi(\infty) = -\frac{\sigma}{2}$$

Also, since $\sigma_x + \sigma_y$ must remain invariant under the transformation $z \to -z$, the real part of $\Phi(z)$ must be even in z. This suggests that the simplest choice for $\Phi(z)$ may be

$$\Phi(z) = \frac{\sigma}{4} - \frac{c}{z^2}$$

Then, trying this *Ansatz* with real c,

$$\Phi'(z) = \frac{2c}{z^3}$$

so that on $z^*z = R^2$, where $\sigma_r = 0 = \tau_{r\theta}$, one has

$$\sigma_r + \sigma_\theta = \sigma_\theta = \sigma - \frac{4c}{R^2}\cos 2\theta = \sigma - \frac{2c}{R^2}(e^{2i\theta} + e^{-2i\theta})$$

$$\sigma_\theta - \sigma_r + 2i\tau_{r\theta} = \sigma_\theta = 2\left(\frac{2ce^{-2i\theta}}{R^2} + \Psi e^{2i\theta}\right)$$

Equating and solving for Ψ,

$$\Psi = -\frac{c}{R^2} + \frac{\sigma e^{-2i\theta}}{2} - \frac{3ce^{-4i\theta}}{R^2}$$

which corresponds to

$$\Psi(z) = -\frac{c}{R^2} + \frac{\sigma R^2}{2z^2} - \frac{3cR^2}{z^4}$$

and

$$\Psi(\infty) = -\frac{c}{R^2} = -\frac{\sigma}{2}$$

yields $2c = \sigma R^2$, so

$$\Psi(z) = -\frac{\sigma}{2}\left(1 - \frac{R^2}{z^2} - \frac{3R^4}{z^4}\right) \tag{8.84}$$

and

$$\Phi(z) = \frac{\sigma}{4}\left(1 - \frac{2R^2}{z^2}\right)$$

Using these Kolosov potentials to solve for the stresses, one has the Kirsch formulae

$$\sigma_r = \frac{\sigma}{2}\left(1 - \frac{R^2}{r^2}\right) + \frac{\sigma}{2}\left(1 - \frac{4R^2}{r^2} + \frac{3R^4}{r^4}\right)\cos 2\theta$$

$$\sigma_\theta = \frac{\sigma}{2}\left(1 + \frac{R^2}{r^2}\right) - \frac{\sigma}{2}\left(1 + \frac{3R^4}{r^4}\right)\cos 2\theta$$

$$\tau_{r\theta} = -\frac{\sigma}{2}\left(1 + \frac{2R^2}{r^2} - \frac{3R^4}{r^4}\right)\sin 2\theta$$

At the hole, $r = R$, and

$$\sigma_\theta = \sigma(1 - 2\cos 2\theta)$$

which attains its maximum value of 3σ (triple the tensile stress applied at $r = \infty$) at $\theta = \pm\pi/2$. These two points are then the initial failure points of the plate if σ is raised to one-third the yield stress of the material.

● EXAMPLES 8

1. Determine the curvature of the image of a unit circle $|z| = 1$ under the mapping $w = f(z)$.

Ans. $\quad dz = iz\,d\theta \quad$ for $\quad z = e^{i\theta}$

$$\arg dz = \frac{\pi}{2} + \arg z$$

$$d(\arg dz) = d(\arg z) = d\theta = \mathrm{Im}\left(\frac{dz}{z}\right)$$

The curvature κ is given by

$$\kappa\,|dw| = d(\arg dw)$$

and $dw = f'(z)\,dz$, so $\arg dw = \arg f'(z) + \arg dz$. Then $d(\arg dw) = d\{\arg f'(z)\} + d\{\arg dz\}$ and $\arg f'(z) = \operatorname{Im}\{\log f'(z)\}$, so $d\{\arg f'(z)\} = \operatorname{Im}(f''/f'\,dz)$. Hence,

$$d(\arg dw) = \operatorname{Im}\left\{\frac{f''}{f'}\,iz\,d\theta + \frac{iz\,d\theta}{z}\right\}$$

$$d(\arg dw) = \operatorname{Re}\left\{1 + z\frac{f''}{f'}\right\}d\theta$$

$$\kappa = \frac{d(\arg dw)}{|dw|} = \frac{\operatorname{Re}(1 + zf''/f')}{|zf'|}$$

2. Show that the angular velocity ω of the tangent to the image of the unit circle $|z| = 1$ under the mapping $w = f(z)$ is given by

$$\omega = 1 + izd_z(\arg f')$$

assuming that the tangent corresponds to a point moving on the unit circle $|z| = 1$ at unit angular velocity.

Ans. $\omega = d_t(\arg dw)$ and $d_t = izd_z$

3. Show that the image of the unit circle $|z| = 1$ is convex (free of dents) if and only if

$$\operatorname{Re}(zf''/f') > -1$$

where $w = f(z)$ is the mapping.

Ans. This is the condition for $\kappa > 0$.

4. Show that the necessary and sufficient condition that every point of the image of $|z| = 1$ under $w = f(z)$ is "visible" from $w = 0$ is

$$\operatorname{Re}\left(\frac{zf'}{f}\right) > 0$$

In this case the image curve is said to be *star-shaped* from $w = 0$.

Ans. For a star-shaped region one must have $d_t(\arg w) > 0$.

5. If $P(z) = az^2 + bz + c$ with $a > 0$, b, c real, find the trajectories of the roots as c increases from zero to infinity while a and b remain constant.

Ans. For $c = 0$ the two roots are $z = 0$ and $z = -b/a$. Since the roots are generally given by

$$z = \frac{-b \pm \sqrt{b^2 - 4ac}}{2a}$$

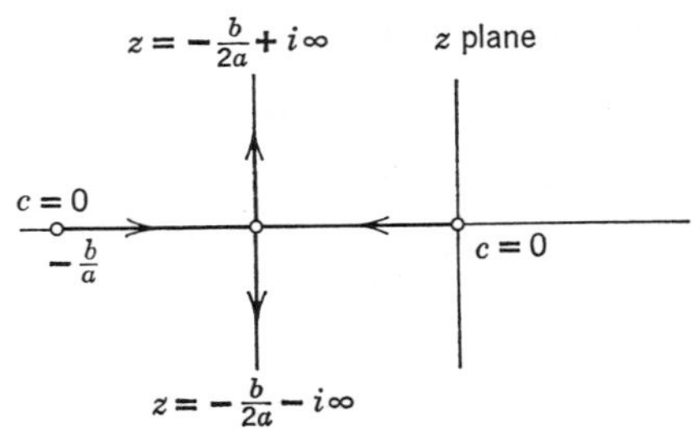

Figure 8.17. Root loci for Example 5.

they will remain real and distinct until $c = b^2/4a$, when they will both become $-b/2a$. As c increases above $b^2/4a$, they will become complex conjugate with real part $-b/2a$ and they will therefore move parallel to the imaginary axis in opposite directions towards $-b/2a \pm i\infty$.

6. Find the trajectories of the roots of $P(z) = az^2 + bz + c$ for $a > 0$, b, c real as b increases from zero to infinity with a and $c > 0$ remaining constant.

Ans. $\partial_b P = (2az + b)\partial_b z + z = 0$

so

$$z^*\partial_b z = -\frac{z^*z}{(2az + b)}$$

and, since $|z|^2 = z^*z$ is real,

$$\mathrm{Re}(z^*\partial_b z) = -|z|^2 \,\mathrm{Re}[(2az + b)^{-1}]$$

But

$$(2az + b) = \pm\sqrt{b^2 - 4ac}$$

so

$$\mathrm{Re}[(2az + b)^{-1}] = 0 \qquad \text{for} \quad b^2 < 4ac$$

Hence, for $0 \leq b^2 < 4ac$, the trajectories are quarter *circles* about the origin of radius $\sqrt{c/a}$. For $b^2 \geq 4ac$,

$$\mathrm{Im}(z^*\partial_b z) = 0$$

so the trajectories lie along the real axis in each direction away from $z = -\sqrt{c/a}$.

7. Find the linear and areal distortion in the mapping $w = z^3$.

Ans. $|d_z w| = 3\,|z|^2$ is the linear distortion factor and $9\,|z|^4$ is the areal distortion factor.

8. Describe the mapping $w = (\frac{1}{2})[z + (1/z)]$.

Ans. Since $d_z w = (\frac{1}{2})[1 - (1/z^2)] = 0$ for $z = \pm 1$, the mapping will fail to be conformal at these points. Considering the circle

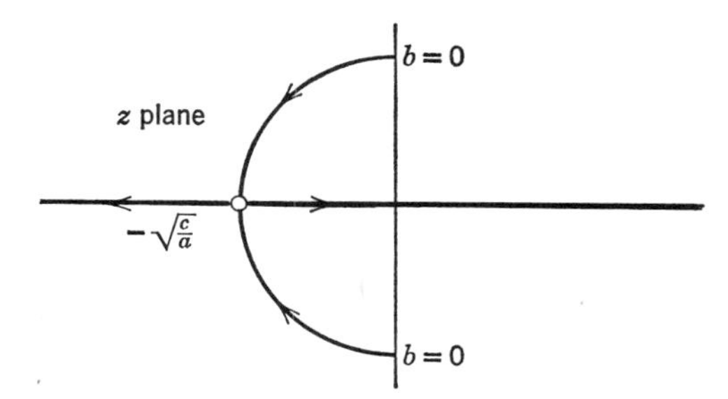

Figure 8.18. Root loci for Example 6.

$|z| = r$ with $z = re^{i\theta}$ and $w = u + iv$, one has

$$2u = \left(r + \frac{1}{r}\right)\cos\theta$$

$$2v = \left(r - \frac{1}{r}\right)\sin\theta$$

so that the image of $|z| = r$ is the ellipse

$$\frac{4u^2}{[r + (1/r)]^2} + \frac{4v^2}{[r - (1/r)]^2} = 1$$

in the w plane. This is also the image of $|z| = 1/r$. As $r \to 1$ the major semiaxis approaches 1 and the minor semiaxis approaches zero, so that the *outside* and the *inside* of the unit circle in the z plane both correspond to the whole w plane cut by a straight segment between $w = -1$ and $w = 1$. The inverse mapping $z = w \pm \sqrt{w^2 - 1}$ is double-valued, with $z = w - \sqrt{w^2 - 1}$ giving a mapping of the cut w plane onto $|z| < 1$ and $z = w + \sqrt{w^2 - 1}$ giving a mapping of the cut w plane onto $|z| > 1$.

9. Map the region remaining when $\operatorname{Im} w > 0 < \operatorname{Re} w$ is removed from $\operatorname{Im} w \geq -a$ onto $\operatorname{Im} z \geq 0$, sending $w = +\infty$ into $z = 0$ and $w = 0$ into $z = 1$.

Ans. The Schwarz–Christoffel mapping is implicitly

$$d_z w = A(z - 1)^{1/2} z^{-1} = \frac{A\sqrt{z - 1}}{z}$$

since the interior angle at $w = 0$ is $3\pi/2$ and at $w = \infty$ (between $\operatorname{Im} w = 0$ and $\operatorname{Im} w = -a$) it is zero. Integration along a vertical path at $w = \infty$ (or about a semicircular arc of infinitesimal radius around $z = 0$) yields the value of the constant A. Thus,

$$ia = \lim_{\varepsilon \to 0} \int_{(1/\varepsilon)-ia}^{1/\varepsilon} dw = A \lim_{\varepsilon \to 0} \int_{\pi}^{0} (\varepsilon e^{i\theta} - 1)^{1/2} i \, d\theta = A\pi$$

so

$$A = \frac{ia}{\pi} \qquad \text{and} \qquad w = \frac{2ia}{\pi} [\sqrt{z-1} + \arctan \sqrt{z-1}]$$

10. Find the Green's function for the semicircular region $1 \geq |z|$, $\text{Re}\, z \geq 0$.

Ans. First one finds the mapping of the given region onto the unit circle via a mapping onto a half-plane. The angles at $z = \pm 1$ must be opened up from $\pi/2$ in the z plane to π in the w plane. If

$$\xi = -\left(\frac{z-1}{z+1}\right)^2$$

the corner angles will be properly opened, but

$$\arg \xi = 2[\arg(z-1) - \arg(z+1)] - \pi$$

indicates that, on the image of the arc, $\arg \xi = 0$, while on the image of the diameter, $\arg \xi = \pi$, so that the semicircular disk is mapped onto the upper half ξ plane. The point $z = a$ becomes $\xi = b = -(a-1)^2/(a+1)^2$. To map the upper half ξ plane onto $|w| \leq 1$, with $\xi = b$ going into $w = 0$, one has

$$w = \frac{\xi - b}{\xi - b^*} = \left[\left(\frac{z-1}{z+1}\right)^2 + \left(\frac{a-1}{a+1}\right)^2\right] \Big/ \left[\left(\frac{z-1}{z+1}\right)^2 + \left(\frac{a^*-1}{a^*+1}\right)^2\right]$$

and the Green's function is

$$g(z,a) = 2q$$

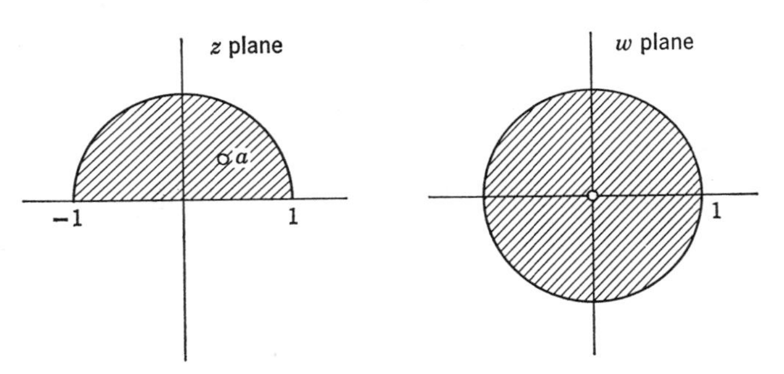

Figure 8.19. Green's function for harmonic Dirichlet problem for interior of semicircle.

11. Verify that the complex flow function

$$f(z) = v_\infty\left(z + \frac{R^2}{z}\right)$$

describes a divergenceless, irrotational (two-dimensional) flow of an ideal fluid past a cylinder of radius R centered at $z = 0$ with (real) velocity v_∞ parallel to the real axis at $z = \infty$.

Ans. The velocity potential ϕ is supposed to be the real part of $f(z)$, so that

$$f(z) = \phi + i\psi$$

and the complex velocity

$$v = f'(z)^* = \partial_x\phi - i\partial_x\psi = \partial_x\phi + i\partial_y\phi = v_1 + iv_2$$

by the Cauchy–Riemann relations, so that the divergence of v is

$$\partial_x v_1 + \partial_y v_2 = \partial_x^2\phi + \partial_y^2\phi = 0$$

since ϕ is harmonic. The curl of v is zero, since

$$\partial_x v_2 - \partial_y v_1 = \partial_x\partial_y\phi - \partial_y\partial_x\phi = 0$$

For the given $f(z)$,

$$v = v_\infty\left(1 - \frac{R^2}{z^*z^*}\right)$$

so at $z = \infty$, $v = v_\infty$, and

$$v^*z = v_\infty\left(z - \frac{R^2}{z}\right)$$

also $\operatorname{Re}(v^*z) = v_\infty \operatorname{Re}[z - (R^2/z)] = 0$ on $z^*z = R^2$, so the direction of v is parallel to the surface of the cylinder.

12. By mapping a circle interior to another and tangent to it at a single point into a circular arc, show that a superposition of uniform flow and vortex flow about the exterior circle can be mapped into the flow past an aerofoil (Joukowsky wing section).

Ans. Take the circular arc in the w plane to pass through $w = \pm 2a$ and $w = 2ib$. Since $(R - 2b)^2 + (2a)^2 = R^2$, the radius of curvature of the arc in the w plane is

$$R = \frac{a^2 + b^2}{b}$$

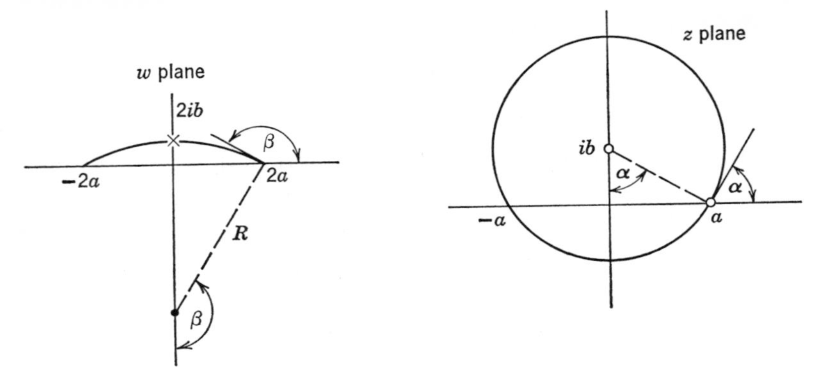

Figure 8.20. Basic mapping for Joukowsky aerofoil.

and the angle β between the arc and the real axis increasing from $w = 2a$ satisfies

$$\sin \beta = \frac{2ab}{a^2 + b^2}$$

Clearly,

$$\frac{w - 2a}{w + 2a} = \left(\frac{z - a}{z + a}\right)^2$$

will take the arc into a circle through $z = \pm a$ in the z plane, since

$$\arg(w - 2a) - \arg(w + 2a) = 2[\arg(z - a) - \arg(z + a)]$$

ensures that the right side will be constant as w traverses the arc. Thus the images of the arc in the w plane must be circular arcs with a common tangent at $z = a$, since the angle 2π required to rotate the tangent at $w = 2a$ from the concave to the convex side of the arc is halved to π in the z plane. The same is true for $w = -2a$ and $z = -a$. By solving for w, the mapping may also be expressed as

$$w = z + \frac{a^2}{z}$$

whence real $z > a$ is seen to correspond to real $w > 2a$ and the angle α between the tangent to the circle in the z plane at $z = a$ and the real axis ($z > a$) is thus half of β (i.e., $\beta = 2\alpha$). The inverse mapping is

$$2z = w \pm \sqrt{w^2 - 4a^2}$$

which is double-valued, so that $w = 2ib$ corresponds to the two points $z = i(b \pm \sqrt{a^2 + b^2})$ whence the center of the circle in the z plane is at $z = ib$ and its radius is $\sqrt{a^2 + b^2}$. The plus sign in the inverse transformation corresponds to mapping the w plane slit along the arc onto the *exterior* of the circle ($w = \infty, z = \infty$) and the minus sign corresponds to mapping the slit w plane onto the *interior* of the circle ($w = \infty, z = 0$). Having established that the mapping of the w plane slit along the arc onto the *exterior* of the circle in the z plane is

$$2z = w + \sqrt{w^2 - 4a^2}$$

one observes that a larger circle in the z plane containing the circle described above and tangent to it at $z = a$ will be mapped into an aerofoil in the w plane with a cusp at $w = 2a$. Supposing the large circle to be surrounded by a superposition of a flow uniform at $z = \infty$ and a vortex, one has for the complex flow function

$$f(z) = v_\infty e^{i\phi}\left[z - c + \frac{\rho^2}{z - c}\right] + i\lambda \log(z - c)$$

where the center of the large circle is at $z = c$, its radius is ρ, and ϕ is the direction of the uniform flow at $z = \infty$. The last (vortex) term is inserted so that by an appropriate choice for λ the stagnation point on the trailing portion may be made to coincide with the cusp. The complex flow function in the w plane is then obtained by replacing z in $f(z)$ by $z = (w + \sqrt{w^2 - 4a^2})/2$. Since λ must be selected so that $F'(2a) = 0$, one has

$$F'(w) = f'(z)\, d_w z$$

$$F'(2a) = 0 = f'(a)\, d_w z\big|_{z=a}$$

$$f'(z) = v_\infty e^{i\phi}\left\{1 - \frac{\rho^2}{(z - c)^2}\right\} + \frac{i\lambda}{z - c}$$

$$f'(a) = v_\infty e^{i\phi}\left[1 - \frac{\rho^2}{(a - c)^2}\right] + \frac{i\lambda}{a - c} = 0$$

$$\lambda = iv_\infty e^{i\phi}\left[1 - \left(\frac{\rho}{a - c}\right)^2\right](a - c)$$

$$\lambda e^{-i[\phi+(\pi/2)]} = v_\infty\left[1 - \left(\frac{\rho}{a - c}\right)^2\right](a - c)$$

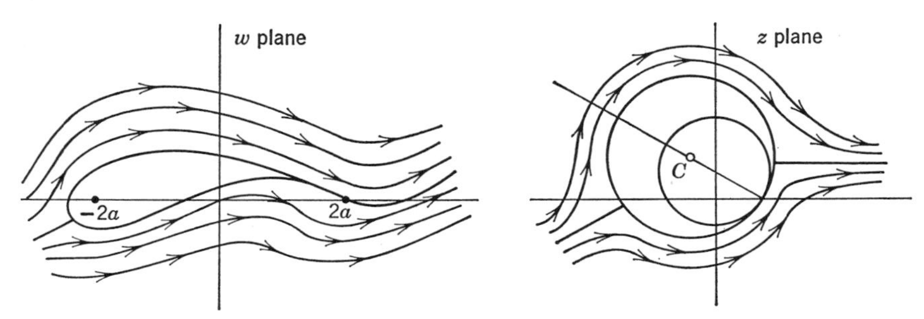

Figure 8.21. Streamline flow past aerofoil as image of flow about cylinder.

so that λ as well as ϕ must be available to match the input parameters on the right side. The pressure on the aerofoil can be calculated from Bernoulli's theorem:

$$p + \frac{v^2}{2} = \text{constant} = p_\infty + \frac{v_\infty^2}{2}$$

13. Find the electric field intensity when a grounded conducting plane strip of width $2c$ is placed in a uniform electric field of intensity E_∞ at angle α.

Ans. If the strip were parallel to the uniform field, say $|\text{Re}\, z| \leq c$, the field could be found by mapping the resultant field when a grounded conducting cylinder of unit radius is inserted into a uniform field k in the w plane. Let the cylinder be $|w| = 1$ and let k be real. Then the complex electropotential function in the w plane is

$$g(w) = -k\left(w - \frac{1}{w}\right)$$

because the corresponding real electrostatic potential is

$$\text{Re}\, g(w) = -k\, \text{Re}(w - w^*) = 0$$

on $w^*w = 1 = |w|^2$, and for large w the electric intensity is

$$-g'(w)^* = k\left(1 - \frac{1}{w^2}\right)^* \cong k$$

which is uniform. Furthermore, by Example 8, this field in the w plane may be mapped into the z plane with $|w| = 1$ going into the strip $|x| \leq c$ by the mapping

$$z = \frac{c}{2}\left(w + \frac{1}{w}\right)$$

The corresponding inverse mapping is

$$w = \frac{z + \sqrt{z^2 - c^2}}{c}$$

where the positive sign before the square root has been chosen to ensure that $z = \infty$ corresponds to $w = \infty$. Hence the complex electropotential function in the z plane becomes

$$f_1(z) = g(w) = g\left(\frac{z + \sqrt{z^2 - c^2}}{c}\right)$$

or

$$f_1(z) = -k\left(\frac{z + \sqrt{z^2 - c^2}}{c} - \frac{c}{z + \sqrt{z^2 - c^2}}\right)$$

$$f_1(z) = -\frac{2k}{c}\sqrt{z^2 - c^2}$$

which for large z is approximately

$$f_1(z) \approx -\frac{2kz}{c}$$

so that, since $E_\infty = -f_1'(z)^*$ for large z, the constant k is identified as

$$k = \frac{cE_\infty}{2}$$

$$f_1(z) = -E_\infty\sqrt{z^2 - c^2}$$

This solves the problem for the case $\alpha = 0$. For $\alpha \neq 0$ the complex electropotential function $f(z)$ must be a linear combination of $f_1(z)$ and z since only such a combination will yield a field uniform for large z. One has

$$f(z) = af_1(z) + bz$$

Also, to yield a field of intensity E_∞ at angle α for large z, $f(z)$ must be approximately

$$f(z) \approx -E_\infty e^{-i\alpha} z$$

Hence,

$$-E_\infty e^{-i\alpha} = -aE_\infty + b$$

Since the strip is grounded, $\text{Re} f(z) = 0$ on the strip and in particular $\text{Re} f(0) = 0$, so that $\text{Re}[af_1(0)] = -E_\infty \, \text{Re}(iac) = 0$

and a must be real. Also, the field must be perpendicular to the strip and in particular $f'(0) = b$ must be imaginary. Thus,

$$a = \cos\alpha \qquad b = iE_\infty \sin\alpha$$

and

$$f(z) = E_\infty(iz\sin\alpha - \sqrt{z^2 - c^2}\cos\alpha)$$

while the field is

$$E = -[f'(z)]^* = E_\infty\left[\frac{z\cos\alpha}{\sqrt{z^2 - c^2}} - i\sin\alpha\right]^*$$

14. Find the distribution of charge along the strip of Example 13.

Ans. If the charge per unit area is σ, the charge per unit length q is

$$q = \oint \sigma\, ds$$

where the contour is an orthogonal section of the surface of the strip. Since the total flux is (Gauss electrostatic theorem)

$$\text{Im} \oint E^*\, dz = 4\pi \oint \sigma\, ds$$

(actually this must hold on *any* closed surface containing charge), one has

$$4\pi\sigma = |f'(z)|$$

for z on the strip surface. Hence,

$$4\pi\sigma = E_\infty \left| \sin\alpha + \frac{z\cos\alpha}{\sqrt{c^2 - z^2}} \right|$$

for z on the strip surface. As expected, the charge density becomes infinite at $z = \pm c$.

15. Find the capacitance per unit length of a conducting elliptic cylinder surrounding a conducting strip. Assume the orthogonal section of the cylinder has the section of the strip extending between the foci of the ellipse.

Ans. Let the equation of the ellipse be ($w = u + iv$)

$$\left(\frac{u}{a}\right)^2 + \left(\frac{v}{b}\right)^2 = 1$$

with the strip section $|u| \leq k = \sqrt{a^2 - b^2}$. The capacitance C per unit length of two concentric circular cylinders (inner one of unit radius, outer one of radius s) is obtained from the potential function

$$\phi = V\frac{\log |z|}{\log s} = 2q \log |z|$$

if the inner cylinder is grounded and the outer cylinder is at voltage V. Since this corresponds to a charge per unit length q, one has

$$C = \frac{q}{V} = \frac{1}{2 \log s}$$

The mapping (see Example 8) $w = (k/2)[z + (1/z)]$ takes $|z| = 1$ into the strip $|u| \le k$ and it also takes $|z| = s$ into the ellipse

$$u = \frac{k}{2}\left(s + \frac{1}{s}\right)\cos\theta \qquad v = \frac{k}{2}\left(s - \frac{1}{s}\right)\sin\theta$$

so that the semiaxes are

$$a = \frac{k}{2}\left(s + \frac{1}{s}\right) \qquad b = \frac{k}{2}\left(s - \frac{1}{s}\right)$$

and, since $s = e^{1/2C}$,

$$\frac{a}{k} = \cosh\left(\frac{1}{2C}\right) \qquad \frac{b}{k} = \sinh\left(\frac{1}{2C}\right)$$

so the capacitance can be expressed as

$$C = \frac{1}{2\,\text{arcosh}(a/k)} = \frac{1}{2\,\text{arcsinh}(b/k)}$$

16. Find the electrostatic potential of an infinite conducting plane from which two parallel strips of negligible width have been removed if the region between the strips is maintained at unit voltage and the region outside the strips is grounded.

Ans. Let the region at unit voltage lie on the real axis with $a < \operatorname{Re} z < b$, while the grounded region is $\operatorname{Re} z < a$ and $\operatorname{Re} z > b$. The harmonic measure of the region at unit voltage is

$$\phi = \frac{1}{\pi}\arg\left(\frac{z-b}{z-a}\right)$$

which approaches 1 as z approaches the portion at unit voltage and 0 as z approaches the grounded portion.

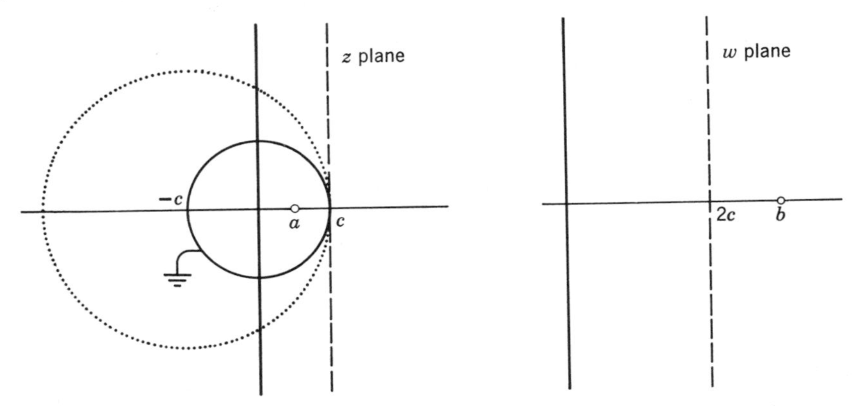

Figure 8.22. Mapping for field and charge density of cylinder containing parallel line charge.

17. Find the field and surface charge density for a grounded conducting cylinder at $|z| = c$ with a parallel line charge q per unit length at $z = a$ (a real; $a < c$).

Ans. Construct the circle $|z + c| = 2c$ tangent to $|z| = c$ at $z = c$ and invert $|z| = c$ and $z = a$ with respect to the constructed circle.

$$w = \frac{4c^2}{z + c}$$

is the inversion and

$$z = \infty \rightarrow w = 0$$

$$z = -c \rightarrow w = \infty$$

$$z = c \rightarrow w = 2c$$

$$z = a \rightarrow w = \frac{4c^2}{a + c} = b$$

$$|z| = c \rightarrow \text{Re } w = 2c$$

$$|z + c| = 2c \rightarrow |w| = 2c$$

so that the cylinder and line charge become a plane and line charge. For the latter the complex electropotential function is

$$g(w) = 2q \log(w - b) - 2q \log[w - (4c - b)]$$

so that the corresponding function in the z plane is

$$f(z) = g\left(\frac{4c^2}{z + c}\right) = 2q \log\left[\frac{c(z - a)}{(az - c^2)}\right]$$

corresponding to field

$$E = -2q\left\{\frac{1}{z^* - a} - \frac{a}{az^* - c^2}\right\}$$

and since

$$\left|\frac{z - a}{z - (c^2/a)}\right|$$

is constant on $|z| = c$, one has there

$$\left|\frac{z - a}{z - (c^2/a)}\right| = \left|\frac{c - a}{c - (c^2/a)}\right| = \frac{a}{c}$$

and $\text{Re}\, f(z) = 0$ on $|z| = c$, as it should. The problem is thus solved by superposing the potential due to the opposite charge at the point inverse to $z = a$ and a constant potential. The surface charge density σ is then given by

$$4\pi\sigma = |f'(z)| = 2q\left|\frac{1}{z - a} - \frac{a}{az - c^2}\right|$$

so that

$$\sigma = \frac{q(c^2 - a^2)}{2\pi c\, |z - a|^2} \qquad \text{for} \quad |z| = c$$

and the cylinder also carries a charge q per unit length according to the Gauss electrostatic theorem. For the case of an uncharged cylinder the electropotential function must be modified by superposing a charge $-q$ per unit length at $z = 0$ so that

$$f(z) = 2q \log\left[\frac{c(z - a)}{z(az - c^2)}\right]$$

18. Find the number of zeros of $w = z^3 + 4z^2 - 2z + 2$ in the right half z plane.

Ans. Construct a closed contour consisting of the imaginary axis I as diameter and a large semicircle C in the right half-plane. Along C the change in arg w is

$$\Delta_C \arg w \approx \Delta_C \arg(z^3) = 3\Delta_C \arg z$$

with approximation becoming exact as the radius of C approaches ∞. Since $\Delta_C \arg z = \pi$ along C, in the limit

$$\Delta_C \arg w = 3\pi$$

As z descends the imaginary axis, $z = iy$ so

$$w = 2(1 - 2y^2) - iy(2 + y^2)$$

and

$$\arg w = \arctan\left[\frac{y(y^2 + 2)}{2(2y^2 - 1)}\right]$$

so that $\arg w$ increases from $-\pi/2$ to $\pi/2$ as y decreases from ∞ to $-\infty$. Thus,

$$\Delta_I \arg w = \pi$$

Hence the total increase in $\arg w$ is

$$(\Delta_C + \Delta_I)\arg w = 4\pi$$

and the number of zeros in the right half-plane is two.

19. If $u(x,y)$ is the real part of an analytic function $f(z)$ for which

$$\lim_{|z|\to\infty} \left|\frac{f(z)}{z}\right| = 0$$

for $\operatorname{Im} z \geq 0$, show that if the values of $u(x,y)$ are given for $y = 0$ [i.e., $u(x,0)$ is given], the *Poisson integral* for the half-plane solves the Dirichlet problem of finding a harmonic $u(x,y)$ taking assigned values on $y = 0$.

Ans. $g(a) = \dfrac{f(a)}{a - z^*}$

is analytic for $\operatorname{Im} a > 0 < \operatorname{Im} z$. Hence, by the Cauchy integral theorem,

$$g(z) = \frac{1}{2\pi i}\oint \frac{g(a)\,da}{a - z} = \frac{1}{2\pi i}\oint \frac{f(a)\,da}{(a - z^*)(a - z)}$$

where the contour consists of $\operatorname{Im} a = 0$ and an infinite semicircle in the upper half a plane. Because of the limit condition on $f(a)$, the integral on the semicircle goes to zero. Thus,

$$g(z) = \frac{f(z)}{z - z^*} = \frac{1}{2\pi i}\int_{-\infty}^{\infty} \frac{f(a)\,da}{|a - z|^2}$$

since a is real in this integral. Thus,

$$f(z) = \frac{y}{\pi}\int_{-\infty}^{\infty} \frac{f(a)\,da}{(x - a)^2 + y^2}$$

and, taking real parts,

$$u(x,y) = \frac{y}{\pi}\int_{-\infty}^{\infty} \frac{u(a,0)\,da}{(x - a)^2 + y^2}$$

which is the Poisson integral for the (upper) half-plane.

20. Solve the harmonic Dirichlet problem for the exterior of a circle of radius R.

Ans. If in (5.9) C is taken to be the circle $|z| = R$ and $|w| > R$, one has, for $f(z)$ analytic when $|z| > R$,

$$f(\infty) - f(w) = \frac{1}{2\pi i}\oint \frac{f(z)\,dz}{z - w}$$

The function

$$g(z) = \frac{Rf(z)}{zw^* - R^2}$$

is also analytic for $|z| > R$, since the zero in the denominator occurs inside $|z| = R$. Also $g(\infty) = 0$, so applying (5.9) to $g(z)$, one has

$$-g(w) = -\frac{Rf(w)}{w^*w - R^2} = \frac{1}{2\pi i}\oint \frac{Rf(z)\,dz}{(zw^* - R^2)(z - w)}$$

or

$$f(w) = -\frac{(w^*w - R^2)}{2\pi i}\int_0^{2\pi} \frac{iRe^{i\phi}f(Re^{i\phi})\,d\phi}{(w^*Re^{i\phi} - R^2)(Re^{i\phi} - w)}$$

and

$$f(w) = \frac{(w^*w - R^2)}{2\pi}\int_0^{2\pi} \frac{f(Re^{i\phi})\,d\phi}{(w^* - Re^{-i\phi})(w - Re^{i\phi})}$$

with $w = re^{i\theta} (r > R)$. Noting that the denominator of the integrand is $|w - Re^{i\phi}|^2$, the real part is

$$u(re^{i\theta}) = \frac{1}{2\pi}\int_0^{2\pi} \frac{(r^2 - R^2)u(Re^{i\phi})\,d\phi}{r^2 + R^2 - 2Rr\cos(\theta - \phi)}$$

the Poisson integral for $r > R$.

21. Construct the most general function $f(z)$ analytic and free of singularities in an annulus $a < |z| < b$ with real part vanishing on $|z| = a$.

Ans. For $g(z)$ analytic and free of singularities for $a^2/b \leq |z| \leq b$ the function

$$f(z) = g(z) - g^*\left(\frac{a^2}{z^*}\right)$$

has the property on $z^*z = a^2$ that

$$\text{Re} f(z) = \text{Re}[g(z) - g^*(z)] = 0$$

22. Construct the most general function $f(z)$ analytic and free of singularities in an annulus $a < |z| < b$ with real part vanishing on $|z| = b$.

Ans. For $h(z)$ analytic and free of singularities for $a < |z| < b^2/a$ the function

$$f(z) = h(z) - h^*\left(\frac{b^2}{z^*}\right)$$

has the property on $z^*z = b^2$ that

$$\text{Re} f(z) = \text{Re}[h(z) - h^*(z)] = 0$$

23. Find the force and moment acting on a body submerged in a non-viscous stationary incompressible irrotational fluid flow (theorem of Blasius).

Ans. According to Bernoulli's theorem $(p + (1/2)\rho\,|v|^2)$ is constant, and since the complex velocity $v = [f'(z)]^*$ as in (8.54), one has

$$p = -\tfrac{1}{2}\rho v^* v = -\tfrac{1}{2}\rho f'(z)^* f'(z)$$

where the additive constant has been set equal to zero, since it will play no role in the determination of the quantities required. The differential (complex) force acting on any element ds of arc is

$$dF = ip\,dz = ip\,dx - p\,dy$$

as may be discerned by considering a small approximately triangular element with sides of length dx and dy and "hypotenuse" of length ds. Thus,

$$dF^* = -ip\,dz^* = \frac{i\rho}{2} f'(z)\,df^*$$

Since the surface of the submerged body consists of streamlines $(d\psi = 0)$,

$$f(z) = \phi + i\psi$$
$$df(z) = d\phi + i\,d\psi = d\phi = df^*(z)$$

on the surface. Hence the total force can be found by integration over the closed contour representing the cross section of the (cylindrical) surface:

$$F^* = \frac{i\rho}{2} \oint [f'(z)]^2\,dz$$

The differential moment $dM = dM^*$ (directed $\perp$ to the z plane) is similarly given by

$$-dM = \text{Re}\left[\frac{\rho}{2} z f'(z)\,df^*\right]$$

so that integrating over the whole contour (surface), one has

$$-M = \operatorname{Re}\left[\frac{\rho}{2}\oint z[f'(z)]^2\, dz\right]$$

These are the Blasius formulae for F and M.

24. Show that a cylinder of radius a inserted in a uniform flow of incompressible nonviscous fluid experiences *no* Blasius force after stationarity is established.

Ans. Here

$$f(z) = v_\infty\left(z + \frac{a^2}{z}\right)$$

and

$$f'(z) = v_\infty\left(1 - \frac{a^2}{z^2}\right)$$

so that

$$F^* = \frac{i\rho v_\infty^2}{2}\oint\left(1 - \frac{a^2}{z^2}\right)^2 dz = 0$$

It should be remembered that there is no viscous drag along the surface of the cylinder since the fluid was supposed to be inviscid. The fact that actual experience with drag of immersed cylinders contradicts the result obtained is sometimes called D'Alembert's paradox. What is M for this ideal case according to Blasius?

● PROBLEMS 8

1. Prove that if κ is the curvature of a closed curve with continuous tangent in the z plane surrounding a simply connected region, then

$$\oint \kappa\, |dz| = 2\pi$$

2. Find the curvature of the image of the unit circle $|z| = 1$ under the Möbius (bilinear) mapping.
3. Find the curvature of the image of the unit circle $|z| = 1$ under the mapping $w = kz^m$.
4. Find the linear and areal distortion of the z plane under the mapping $w = z^3$.
5. Does the maximum modulus theorem imply that $|f(z)|$ cannot have a maximum value among all those values inside a region of analyticity

for which $z = g(t)$ where $g(t)$ is a given (complex) function of the real variable t?

6. Determine the directions at $z = 0$ for which the absolute value of e^{-cz^2} decreases most rapidly as $|z|$ increases. [*Hint:* The absolute values are constant along a family of hyperbolas.] The orthogonal trajectories are called the curves of steepest descent.
7. Find the point of the region determined by $0 \leq \operatorname{Re} z$; $0 \leq \operatorname{Im} z$; $\operatorname{Re} z + \operatorname{Im} z \leq 3$ for which e^z has a maximum modulus.
8. Find the maximum value of $|e^{z^2}|$ in the region $|z^2 - a| = R$.
9. Find the number of roots of $z^3 + z + 1 = 0$ in the first quadrant.
10. Find the number of zeros of $z^4 - 10z^3 + 33z^2 - 46z + 30 = 0$ in the right half-plane.
11. Find the number of roots of $z^4 - 2z^3 - 2z + 26 = 0$ in the sector within 45° of the imaginary axis.
12. Find the number of zeros of $z^{19} + z^{17} - z^2 - 1 = 0$ in the upper half-plane.
13. Show that the linear distortion in a Möbius (bilinear) mapping (8.11) $|dw/dz| = |ad - bc|/|cz + d|^2$.
14. Find the areal distortion in the wedge mapping $w = z^{\pi/\alpha}$.
15. Find the Schwarz–Christoffel mapping of $|\operatorname{Re} z| \leq a$; $\operatorname{Im} z \geq 0$ onto $\operatorname{Im} w \geq 0$.
16. Find the Schwarz–Christoffel mapping of the portion of the upper half z plane cut from $z = ia$ to $z = -\infty + ia$ onto the upper half w plane, with $z = -\infty$ corresponding to $w = 0$ and $z = ia$ to $w = -1$.
17. Find the Schwarz–Christoffel mapping of $\operatorname{Re} z \geq 0 \leq \operatorname{Im} z$ from which $\operatorname{Re} z > h$ and $\operatorname{Im} z > g$ has been removed onto the upper half w plane with $z = 0$ to $w = -1$, $z = h + ig$ to $w = a$, and $z = \infty$ to $w = 0$.
18. Map $|z - i| \leq 2$ onto $\operatorname{Im} w \geq 0$.
19. Map the interior of the rectangle $|\operatorname{Im} z| \leq a$, $|\operatorname{Re} z| \leq b$ onto $\operatorname{Im} w \geq 0$.
20. Find the Green's function (electrostatic potential of a line charge) for a (grounded) circular cylinder $|z| = a$.
21. Find the electrostatic potential inside a grounded circular cylinder $|z| = a$ containing line charges q_k at locations $z = z_k$.
22. Derive the Poisson integral

$$u(re^{i\theta}) = \frac{1}{2\pi} \int_0^{2\pi} \frac{(R^2 - r^2)u(Re^{i\phi})\,d\phi}{R^2 + r^2 - 2Rr\cos(\theta - \phi)}$$

for the electrostatic potential inside a dielectric cylinder (radius R) containing no charges if the surface of the cylinder is maintained at a given potential. [*Hint:* Apply the Cauchy integral theorem to $f(z)/(za^* - R^2)$, where a is a point inside $|z| = R$ at which the potential $\operatorname{Re} f(a)$ is to be calculated.]

23. By superposing the potentials for Problems 21 and 22 derive the case of a dielectric cylinder containing parallel line charges (Poisson–Jensen theorem).
24. Find the modified flow function to describe the flow in a corner of angle α after a cylinder $|z| = a$ has been inserted at the origin.
25. Find the velocity field described by the flow function $f(z) = z^2$.
26. Find the velocity field described by the flow function $f(z) = \sqrt{z}$.
27. Find the velocity field for flow at the end of a semi-infinite rectangular channel.
28. Find the velocity field of four vortices of strength C at $z = \pm(1 + i)$ and $-C$ at $z = \pm(-1 + i)$.
29. Find the flow function

$$f(z) = v_\infty\left\{\frac{2(R^2 - a^2)}{z + \sqrt{z^2 - 4a^2}} + z\right\}$$

about the ellipse

$$\frac{x^2}{[R + (a^2/R)]^2} + \frac{y^2}{[R - (a^2/R)]^2} = 1$$

which reduces to uniform flow at $z = \infty$.
30. Find the flow function and velocity for a dipole inside a circle.
31. Show how the Taylor expansion of a flow function analytic at $z = \infty$ can be interpreted as a superposition of multipoles.
32. Find the Green's function for the Dirichlet problem for the Laplace equation for $0 \leq \operatorname{Re} z \leq 1$.
33. Find the Green's function for the Dirichlet problem for the Laplace equation for $|z| \geq R$.
34. Prove that the (arithmetic) mean of the values of an analytic function on the circumference of a circle is identical with its value at the center.
35. Prove that if a function $f(z)$ is regular and has no zeros for $|z| \leq a$, then the average of its argument $[\arg f(z)]$ on the circumference is $\arg f(0)$.
36. Show that the (harmonic measure) potential of two parallel conducting strips at unit voltage coplanar with and insulated from the rest of a conducting, grounded plane is

$$\phi = \frac{1}{\pi}\arg\left\{\left(\frac{z - b}{z - a}\right)\left(\frac{z - c}{z - d}\right)\right\}$$

37. Show that if $f(z)$ is analytic throughout a region R and on its boundary and $|f(z)| \neq 0$ inside R, then the *minimum* value of $|f(z)|$ must occur on the boundary.
38. Show from (8.45) that a Green's function for the *Neumann problem* (of finding the values of a harmonic function u within a region from the

given values of $\partial_n u$, the normal derivative of u on the boundary C) may be constructed by requiring the following:

(1) $\partial_n g = 4\pi/L$ on C (L = perimeter of C);
(2) $g(z,a)$ is harmonic except at $z = a$;
(3) $g(z,a) \approx 2 \log |z - a|$ for $z \approx a$.

Notice that although u as a harmonic function must, by (8.47), have a zero average for its normal derivative $\partial_n u$ on C, the average of $\partial_n g$ on C is, by (1), 4π. Note also that (8.45) results directly in an integral equation for u.

39. Find a function real and harmonic in $a < |z| < b$ taking the values

$$[a^2 - (a^2/b)^2]\cos 2\theta \quad \text{on } |z| = b$$

and

$$e^{a \cos \theta} \cos(a \sin \theta) - e^{-(b^2/a)\cos \theta} \cos[(b^2/a)\sin \theta] \quad \text{on } |z| = a$$

40. Find a function real and harmonic in $a < |z| < b$ taking the values zero on $|z| = a$ and $2 \log(b/a)$ on $|z| = b$.
41. If $f(z) = g(z) + h(z) - g^*(a^2/z^*) - h^*(b^2/z^*)$ with $g(z)$ and $h(z)$ as in Examples 21 and 22, find $\operatorname{Re} f(z)$ on $|z| = a$ and on $|z| = b$.
42. Show that the radial and tangential displacements corresponding to (8.84) are

$$u_r = \frac{\sigma}{4Gr}\left\{(1 - 2\nu)r^2 + R^2 + \left[4(1 - \nu)R^2 + r^2 - \frac{R^4}{r^2}\right]\cos 2\theta\right\}$$

$$u_\theta = -\frac{\sigma}{4Gr}\left[2(1 - 2\nu)R^2 + r^2 - \frac{R^4}{r^2}\right]\sin 2\theta$$

43. Show that the Kolosov potentials for the infinite plate with a circular hole of radius R under biaxial tension σ at infinity are given by

$$\Phi(z) = \frac{\sigma}{2} \qquad \Psi(z) = \frac{\sigma R^2}{z^2}$$

Find the stresses σ_r, σ_θ, $\tau_{r\theta}$ and the displacements.

44. Show that the infinite plate with a circular hole under pressure p with zero stresses at infinity corresponds to the Kolosov potentials

$$\Phi(z) = 0 \qquad \Psi(z) = \frac{pR^2}{z^2}$$

Find the stresses and displacements.

45. Show that the Kolosov potentials for a cylindrical shell of outer radius b and pressure P_b and inner radius a and pressure P_a are

$$\Phi(z) = -\frac{P_b b^2 - P_a a^2}{2(b^2 - a^2)}$$

$$\Psi(z) = -\frac{(P_b - P_a)b^2 a^2}{(b^2 - a^2)z^2}$$

Find the stresses and displacements.

46. The inside and outside of a cylindrical shell of radii a and b $(b > a)$ are maintained at temperatures T_a and T_b. Find the temperature at any radial distance inside the shell.

47. Show that the Neumann problem for Laplace's equation in a unit circle has a solution only if

$$\int_0^{2\pi} f(\theta)\, d\theta = 0$$

where $f(\theta)$ is the value of $\partial_n u$ on the circle. Also show that for $|z| < 1$, if the above is true, then with $|w| = 1$

$$u(z) = \text{Re}\left\{\frac{1}{\pi i}\int \frac{dz}{z} \oint \frac{f(\theta)\, dw}{w - z}\right\}$$

48. Show that the potential due to two (metal) half-planes (parallel coplanar) maintained at potentials V and $-V$ with a gap of distance $2a$ between them can be found by using the Schwarz–Christoffel mapping:

$$w = a \sin\left(\frac{\pi z}{2V}\right)$$

taking half the uniform field between $\text{Re}\, z = \pm V$, $\text{Im}\, z \geq 0$ into the upper half w plane.

49. In Problem 48 $-iw$ is the w plane rotated through a quadrant. Then the upper half $-iw$ plane is half the field of a vertical (metal) strip. By the Schwarz–Christoffel mapping,

$$-iw = \sin\left(\frac{\pi\zeta}{2b}\right)$$

the upper half iw plane is mapped onto the portion of the upper half ζ plane between $\text{Re}\, \zeta = \pm b$, while if

$$a = \sinh\left(\frac{\pi c}{2b}\right)$$

the metal strip becomes another vertical one in the ζ plane on the imaginary axis with $|\text{Im}\,\zeta| \le c$. In this way, show that the field of an array of parallel vertical metal strips of length $2c$ and (horizontal) separation $2b$ may be described by the complex electropotential

$$f(w) = \frac{2V}{\pi} \arcsin\left[\frac{\sinh(\pi w/2ib)}{\sinh(\pi c/2b)}\right]$$

50. Show that the mapping for which

$$d_z w = \frac{1}{\sqrt{1 - z^4}}$$

represents $|z| \le 1$ on a square in the w plane.

51. Show that the mapping for which

$$d_z w = \frac{\sqrt{1 - z^4}}{z^2}$$

maps $|z| \ge 1$ onto the exterior of a square.

52. Show that the Schwarz–Christoffel mapping for which

$$d_z w = (1 - z^2)^{-2/3}$$

maps an equilateral triangle.

53. For stationary current flow in a sheet the (complex) current density J (current crossing unit length in the sheet) is related to the potential ϕ by $J^* = -\sigma\partial^*\phi$ (where σ is conductivity). In terms of the complex electrokinetic potential $f(z)$ this yields $J^* = \sigma f'(z)$. Show that for the case of a circular metal disk of conductivity σ_1 and radius a embedded in an infinite plane metal sheet of conductivity σ_2 with a point source of current j at $z = c = c^* > a$ one has

$$f(z) = \frac{j}{2\pi\sigma_2}\left[\log(z - c) + \left(\frac{\sigma_1 - \sigma_2}{\sigma_1 + \sigma_2}\right)\log\left(\frac{cz}{cz - a^2}\right)\right]$$

54. There is a single point source of current in an infinite plane metal sheet. Show that the potential difference between any two points is half the potential difference between the same two points if a circular disk whose circumference contains them is removed from the sheet.

55. Find the magnetic intensity due to a set of parallel wires carrying currents j_k located at $z = z_k$.

56. Find the magnetic intensity of a wire of radius a carrying current j in a uniform transverse magnetic field H.

57. Show that if an incompressible nonviscuous fluid flows about an immersed cylinder of radius a with a complex flow function

$$f(z) = v_\infty\left(z + \frac{a^2}{z}\right) + \frac{i\Gamma}{2\pi}\log\left(\frac{z}{a}\right)$$

the cylinder experiences a transverse Blasius force given by

$$f = i\rho v_\infty \Gamma$$

which vanishes if the circulation Γ approaches zero. The presence of such a transverse force on a rotating body in a uniform flow is called *Magnus effect*. It is always directed from the point where the circulation maximally opposes the uniform flow to the point where it maximally reinforces it.

CHAPTER 9:

Laplace and Fourier Transforms

9.1 LAPLACE INVERSION FORMULAE

We have seen that analytic functions possess Taylor expansions (in series of nonnegative powers) about any point where they are nonsingular and that these converge in the interior of a circle passing through the nearest singularity. There exists also an integral analog to this power series expansion. Thus if $g(z)$ is analytic and free of singularity in a neighborhood of $z = 0$, one has

$$g(z) = \sum_{k=0}^{\infty} f_k z^k \tag{9.1}$$

convergent for $|z| < R$, some radius of convergence. The exponential mapping $z = e^{-s}$ takes the interior of the circle $|z| < R$ into the half-plane:

$$\operatorname{Re} s > -\log R \tag{9.2}$$

with $z = 0$ corresponding to an indefinitely large positive value of the real part of s. The resulting series is

$$\psi(s) = \sum_{k=0}^{\infty} f_k e^{-sk} = g(e^{-s}) \tag{9.3}$$

An analytic function of s convergent in the region (9.2) is obtained by replacing the discrete variable k, which takes only integer values in (9.3), by a continuous variable t, which ranges from $t = 0$ to $t = \infty$ in an integration which replaces the summation in (9.3). Thus,

$$F(s) = \int_0^\infty f(t)e^{-st}\,dt \tag{9.4}$$

is the analog and is called the *Laplace transform* of $f(t)$. This integral will converge to an analytic function of $F(s)$ for a sufficiently large value of the real part of s if $f(t)$ does not become infinite too rapidly as $t \to \infty$. The boundary of the region of convergence of (9.4) is the line $\operatorname{Re} s = \sigma$, and σ is called the *abscissa of convergence* of the Laplace transform. Notice that the knowledge of $F(s)$ even for a single value of s requires a knowledge of $f(t)$ over every subinterval of $(0, \infty)$.

In the case of (9.1) we have also previously derived a formula for the coefficients

$$f_k = \frac{1}{2\pi i}\oint \frac{g(z)\,dz}{z^{k+1}} \tag{9.5}$$

With a change of variable $z = e^{-s}$ this would become

$$f_k = \frac{1}{2\pi i}\oint e^{sk}\psi(s)\,ds \tag{9.6}$$

provided the contour were transversed in a clockwise (rather than the usual counterclockwise) direction. Because of the analogy between (9.3) and (9.4) we might expect a similar formula to express $f(t)$ as an integral of $e^{st}F(s)$.

To obtain such a formula let us recall that $F(z)$ is analytic and free of singularities in the half-plane $\operatorname{Re} s > \sigma$ so that a closed contour consisting of a straight line parallel to the imaginary axis close to the abscissa of convergence and a semicircular arc joining the ends of this line and cutting the real positive axis could be used with the Cauchy integral theorem to yield (the straight line is just inside the half-plane of convergence)

$$F(s) = \frac{1}{2\pi i}\oint \frac{F(z)\,dz}{s - z} \tag{9.7}$$

where the direction of integration is again clockwise. When the contour is expanded so that the radius of the semicircular part becomes infinite, the integral along the curved part will approach zero if the limit of $|z|^\varepsilon\,|F(z)|$ exists as $|z| \to \infty$ for ε a small fixed positive number. Under these circumstances (9.7) becomes

$$F(s) = \frac{1}{2\pi i}\int_{c-i\infty}^{c+i\infty} \frac{F(z)\,dz}{s - z} \tag{9.8}$$

where c is a real number slightly larger than the abscissa of convergence of (9.4). Also, one has

$$\frac{1}{s - z} = \int_0^\infty e^{-(s-z)u}\, du \tag{9.9}$$

so that substituting this into (9.8) and interchanging the order of the integrations, one has

$$F(s) = \int_0^\infty e^{-su} \left[\frac{1}{2\pi i} \int_{c-i\infty}^{c+i\infty} e^{zu} F(z)\, dz \right] du \tag{9.10}$$

On comparing this result with (9.4), one has (replacing u by t and z by s)

$$f(t) = \frac{1}{2\pi i} \int_{c-i\infty}^{c+i\infty} e^{st} F(s)\, ds \tag{9.11}$$

the *inverse Laplace transform* of $F(s)$. Together (9.4) and (9.11) are called the *Laplace inversion formulae*.

An example of (9.11) is afforded by

$$F(s) = \frac{se^{-s}}{s^2 + 1}$$

It is desired to find the (inverse) function $f(t)$ with Laplace transform $F(s)$. The inversion integral is

$$\frac{1}{2\pi i} \int_{c-i\infty}^{c+i\infty} \frac{se^{-s} e^{st}\, ds}{s^2 + 1}$$

where $c > 0$, since the only singularities of $F(s)$ in the finite s plane lie at $s = \pm i$ on the imaginary axis and the right half-plane is the region of convergence of $F(s)$. For $t < 1$ the exponential $e^{s(t-1)}$ will approach zero for $\operatorname{Re} s \to +\infty$, so the inversion integral can be made part of a contour integral formed of the vertical line of integration and a large semicircular arc in the *right* half-plane. The integral on the latter will approach zero (for $t < 1$) as its radius becomes infinite, and since the contour encloses *no singularities* of

$$\frac{se^{2(t-1)}}{s^2 + 1}$$

the inversion integral must be zero for $t < 1$. More rigorously the argument would follow Jordan's lemma (Example 14, Chapter 6). For $t > 1$ the exponential $e^{s(t-1)}$ will approach zero for $\operatorname{Re} s \to -\infty$, so the inversion integral can be made part of a contour integral with a large semicircular arc in the *left* half-plane. Along the latter the integral will approach zero, so that the inversion integral is equal to the contour integral. As the radius becomes infinite the contour encloses all finite singularities of

$$\frac{se^{s(t-1)}}{s^2 + 1}$$

These are two simple poles with residues $e^{i(t-1)}/2$ at $s = i$ and $e^{-i(t-1)}/2$ at $s = -i$. Hence the result of the integration is

$$\frac{1}{2\pi i}\int_{c-i\infty}^{c+i\infty} e^{st}F(s)\,ds = \begin{cases}\cos(t-1) & \text{for} \quad t > 1\\ 0 & \text{for} \quad t < 1\end{cases}$$

which is also easily verified to yield $F(s) = se^{-s}/(s^2 + 1)$ by direct integration.

9.2 PROPERTIES OF LAPLACE TRANSFORM

The path of integration in (9.11) may be made part of a closed contour by connecting the "ends" by a semicircular arc of "infinite" radius, and the integral on the right may be evaluated by the residue theorem. We note, however, that for $t > 0$ the arc must be swung "about" the left half s plane to make the integral of $e^{st}F(s)$ approach zero as the arc radius approaches infinity (exponential attenuation).

This yields exactly the result (9.11) since u in (9.9) is positive. If, however, $t < 0$, the integral on the right of (9.11) can be augmented by that along a large semicircular arc in the right half s plane which latter will approach zero as the arc radius approaches infinity. Now with $t < 0$ the contour lies inside the region of convergence of (9.4) so there are no singularities within it and we have

$$0 = \frac{1}{2\pi i}\int_{c-i\infty}^{c+i\infty} e^{st}F(s)\,ds \qquad \text{for} \quad t < 0 \tag{9.12}$$

Notice that the contour is completed by an arc such that $t(\text{Re } s) < 0$ for large $|s|$ in each case. Because of (9.12) it is generally assumed that the function being transformed is zero for $t < 0$.

The Laplace transform of an integer power m of t is

$$\int_0^\infty e^{-st}t^m\,dt = \frac{m!}{s^{m+1}} \tag{9.13}$$

as is readily verified by integration by parts.

The Laplace transform of $e^{i\omega t}$ is

$$\int_0^\infty e^{-st}e^{i\omega t}\,dt = \frac{1}{s - i\omega} \tag{9.14}$$

for real ω. Thus the real and imaginary parts of (9.14) yield

$$\int_0^\infty e^{-st}\cos\omega t\,dt = \frac{s}{s^2+\omega^2} \tag{9.15}$$

$$\int_0^\infty e^{-st}\sin\omega t\,dt = \frac{\omega}{s^2+\omega^2} \tag{9.16}$$

The Laplace transform of the unit step function $u(t-a)$ defined by

$$u(t-a) = \begin{cases} 1 & \text{for} \quad t \geq a > 0 \\ 0 & \text{for} \quad t < a \end{cases} \tag{9.17}$$

is

$$\int_0^\infty e^{-st}u(t-a)\,dt = \int_a^\infty e^{-st}\,dt = \frac{e^{-sa}}{s} \tag{9.18}$$

The differential of the unit step is

$$du(t-a) = \begin{cases} 1 & \text{for} \quad t = a \\ 0 & \text{for} \quad t \neq a \end{cases} \tag{9.19}$$

since the difference between the two values of u for $t > a$ and $t < a$ is unity and the difference between two values of u for two points t_1 and t_2 on the same side of a is zero. One may then integrate the exponential e^{-st} with respect to du, obtaining

$$\int_0^\infty e^{-st}\,du = e^{-sa} \tag{9.20}$$

This is called a *Stieltjes integral* of e^{-st}, since we are integrating with respect to a function of t rather than with respect to t itself as in a Riemann integral. In the integration of (9.20) no value of t except $t = a$ survives because du is zero except for $t = a$. It is customary in physical literature to pretend that the derivative du/dt exists and to denote it by

$$\frac{du}{dt} = \delta(t-a) \tag{9.21}$$

which is called the *Dirac delta function*. Clearly, the derivative of u does exist and is zero in all intervals *not* containing $t = a$, since du is zero for $t \neq a$. Just as clearly, the change in u in an infinitesimal interval containing $t = a$ is unity and the derivative does not exist at $t = a$. Physically one may visualize $\delta(t-a)$ as a limiting case of a rectangular pulse of unit area with base centered at $t = a$ as the height increases indefinitely. The association of $\delta(t - a)$ with unit area follows from the formal integration

$$\int_{-\infty}^{\infty} \delta(t - a)\, dt = \int_{-\infty}^{\infty} du = 1 \tag{9.22}$$

since the range of the latter Stieltjes integration includes the one point ($t = a$) at which du is not zero. Similarly, one has

$$\int_{-\infty}^{\infty} \delta(t - a) f(t)\, dt = \int_{-\infty}^{\infty} f(t)\, du = f(a) \tag{9.23}$$

All the various formal properties of the delta function (which is actually undefined at $t = a$) can be deduced from the properties of the unit step function (9.17). Functions such as the delta function are now called *generalized functions*, and their properties all follow from the theory of Stieltjes integration.

The Laplace transform of the delta function is then

$$\int_{0}^{\infty} e^{-st} \delta(t - a)\, dt = \int_{0}^{\infty} e^{-st}\, du = e^{-sa} \tag{9.24}$$

supposing that $a > 0$. If a were negative, the delta function would be zero for t positive and the right side of (9.24) would be replaced by a zero.

Many simple properties of the Laplace transform follow directly from (9.4). The notation

$$F(s) \rightleftarrows f(t) \tag{9.25}$$

will be used to denote the correspondence between a function $f(t)$ (defined as zero for $t < 0$) and its Laplace transform $F(s)$. Then simple substitution into (9.4) yields

$$F(ms) \rightleftarrows \frac{1}{m} f\left(\frac{t}{m}\right) \qquad \text{for} \quad m > 0 \tag{9.26}$$

$$F(s - a) \rightleftarrows e^{at} f(t) \tag{9.27}$$

$$e^{-as} F(s) \rightleftarrows \begin{cases} f(t - a) & \text{for} \quad t \geq a \geq 0 \\ 0 & \text{for} \quad t < a \end{cases} \tag{9.28}$$

By repeated differentiation of (9.4), one has

$$d_s^n F(s) \rightleftarrows (-t)^n f(t) \tag{9.29}$$

Integration of (9.4) yields

$$\int_{s}^{\infty} F(s)\, ds \rightleftarrows \frac{f(t)}{t} \tag{9.30}$$

The Laplace transform of the nth derivative of $f(t)$ is

$$\int_0^\infty e^{-st} d_t^n f(t)\, dt = \int_0^\infty e^{-st}\, d(d_t^{n-1} f) \tag{9.31}$$

which by integration by parts yields

$$d_t^{n-1} f(0) + s \int_0^\infty e^{-st} d_t^{n-1} f\, dt \tag{9.32}$$

Similarly, repeated integration by parts of (9.31) yields

$$s^n F(s) - \sum_{k=0}^{n-1} f^{(k)}(0) s^{n-k-1} \rightleftarrows d_t^n f = f^{(n)}(t) \tag{9.33}$$

If this is applied to the derivative of the function

$$g(t) = \int_0^t f(u)\, du \tag{9.34}$$

one has

$$G(s) = \frac{F(s)}{s} \rightleftarrows g(t) \tag{9.35}$$

The product of two Laplace transforms is *not* the Laplace transform of the product of the functions, but is the Laplace transform of the *convolution* $h(t)$ of the two functions defined by

$$h(t) = \int_0^t f(\tau) g(t - \tau)\, d\tau \tag{9.36}$$

which may be verified as follows:

$$\int_0^\infty \int_0^\infty e^{-s(t+u)} g(u) f(t)\, du\, dt = G(s)F(s) \tag{9.37}$$

by multiplication. Here, the values of u and t range over the first quadrant of u, t plane. Make the transformation

$$\begin{cases} v = t + u \\ u = u \end{cases} \tag{9.38}$$

with Jacobian

$$J\left(\frac{v,u}{t,u}\right) = \begin{vmatrix} \partial_t v & \partial_t u \\ \partial_u v & \partial_u u \end{vmatrix} = \begin{vmatrix} 1 & 0 \\ 1 & 1 \end{vmatrix} = 1 \tag{9.39}$$

so that $du\, dt = du\, dv$. Then (9.37) becomes $H(s)$, given by

$$\int_0^\infty e^{-sv} \int_0^v g(u) f(v - u)\, du\, dv = H(s) \tag{9.40}$$

since u cannot become larger than v, because then t would be negative. Equation (9.40) shows that

$$H(s) = F(s)G(s) \rightleftarrows h(t) = \int_0^t g(\tau)f(t-\tau)\,d\tau \tag{9.41}$$

There are important asymptotic relationships between the initial and final values of a function and the values of the Laplace transform of its derivative for large and small values of s. Thus (9.33) shows that

$$sF(s) - f(0) \rightleftarrows f'(t) = d_t f$$

and (9.4) applied to $f'(t)$ shows that

$$\lim_{s\to\infty} sF(s) = f(0) \tag{9.42}$$

As $s \to 0$

$$\lim_{s\to 0} sF(s) = f(\infty) \tag{9.43}$$

since the limit of the transform of $f'(t)$ is $f(\infty) - f(0)$.

9.3 GAMMA FUNCTION

The gamma function $\Gamma(z)$ is a generalization of $(n-1)!$ to which the function reduces when $z = n$ (a positive integer) rather than a general complex number. It may be defined by the (Hankel) contour integral

$$\frac{1}{\Gamma(z)} = \frac{1}{2\pi i}\int_L e^{\sigma}\sigma^{-z}\,d\sigma \tag{9.44}$$

where L is a contour "encircling" the negative real σ axis. The integrand is *uniquely defined* on the principal branch of the $\log\sigma$ (i.e., the branch for which $|\arg\sigma| < \pi$) and is analytic in z.

$$e^{\sigma}\sigma^{-z} = e^{\sigma}e^{-z\log\sigma} = e^{\sigma - z\log\sigma}$$

Since the contour never comes closer to $\sigma = 0$ than $\varepsilon > 0$, the integral is uniformly convergent for all finite values of z, so it follows that $1/\Gamma(z)$ has no singularities in the finite z plane and must be an entire function.

Since $e^{\lambda\sigma}$ is an entire function of σ,

$$\int_L e^{\lambda\sigma}\,d\sigma = 0 \tag{9.45}$$

and

$$d_\lambda^n \int_L e^{\lambda\sigma}\, d\sigma = \int_L \sigma^n e^{\lambda\sigma}\, d\sigma = 0 \tag{9.46}$$

for any value of λ (including $\lambda = 1$), so it is readily seen that $1/\Gamma(z)$ is zero for $z = -n$ where n is any nonnegative integer.

Furthermore, integration by parts yields

$$\frac{1}{\Gamma(z)} = \frac{1}{2\pi i}\int_L e^\sigma \sigma^{-z}\, d\sigma = \frac{z}{2\pi i}\int_L e^\sigma \sigma^{-(z+1)}\, d\sigma \tag{9.47}$$

so that

$$\Gamma(z+1) = z\Gamma(z) \tag{9.48}$$

Since the residue of e^σ/σ at $\sigma = 0$ is unity,

$$\frac{1}{\Gamma(1)} = \frac{1}{2\pi i}\int_L e^\sigma \frac{d\sigma}{\sigma} = 1 \tag{9.49}$$

so that (9.48) implies

$$\Gamma(n+1) = n\Gamma(n) = n(n-1)\Gamma(n-1) = n! \tag{9.50}$$

for n a positive integer, and this suggests defining $0! = 1 = \Gamma(1)$, which is customary.

Since the integrand of (9.44) has only a branch point at $\sigma = 0$, the contour may be connected by quarter circular arcs to a line parallel to the imaginary σ axis and infinitesimally to the right of it. Since the integral along the curved portion approaches zero as it becomes infinite, the remaining integral along the straight portion yields

$$\frac{1}{\Gamma(z)} = \frac{1}{2\pi i}\int e^{st}(st)^{-z}\, d(st) = \frac{t^{1-z}}{2\pi i}\int_{c-i\infty}^{c+i\infty} e^{st}s^{-z}\, ds \tag{9.51}$$

where st has been substituted for σ. Thus,

$$\frac{t^{z-1}}{\Gamma(z)} = \frac{1}{2\pi i}\int_{c-i\infty}^{c+i\infty} e^{st}s^{-z}\, ds \tag{9.52}$$

can be identified with (9.11) as the inverse Laplace transform. Accordingly, (9.4) yields

$$s^{-z} = \int_0^\infty e^{-st}\frac{t^{z-1}}{\Gamma(z)}\, dt \tag{9.53}$$

or

$$\frac{\Gamma(z)}{s^z} = \int_0^\infty e^{-st}t^{z-1}\, dt \tag{9.54}$$

for Re $z > 0$. For $s = 1$ this becomes

$$\Gamma(z) = \int_0^\infty e^{-t} t^{z-1}\, dt \tag{9.55}$$

which is sometimes taken as a definition of $\Gamma(z)$.

One can decompose (9.55) into a sum of two integrals:

$$\Gamma(z) = \int_0^1 e^{-t} t^{z-1}\, dt + \int_1^\infty e^{-t} t^{z-1}\, dt \tag{9.56}$$

The last integral is uniformly convergent for all finite z and therefore is an entire function of z. The first integral can be evaluated by termwise integration of

$$e^{-t} t^{z-1} = \sum_{n=0}^{\infty} \frac{(-1)^n t^{z+n-1}}{n!}$$

yielding

$$\int_0^1 e^{-t} t^{z-1}\, dt = \sum_{n=0}^{\infty} \frac{(-1)^n}{(z+n)n!} \tag{9.57}$$

whence $\Gamma(z)$ is seen to have simple poles at $z = -n$ with corresponding residues $(-1)^n/n!$

The convolution theorem (9.41) can be used to establish the identity

$$B(p,q) = \frac{\Gamma(p)\Gamma(q)}{\Gamma(p+q)} \tag{9.58}$$

for the beta function defined by

$$B(p,q) = \int_0^1 u^{p-1}(1-u)^{q-1}\, du \tag{9.59}$$

since, with $\tau = ut$,

$$t^{p+q-1} B(p,q) = \int_0^t \tau^{p-1}(t-\tau)^{q-1}\, d\tau$$

and the Laplace transform of this yields

$$\frac{\Gamma(p+q)}{s^{p+q}} B(p,q) = \frac{\Gamma(p)}{s^p} \frac{\Gamma(q)}{s^q}$$

whence (9.58) follows.

The value of $\Gamma'(1)$ may be determined as follows: From (9.55) differentiation yields

$$\Gamma'(z) = \int_0^\infty e^{-t} t^{z-1} \log t\, dt$$

so

$$\Gamma'(1) = \int_0^\infty e^{-t} \log t \, dt = \int_0^1 e^{-t} \log t \, dt + \int_1^\infty e^{-t} \log t \, dt$$

The integral over (0,1) is equivalent to

$$\int_0^1 (e^{-t} - 1) \frac{dt}{t}$$

as may be seen by integrating the latter by parts. Thus,

$$\Gamma'(1) = \int_0^1 (e^{-t} - 1) \frac{dt}{t} + \int_1^\infty e^{-t} \frac{dt}{t}$$

so that, since $e^{-t} = \lim_{n\to\infty}[1 - (t/n)]^n$,

$$-\Gamma'(1) = \lim_{n\to\infty} \left[\int_0^1 \left\{1 - \left(1 - \frac{t}{n}\right)^n\right\} \frac{dt}{t} - \int_1^n \left(1 - \frac{t}{n}\right)^n \frac{dt}{t} \right]$$

and

$$-\Gamma'(1) = \lim_{n\to\infty} \left[\int_0^n \left\{1 - \left(1 - \frac{t}{n}\right)^n\right\} \frac{dt}{t} - \int_1^n \frac{dt}{t} \right]$$

$$-\Gamma'(1) = \lim_{n\to\infty} \left[\int_0^1 \left(\frac{1 - u^n}{1 - u}\right) du - \int_1^n \frac{dt}{t} \right]$$

Hence,

$$-\Gamma'(1) = \lim_{n\to\infty} \left\{ \sum_{k=1}^n \frac{1}{k} - \log n \right\} = \gamma \tag{9.60}$$

where $\gamma \simeq 0.5772157$, is the Euler–Mascheroni constant.

An expansion for the logarithmic derivative of the gamma function may be obtained from (9.58):

$$\frac{\Gamma(z - \varepsilon)\Gamma(\varepsilon)}{\Gamma(z)} = \int_0^1 u^{\varepsilon-1}[(1 - u)^{z-\varepsilon-1}] \, du$$

Whence,

$$\frac{\Gamma(z - \varepsilon)\Gamma(\varepsilon)}{\Gamma(z)} - \frac{1}{\varepsilon} = \int_0^1 u^{\varepsilon-1}[(1 - u)^{z-\varepsilon-1} - 1] \, du$$

and using

$$u^\varepsilon \simeq 1 + \varepsilon \log u$$

$$(1 - u)^{-\varepsilon} \simeq 1 - \varepsilon \log(1 - u)$$

one has

$$\lim_{\varepsilon\to 0} \left[\frac{\Gamma(z - \varepsilon)\Gamma(\varepsilon)}{\Gamma(z)} - \frac{1}{\varepsilon} \right] = \int_0^1 u^{-1}[(1 - u)^{z-1} - 1] \, du$$

For z real and greater than unity,

$$\Gamma(z-\varepsilon) = \Gamma(z) - \varepsilon\Gamma'(z) + \frac{\varepsilon^2\Gamma''(z)}{2} + \cdots$$

and for $\varepsilon > 0$,

$$\Gamma(\varepsilon) = \frac{1}{\varepsilon} + C_0 + C_1\varepsilon + C_2\varepsilon^2 + \cdots$$

since $\Gamma(\varepsilon)$ has a simple pole of unit residue at the origin. Hence,

$$\lim_{\varepsilon\to 0}\left[\frac{\Gamma(z-\varepsilon)\Gamma(\varepsilon)}{\Gamma(z)} - \frac{1}{\varepsilon}\right] = C_0 - \frac{\Gamma'(z)}{\Gamma(z)}$$

and taking

$$u^{-1} = \frac{1}{1-(1-u)} = \sum_{k=0}^{\infty}(1-u)^k$$

in the integral to which the limit is equal, one has

$$\sum_{k=0}^{\infty}\int_0^1 [(1-u)^{k+z-1} - (1-u)^k]\, du = \sum_{k=0}^{\infty}\int_0^1 (v^{k+z-1} - v^k)\, dv$$

so that

$$C_0 - \frac{\Gamma'(z)}{\Gamma(z)} = \sum_{k=0}^{\infty} \frac{1}{k+z} - \frac{1}{k+1}$$

and

$$C_0 = \lim_{z\to 1}\frac{\Gamma'(z)}{\Gamma(z)} = \frac{\Gamma'(1)}{\Gamma(1)} = -\gamma$$

so that

$$\frac{\Gamma'(z)}{\Gamma(z)} = -\gamma - \sum_{k=0}^{\infty}\left(\frac{1}{k+z} - \frac{1}{k+1}\right) \tag{9.61}$$

which may be rearranged as

$$\frac{\Gamma'(z)}{\Gamma(z)} = -\gamma - \frac{1}{z} + \sum_{n=1}^{\infty}\left(\frac{1}{n} - \frac{1}{n+z}\right) \tag{9.62}$$

Indefinite integration then yields

$$-\log\Gamma(z) = \gamma z + \log z - \sum_{n=1}^{\infty}\left\{\frac{z}{n} - \log\left(1+\frac{z}{n}\right)\right\} \tag{9.63}$$

where the portion of the integration constant not in the summation has been found to be zero by letting $z \to 0$. Whence the product

$$\frac{1}{\Gamma(z)} = ze^{\gamma z}\prod_{n=1}^{\infty}\left(1+\frac{z}{n}\right)e^{-z/n} \tag{9.64}$$

is convergent for all finite z.

By taking $\sigma = xe^{i\pi}$ on the upper portion of L and $\sigma = xe^{-i\pi}$ on the lower portion of L, it follows from the definition of $\Gamma(z)$ that

$$\frac{1}{\Gamma(z)} = \frac{e^{i\pi z}}{2\pi i}\int_0^\infty e^{-x}x^{-z}\,dx - \frac{e^{-i\pi z}}{2\pi i}\int_0^\infty e^{-x}x^{-z}\,dx$$

or

$$\frac{1}{\Gamma(z)} = \frac{\sin \pi z}{\pi}\int_0^\infty e^{-x}x^{-z}\,dx = \frac{\sin \pi z}{\pi}\Gamma(1-z)$$

so that [functional equation for $\Gamma(z)$]

$$\frac{1}{\Gamma(z)\Gamma(1-z)} = \frac{\sin \pi z}{\pi} \tag{9.65}$$

and since each side is an entire function of z, the result must hold for all finite values of z. In particular, for $z = \frac{1}{2}$,

$$\Gamma(\tfrac{1}{2}) = \sqrt{\pi}$$

9.4 METHOD OF STEEPEST DESCENTS

The (saddle-point) method of steepest descents is concerned with asymptotic approximations (valid for large s) to integrals of the form

$$I(s) = \int_A^B e^{sf(z)}\,dF(z) \tag{9.66}$$

Here it will be supposed that $f(z)$ is regular (analytic and free of singularities) in a region including the path of integration (and any deformations considered thereof). Also, s will be taken to be real and positive and $F(z) = z$.

For a fixed value of s the modulus of the integrand is determined by the real part of $f(z)$, and one may imagine a surface constructed "over" the complex z plane with each point at a height "above" that plane equal to the corresponding value of the real part of $f(z)$. This *surface cannot possess any finite relative maxima* (peaks) or *relative minima*, since these would require the nonvanishing second derivatives of the real part of $f(z)$ with respect to x and with respect to y to have the same sign at such a point, in contradiction to Laplace's equation, which states that the second partial derivative of $\operatorname{Re} f(z)$ with respect to x is the negative of the second partial derivative with respect to y. The surface can, however, possess points at which its tangent plane is horizontal. These will be located at the zeros of $f'(z)$, and at such points the first partial derivatives of $\operatorname{Re} f(z)$ with respect to x and y will be zero. If $f''(z) \neq 0$ at such a point, the level lines [along which $\operatorname{Re} f(z)$ is constant] passing through the point will divide the surface into at least four portions with two opposite regions *uphill* from the point at which $f'(z) = 0$ separating two opposite regions *downhill* from the same point. Such a point or the corresponding value of z for which $f'(z) = 0 \neq f''(z)$ is called a *saddle point* because of the shape of the surface $\text{Re} f(z)$ in its neighborhood. The shape is similar to that of a pass between mountains leading from one valley to another.

Because $f(z)$ has no singularities in the region considered, the path of integration between A and B may be varied without changing the value of $I(s)$. Along different paths the contributions to the (same) total value will, however, be differently distributed. If a path can be found so that the integrand only has appreciable values in the neighborhood of a point, the integral may be approximated by a much simpler one. In particular, if the path of integration can be deformed into one passing through a saddle point along the directions of steepest descent (between the two valleys), then the value of the modulus of the integrand will be most quickly reduced as points further from the saddle point are traversed. Ideally the points A and B should be downhill in opposite directions from the saddle point, although one of them might be located at the saddle point.

Suppose that these conditions are fulfilled and that $f(z)$ has a (suitably located) saddle point at $z = a$ so that $f'(a) = 0 \neq f''(a)$. Then

$$I(s) = \int_A^B e^{sf(z)}\, dz = e^{sf(a)} \int_A^C e^{s[f(z)-f(a)]}\, dz$$

and it is more convenient to consider

$$I(s)e^{-sf(a)} = \int_A^B e^{s[f(z)-f(a)]}\, dz$$

with $w = f(z)$, $b = f(a)$, and

$$w - b = f(z) - f(a) = u + iv$$

and to imagine that the path of integration has already been deformed so as to pass through $z = a$. As one moves away from a in the z plane there are two opposite directions such that $\text{Re}[f(z) - f(a)]$ *decreases faster in these directions than in any other opposite directions.* These will be the (two opposite) directions of steepest descent along the surface $u = \text{Re}[f(z) - f(a)]$. They will form the orthogonal trajectory in the z plane to the $u =$ constant lines, and therefore they will be along a $v =$ constant line. The latter will pass through $z = a$, and since $v = 0$ for $z = a$, it will *remain zero* along this line. Thus the value of $[f(z) - f(a)]$ will remain real and will in fact decrease as z moves away from a in either opposite direction along the *path of steepest descent*. Along this path, u will *decrease faster* than along any other path through $z = a$, and since the value of u was zero at $z = a$, it will decrease through *negative* values as z moves away from a.

Suppose the path of integration is along the path of steepest descent including the point a and extending between two points (separated by a) for which $\text{Re}[f(z) - f(a)]$ has decreased by the *same* amount T from the elevation (on the u surface) at the saddle point. Suppose further that the remaining portions of the integration path are (rectifiable, non-self-intersecting) continuous curves (AC and DB) joining the two points to the points A and B respectively, in such a way that u never exceeds $-T$ on the latter portions. Then

$$e^{-sf(a)}I(s) = \int_A^C + \int_D^B + \int_C^D e^{s[f(z)-f(a)]}\,dz \tag{9.67}$$

and, for AC and DB deformed to straight-line segments,

$$\left|\int_A^C\right| \le e^{-sT}\,|C - A|$$

$$\left|\int_D^B\right| \le e^{-sT}\,|B - D|$$

Calling τ the modulus of u ($\text{Re}[f(z) - f(a)]$) along the path of steepest descent, one has

$$\tau = -u = f(a) - f(z) = b - w$$

with the Taylor series

$$-\tau = \frac{(z - a)^2 f''(a)}{2} + \cdots$$

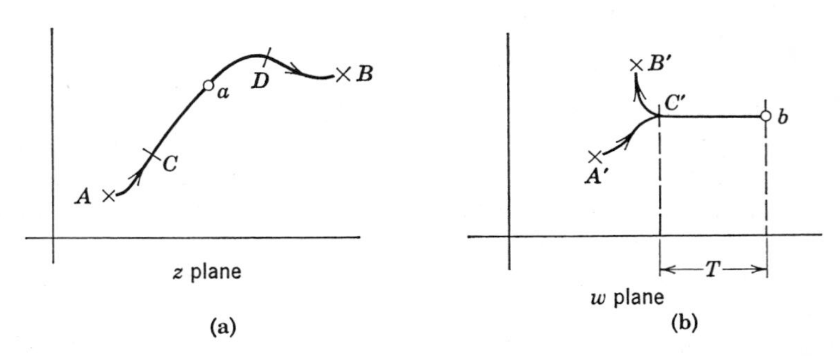

Figure 9.1. (a) Integration path AB including portion CD along steepest descent path through saddle point $z = a$. (b) Image of integration path in w plane: $w = f(z)$; $b = f(a)$; $A' = f(A)$; $B' = f(B)$; $C' = f(C) = f(D)$.

since $f'(a) = 0$. There will be *two* inverse functions

$$z = g_+(\tau) \qquad \text{and} \qquad z = g_-(\tau)$$

corresponding to the branch point at $z = a$. Thus,

$$g_+(\tau) = a + i\sqrt{\tau}\sqrt{\frac{2}{f''(a)}} + \cdots \tag{9.68}$$

$$g_-(\tau) = a - i\sqrt{\tau}\sqrt{\frac{2}{f''(a)}} + \cdots \tag{9.69}$$

The two branches correspond to the two portions of the path of steepest descent on opposite sides of the saddle point $z = a$. In Figure 9.1 these w plane paths have been shown as coincident for $0 \leq \tau \leq T$, which they would be if all the derivatives of $f''(z)$ were zero at $z = a$. If this is not the case, there will be a slight separation between the two paths which will, however, still possess a cusp at $w = b$. As long as $f''(a) \neq 0$, the discussion remains essentially the same.

Corresponding to $g_+(\tau)$ and $g_-(\tau)$ one has the derivatives

$$g_+'(\tau) = \frac{i}{\sqrt{\tau}}\frac{1}{\sqrt{2f''(a)}} + \cdots$$

$$g_-'(\tau) = -\frac{i}{\sqrt{\tau}}\frac{1}{\sqrt{2f''(a)}} + \cdots$$

and

$$\int_C^D e^{s[f(z)-f(a)]}\,dz \cong i\sqrt{\frac{2}{f''(a)}}\int_0^T e^{-s\tau}\frac{d\tau}{\sqrt{\tau}}$$

Since

$$\int_0^T e^{-s\tau}\frac{d\tau}{\sqrt{\tau}} = \int_0^\infty e^{-s\tau}\frac{d\tau}{\sqrt{\tau}} - \int_T^\infty e^{-s\tau}\frac{d\tau}{\sqrt{\tau}}$$

while

$$\int_0^\infty e^{-s\tau}\frac{d\tau}{\sqrt{\tau}} = \frac{\Gamma(\frac{1}{2})}{\sqrt{s}}$$

and

$$\left|\int_T^\infty e^{-s\tau}\frac{d\tau}{\sqrt{\tau}}\right| \leq \frac{e^{-sT}}{s\sqrt{T}}$$

one may write

$$\left|e^{-sf(a)}I(s) - \frac{i\Gamma(\frac{1}{2})}{\sqrt{s}}\sqrt{\frac{2}{f''(a)}}\right| \leq e^{-sT}\left(|C - A| + |B - D| + \frac{\sqrt{2}}{s\sqrt{T\,|f''(a)|}}\right)$$

which for fixed (small) T approaches zero as $s \to \infty$. Thus, recalling that $\Gamma(\frac{1}{2}) = \sqrt{\pi}$, one has the asymptotic approximation

$$I(s) \sim \frac{ie^{sf(a)}}{\sqrt{s}} \sqrt{\frac{2\pi}{f''(a)}} \tag{9.70}$$

valid for large s.

9.5 METHOD OF STATIONARY PHASE

The method of stationary phase is closely related to that of steepest descents and is concerned with the approximate evaluation of integrals of the form

$$I(\omega) = \int_a^b e^{i\omega\phi(t)} f(t)\, dt \tag{9.71}$$

where $\phi(t)$ and $f(t)$ are real functions. It is often loosely argued in a physical fashion that the high-frequency components cancel in a manner similar to interference phenomena in optics and that thus all contributions from the integrand except in the neighborhood of points (of stationary phase) for which $\phi'(t) = 0$ may be neglected. The significance of the zeros of $\phi'(t)$ may be more clearly exhibited by a functional inversion argument which reduces $I(\omega)$ to a Fourier transform. Thus if

$$\tau = \phi(t) \qquad\qquad t = \psi(\tau)$$

are inverse functions [with $\psi(t)$ differentiable], $I(\omega)$ may be written

$$I(\omega) = \int_\alpha^\beta e^{i\omega\tau} f[\psi(\tau)]\psi'(\tau)\, d\tau \tag{9.72}$$

if $\beta = \phi(b) > \alpha = \phi(a)$ and $I(\omega)$ is recognized as the Fourier transform of

$$g(\tau) = \begin{cases} f[\psi(\tau)]\psi'(\tau) & \text{for} \quad \alpha \le \tau \le \beta \\ 0 & \text{otherwise} \end{cases}$$

This represents a spectral decomposition of $I(\omega)$ in which the "Fourier components" corresponding to large $\psi'(\tau)$ predominate if $f(t)$ does not (fortuitously) have pronounced minima at the points of stationary phase. By virtue of the theorem on derivatives of inverse functions

$$\phi'(t)\psi'(\tau) = 1 \tag{9.73}$$

large values of $\psi'(\tau)$ correspond to *small* values of $\phi'(t)$, which are therefore associated with the predominant contributions to $I(\omega)$. Also, since $dt = \psi'(\tau)\, d\tau$, one concludes that $\phi'(t) \ll 1$ implies $\psi'(\tau) \gg 1$ implies $dt \gg d\tau$ so that for any portion dt of the *original* range of integration a much smaller infinitesimal $d\tau$ of the *new* range is required for the same contribution to the integral.

Although the Fourier viewpoint is conceptually useful in understanding the method of stationary phase, the practical method of evaluation consists of the following:

(1) Determination of the points $\{s\}$ of stationary phase for which $\phi'(t) = 0$.

(2) Expansion of $\phi(t)$ in a Taylor series about each such point. The first nonconstant term in each such series will then be of degree not less than quadratic, since the first derivative is zero there. For $\phi'(s) = 0$ one then has [for $\phi''(s) \neq 0$]

$$\phi(t) \cong \phi(s) + \phi''(s)\frac{(t-s)^2}{2} \tag{9.74}$$

(3) Approximate evaluation of (9.71) by substitution of (9.74) into (9.71), replacing $f(t)$ by $f(s)$, and supposing ω to have been chosen so large that the integration only yields appreciable values for a small neighborhood of $t = s$. Thus, for only *one* zero of $\phi(t)$,

$$I(\omega) \cong e^{i\omega\phi(s)}f(s)\int_{-\infty}^{\infty} e^{i\omega\phi''(s)(t-s)^2/2}\,dt \tag{9.75}$$

it being argued that extension of the integration range from a small neighborhood of $t = s$ to $(-\infty, \infty)$ does *not* introduce appreciable error for sufficiently large ω (see Example 21). The approximation (9.75) is then easily evaluated from knowledge of the (Fresnel) integral (see Example 20) on the right. Hence,

$$I(\omega) \cong e^{i\omega\phi(s)}f(s)(1+i)\sqrt{\frac{\pi}{\omega\phi''(s)}} \tag{9.76}$$

Naturally, if there were more than one zero of $\phi'(t)$ in $a \leq t \leq b$, similar terms would have to be added for each zero.

9.6 STIRLING ASYMPTOTIC FORMULA FOR $\Gamma(z)$

Applying the (saddle-point) method of steepest descents to the (Hankel) definition of $\Gamma(z)$, one can obtain the Stirling formula for the asymptotic behavior of $\Gamma(z)$ for large $|z|$. Thus,

$$\frac{1}{\Gamma(z)} = \frac{1}{2\pi i}\int_L e^{\sigma}\sigma^{-z}\,d\sigma = \frac{e^{-i\pi z/2}}{2\pi}\int e^{is}s^{-z}\,ds$$

where the path of integration has been deformed to one just to the right of the imaginary axis $\sigma = is$. The integrand of the latter integral can be written

$$e^{is - z \log s} = e^{\phi(s)}$$

with

$$\begin{aligned}
\phi(s) &= is - z \log s \\
\phi'(s) &= i - z/s \\
\phi''(s) &= z/s^2 \\
\phi(-iz) &= z - z \log z + i\pi z/2 \\
\phi'(-iz) &= 0 \\
\phi''(-iz) &= -1/z
\end{aligned}$$

Hence, $\phi(s) \approx \phi(-iz) + (s + iz)^2\phi''(-iz)/2$,

$$\frac{1}{\Gamma(z)} \sim \frac{e^z}{2\pi z^z} \int_{-\infty}^{\infty} e^{(s+iz)^2/2z}\, ds = \frac{e^z}{z^z}\sqrt{z/2\pi}$$

Thus,

$$\Gamma(z) \sim z^z e^{-z}\sqrt{2\pi/z} \tag{9.77}$$

which is *Stirling's asymptotic formula* for $\Gamma(z)$ for large $|z|$.

9.7 LEGENDRE DUPLICATION FORMULA

By means of Liouville's theorem *the Legendre duplication formula*

$$\Gamma(2z)\sqrt{\pi} = 2^{2z-1}\Gamma(z)\Gamma(z + \tfrac{1}{2}) \tag{9.78}$$

may now be established. Consider

$$\psi(z) \equiv \frac{\Gamma(2z)}{2^{2z}\Gamma(z)\Gamma[z + (\frac{1}{2})]}$$

which has no singularities in the finite z plane, since the simple poles of $\Gamma(2z)$ at $z = -n/2$ for $n = (0,1,2,\ldots)$ are canceled by the simple poles of $\Gamma(z)$ at $z = -n$ for $n = (0,1,2,3,\ldots)$ and the simple poles of $\Gamma[z + (\frac{1}{2})]$ at $z = -(\frac{1}{2}) - n$ for $n = (0,1,2,\ldots)$. Thus $\psi(z)$ is an entire function. Furthermore, by the Stirling formula for large $|z|$,

$$\psi(z) \sim \frac{1}{2\sqrt{\pi}}$$

so that $\psi(z)$ is bounded as $|z| \to \infty$, and hence by Liouville's theorem it must be a constant (viz. $1/2\sqrt{\pi}$) and the Legendre duplication formula follows.

9.8 LAPLACE TRANSFORMS OF SERIES

Two theorems on Laplace transformation of series are especially useful and we shall quote them without proof, citing references in which such demonstrations may be found. The first is a modification of Hardy's theorem.* The series

$$t^{\alpha} \sum_{n=0}^{\infty} a_n t^n \qquad 0 \leq \alpha < 1$$

may be termwise Laplace transformed provided the result converges for some $|s| > 0$. Thus,

$$\int_0^{\infty} e^{-st} \sum_{n=0}^{\infty} a_n t^{n+\alpha}\, dt = \sum_{n=0}^{\infty} a_n \frac{\Gamma(n+\alpha+1)}{s^{n+\alpha+1}} \tag{9.79}$$

provided the series in $1/s$ converges.

It should not be supposed that the resulting series for all such functions converge, as the following example shows:

$$\sum_{n=0}^{\infty} (-1)^n \frac{t^{4n}}{n!} = e^{-t^4} \tag{9.80}$$

The result of transforming this is

$$\sum_{n=0}^{\infty} (-1)^n \frac{(4n)!}{n!\, s^{4n+1}} \tag{9.81}$$

which *does not converge* for any finite s, even though the original series converged for all finite t values.

An *asymptotic expansion* of a function $F(s)$ in terms of a series of functions F_k/s^k is an approximation for $F(s)$ such that

$$\lim_{s \to \infty} \frac{F(s) - \sum_{k=0}^{n} F_k s^{-k}}{s^{-n}} = 0 \tag{9.82}$$

This implies that the error in approximating $F(s)$ by $(n + 1)$ terms $(0 \leq k \leq n)$ of the series will be proportional to s^{-n-1} for values of s sufficiently large. It is understood in the limiting process in (9.82) that n is fixed. This means that the approximation (for which the symbol $\sim$ is used) may be useful even though the infinite series

$$F(s) \sim \sum_{k=0}^{\infty} F_k s^{-k}$$

diverges. The latter is the case for (9.81).

* E. C. Titchmarsh, *Theory of Functions*. Oxford: Oxford University Press, 1939, p. 47.

The second result mentioned above is a modification of Watson's lemma.* A function $f(t)$ possessing an expansion of the form

$$f(t) = t^{\alpha} \sum_{n=0}^{\infty} a_n t^n \qquad 0 \leq \alpha < 1$$

convergent in some neighborhood of the origin, may be termwise Laplace transformed to generate an asymptotic expansion for the Laplace transform $F(s)$ of $f(t)$, provided there exist positive numbers p and q such that $|f(t)| < pe^{qt}$. The asymptotic expansion will be valid for s interior to the first and fourth quadrant and of large modulus. Heuristically one may expect that for large positive real part of s the exponential factor e^{-st} will "erase" all values of the integrand except those near $t = 0$.

9.9 FOURIER SERIES AND TRANSFORMS

The result of restricting the values of z to a circle of radius $r < R$ (the radius of convergence) in (9.1) is to generate a *Fourier series* from the power series (9.1):

$$g(re^{i\theta}) = \sum_{k=0}^{\infty} f_k r^k e^{ik\theta} = \sum_{k=0}^{\infty} \psi_k e^{ik\theta} \tag{9.83}$$

with $\psi_k = r^k f_k$. For fixed r we may write

$$\psi(\theta) = \sum_{k=0}^{\infty} \psi_k e^{ik\theta} \tag{9.84}$$

where the coefficients are given by (Fourier coefficient formula)

$$\psi_k = \frac{1}{2\pi} \int_0^{2\pi} e^{-ik\theta} \psi(\theta)\, d\theta \tag{9.85}$$

which follows from the orthogonality property of the exponential

$$\int_0^{2\pi} e^{i(m-k)\theta}\, d\theta = 2\pi\delta_{mk} \tag{9.86}$$

If r were permitted to become equal to R (the radius of convergence), the convergence properties of (9.84) would become more complicated, since the circle of convergence contains at least one singularity. Generally, if $\psi(\theta)$ were of *bounded variation*, i.e., if

$$\int_0^{2\pi} |d\psi(\theta)|$$

* E. T. Copson, *Theory of Functions of a Complex Variable*. Oxford: Oxford University Press, 1935, p. 218.

were bounded, (9.84) would converge to

$$\lim_{\varepsilon \to 0} \frac{\psi(\theta + \varepsilon) + \psi(\theta - \varepsilon)}{2}$$

for each value of θ. For any value of θ for which $\psi(\theta)$ is continuous, the limit reduces simply to $\psi(\theta)$.

Similar results would hold for the values of a function analytic and free of singularities in an annulus when z is restricted to a circle interior to the annulus and concentric with it.

The Laurent expansion for fixed r would become a Fourier series

$$\psi(\theta) = \sum_{-\infty}^{\infty} \psi_k e^{ik\theta} \tag{9.87}$$

with (9.85) again.

The continuous analog of this would be as in (9.4):

$$\Psi(\theta) = \int_{-\infty}^{\infty} \psi(t) e^{i\theta t}\, dt \tag{9.88}$$

or replacing θ by $-\omega$ and redefining Ψ,

$$\Psi(\omega) = \int_{-\infty}^{\infty} \psi(t) e^{-i\omega t}\, dt \tag{9.89}$$

This is called the *Fourier transform* of $\psi(t)$.

9.10 FOURIER INVERSION FORMULAE

It is possible to find an inversion formula for $\psi(t)$ in terms of $\Psi(\omega)$. This turns out to be

$$\psi(t) = \frac{1}{2\pi} \int_{-\infty}^{\infty} \Psi(\omega) e^{i\omega t}\, d\omega \tag{9.90}$$

as may be derived from the inversion formula for the *bilateral Laplace transform* as follows: The bilateral Laplace transform of $f(t)$ is

$$F(s) = \int_{-\infty}^{\infty} e^{-st} f(t)\, dt \tag{9.91}$$

and unlike (9.4) it is the integral analog of an exponentially mapped ($z = e^{-s}$) Laurent series rather than (9.1). Since this can be decomposed into two (unilateral) Laplace transforms,

$$F(s) = \int_{0}^{\infty} e^{-st} f(t)\, dt + \int_{0}^{\infty} e^{st} f(-t)\, dt \tag{9.92}$$

in which the first integral converges for large positive real part of s and the second integral converges for large negative real part of s, the region of convergence of (9.91) is seen to be a strip parallel to the imaginary s axis. This is also clear from the fact that the exponential mapping $z = e^{-s}$ changes the annulus of convergence into such a strip.

The inversion formula for (9.91) is formally identical with (9.11) (see Example 25) where c is now a real number within the strip. This is then an analog of the Laurent coefficient formula (6.8). If one writes

$$f(t) = \frac{1}{2\pi i}\int_{c-i\infty}^{c+i\infty} e^{st}F(s)\,ds \tag{9.93}$$

where $F(s)$ is given by (9.91), upon substituting $s = c + i\omega$, one has

$$e^{-ct}f(t) = \frac{1}{2\pi}\int_{-\infty}^{\infty} e^{i\omega t}F(c + i\omega)\,d\omega \tag{9.94}$$

Defining

$$\begin{aligned} g(t) &= e^{-ct}f(t) \\ G(\omega) &= F(c + i\omega) \end{aligned} \tag{9.95}$$

(9.94) becomes

$$g(t) = \frac{1}{2\pi}\int_{-\infty}^{\infty} e^{i\omega t}G(\omega)\,d\omega \tag{9.96}$$

and (9.91) becomes

$$F(c + i\omega) = \int_{-\infty}^{\infty} e^{-i\omega t}e^{-ct}f(t)\,dt$$

or

$$G(\omega) = \int_{-\infty}^{\infty} e^{-i\omega t}g(t)\,dt \tag{9.97}$$

The results (9.96) and (9.97) are called the *Fourier inversion formulae*, and the functions $g(t)$ and $G(\omega)$ are said to be *Fourier transforms* of one another. We note that these formulae are much more symmetrical than the Laplace inversion formulae (9.4) and (9.11).

Let us apply (9.96) to the case where $G(\omega) = \delta(\omega - \omega')$:

$$g(t) = \frac{1}{2\pi}\int_{-\infty}^{\infty} e^{i\omega t}\delta(\omega - \omega')\,d\omega = \frac{e^{i\omega' t}}{2\pi} \tag{9.98}$$

so formal substitution into (9.97) yields

$$\delta(\omega - \omega') = \frac{1}{2\pi}\int_{-\infty}^{\infty} e^{-i(\omega-\omega')t}\,dt \tag{9.99}$$

which does not exist in a rigorous sense, as previously explained. Nevertheless, it can be used as a formal idealization of

$$\frac{1}{2\pi}\int_{-T}^{T} e^{-i(\omega-\omega')t}\,dt = \frac{\sin[(\omega'-\omega)T]}{\pi(\omega'-\omega)}$$

for very large T, since the latter only has appreciable values for $|\omega' - \omega| < \pi/T$ and

$$\frac{1}{\pi}\int_{-\infty}^{\infty} \frac{\sin[(\omega'-\omega)T]}{\omega'-\omega}\,d\omega = 1 \tag{9.100}$$

in any case.

Consider the integral

$$I = \int_{-\infty}^{\infty} g_1(t)g_2^*(t)\,dt \tag{9.101}$$

where $g_1(t)$, $g_2(t)$ and the corresponding $G_1(\omega)$, $G_2(\omega)$ satisfy (9.96) and (9.97). Then replacing $g_1(t)$ by (9.96) yields

$$I = \frac{1}{2\pi}\int_{-\infty}^{\infty}\int_{-\infty}^{\infty} e^{i\omega t}G_1(\omega)\,d\omega\, g_2^*(t)\,dt$$

or

$$I = \frac{1}{2\pi}\int_{-\infty}^{\infty} G_1(\omega)\int_{-\infty}^{\infty} e^{i\omega t}g_2^*(t)\,dt\,d\omega$$

which yields

$$I = \int_{-\infty}^{\infty} g_1(t)g_2^*(t)\,dt = \frac{1}{2\pi}\int_{-\infty}^{\infty} G_1(\omega)G_2^*(\omega)\,d\omega \tag{9.102}$$

the *generalized Parseval relation*. If $g(t) \equiv g_1(t) \equiv g_2(t)$, then $G_1(\omega) \equiv G_2(\omega) \equiv G(\omega)$ and (9.102) becomes

$$\int_{-\infty}^{\infty} |g(t)|^2\,dt = \frac{1}{2\pi}\int_{-\infty}^{\infty} |G(\omega)|^2\,d\omega \tag{9.103}$$

the *Parseval relation*.

1 PROPERTIES OF FOURIER TRANSFORMS

The Fourier transform of the integral of a function may be obtained as follows: Let

$$G(\omega) = \int_{-\infty}^{\infty} e^{-i\omega t}g(t)\,dt$$

$$g(t) = \frac{1}{2\pi}\int_{-\infty}^{\infty} e^{i\omega t}G(\omega)\,d\omega$$

The transform of $g(t) + k$ with k any constant is given by

$$G(\omega) + 2\pi k\delta(\omega) = \int_{-\infty}^{\infty} e^{-i\omega t}[g(t) + k]\, dt$$

and the inverse transform is

$$g(t) + k = \frac{1}{2\pi}\int_{-\infty}^{\infty} e^{i\omega t}[G(\omega) + 2\pi k\delta(\omega)]\, d\omega$$

Differentiating this, one has

$$g'(t) = \frac{1}{2\pi}\int_{-\infty}^{\infty} e^{i\omega t} i\omega[G(\omega) + 2\pi k\delta(\omega)]\, d\omega$$

Hence, if $f(t) = g'(t) = [g(t) + k]'$ and $F(\omega)$ is the transform of $f(t)$, one has

$$F(\omega) = i\omega[G(\omega) + 2\pi k\delta(\omega)]$$

so that

$$G(\omega) = \frac{F(\omega)}{i\omega} - 2\pi k\delta(\omega) \tag{9.104}$$

If, in particular, $g(t) = \int_{-\infty}^{t} f(t)\, dt$ so that $g(-\infty) = 0$ and $g(\infty) = F(0)$, then the transform of $G(\omega)$ is

$$g(t) = \frac{1}{2\pi i}\int_{-\infty}^{\infty} e^{i\omega t}\frac{F(\omega)}{\omega}\, d\omega - k$$

and

$$g(t) = \frac{1}{\pi}\int_{0}^{\infty} \left| \frac{e^{i\tau}F(\tau/t) - e^{-i\tau}F(-\tau/t)}{2i\tau} \right| d\tau - k$$

Hence,

$$g(\infty) = \frac{F(0)}{\pi}\int_{0}^{\infty} \frac{\sin \tau}{\tau}\, d\tau - k = \frac{F(0)}{2} - k = F(0)$$

and

$$k = -\frac{F(0)}{2} \tag{9.105}$$

so that

$$G(\omega) = \frac{F(\omega)}{i\omega} + \pi F(0)\delta(\omega) \tag{9.106}$$

For

$$g(t) = u(t) = \int_{-\infty}^{t} \delta(t)\, dt = \begin{cases} 1 & \text{for } t > 0 \\ 0 & \text{for } t < 0 \end{cases}$$

$$f(t) = \delta(t) \qquad \text{and} \qquad F(\omega) = \int_{-\infty}^{\infty} e^{-i\omega t}\delta(t)\,dt = 1$$

so that

$$G(\omega) = U(\omega) = \frac{1}{i\omega} + \pi\delta(\omega) \tag{9.107}$$

in a formal sense.

Many of the simple properties of Laplace transforms have analogs with Fourier transforms. Using again the correspondence notation

$$G(\omega) \leftrightarrows g(t) \tag{9.108}$$

one has

$$G(m\omega) \leftrightarrows \frac{1}{m}\, g\left(\frac{t}{m}\right) \tag{9.109}$$

$$G(\omega - \omega') \leftrightarrows e^{i\omega' t} g(t) \tag{9.110}$$

$$e^{-i\omega a} G(\omega) \leftrightarrows g(t - a) \tag{9.111}$$

$$d_\omega^n G(\omega) \leftrightarrows (-it)^n g(t) \tag{9.112}$$

$$\frac{1}{2\omega}\int_{-\omega}^{\omega} G(\omega)\,d\omega = \int_{-\infty}^{\infty} g(t)\,\frac{\sin \omega t}{\omega t}\,dt \tag{9.113}$$

$$(i\omega)^n G(\omega) \leftrightarrows d_t^n g(t) \tag{9.114}$$

The Fourier transform of the function $h(t)$ defined by

$$h(t) = \int_{-\infty}^{\infty} f(\tau) g(t - \tau)\,d\tau \tag{9.115}$$

is

$$H(\omega) = \int_{-\infty}^{\infty} e^{-i\omega t} h(t)\,dt = \int_{-\infty}^{\infty} f(\tau) \int_{-\infty}^{\infty} e^{-i\omega t} g(t - \tau)\,dt\,d\tau$$

or

$$H(\omega) = \int_{-\infty}^{\infty} e^{-i\omega\tau} f(\tau)\,d\tau \int_{-\infty}^{\infty} e^{-i\omega s} g(s)\,ds$$

so that

$$H(\omega) = F(\omega) G(\omega) \tag{9.116}$$

the *convolution theorem* for Fourier transforms. Equation (9.115) is the appropriate definition of *convolution* (*Faltung*, resultant) for Fourier transforms. We note that as in the case of Laplace transforms the product of two Fourier transforms is *not* the Fourier transform of the product of the functions.

It is important in dealing with Laplace and Fourier transforms to recognize that changing a function even in a small interval will affect the value of the transform for (almost) all values of the transformed variable. This may be verified by applying (9.4) or (9.97) to a function vanishing outside a small range and constant within it.

9.12 BANDWIDTH-DURATION AND UNCERTAINTY PRINCIPLES

If the variables t and ω are interpreted as time and frequency, $f(t)$ may be thought of as a time (-dependent) signal and $F(\omega)$ as its corresponding frequency spectrum (a term sometimes used for $|F(\omega)|$) which describes the extent to which various frequencies participate in the Fourier representation of the function $f(t)$. If it is also supposed that $f(t)$ is so normalized that

$$\int_{-\infty}^{\infty} |f(t)|^2 \, dt = 1 \tag{9.117}$$

$|f(t)|^2 \, dt$ can be interpreted as a probability that a nonzero signal occurs between t and $(t + dt)$. Then via the Parseval relation (9.103),

$$\frac{1}{2\pi} \int_{-\infty}^{\infty} |F(\omega)|^2 \, d\omega = 1 \qquad \omega = 2\pi\nu$$

and $|F(\omega)|^2 \, d\nu$ can be correspondingly interpreted as a probability that the signal contains frequencies between ν and $\nu + d\nu$, and the origin of time may be so chosen that

$$\langle t \rangle = \int_{-\infty}^{\infty} t \, |f(t)|^2 \, dt = 0 \tag{9.118}$$

The "average (angular) frequency" of the signal may be defined as

$$\langle \omega \rangle = \frac{1}{2\pi} \int_{-\infty}^{\infty} \omega \, |F(\omega)|^2 \, d\omega \tag{9.119}$$

while the "statistical duration" (standard deviation of times of signal occurrence) of the signal is taken to be σ_t, with

$$\sigma_t^2 = \int_{-\infty}^{\infty} t^2 \, |f(t)|^2 \, dt = \frac{1}{2\pi} \int_{-\infty}^{\infty} |F'(\omega)|^2 \, d\omega \tag{9.120}$$

The latter integral follows from an application of the Parseval relation to the derivative of the Fourier transform $F(\omega)$ of $f(t)$:

$$F'(\omega) = -i \int_{-\infty}^{\infty} e^{-i\omega t} t f(t) \, dt$$

The "statistical (angular) bandwidth" σ_ω is

$$\sigma_\omega^2 = \frac{1}{2\pi}\int_{-\infty}^{\infty}(\omega - \langle\omega\rangle)^2\,|F(\omega)|^2\,d\omega \tag{9.121}$$

In terms of these definitions the *bandwidth–duration principle*

$$2\sigma_t\sigma_\omega \geq 1 \tag{9.122}$$

can be established by using a form of the *Schwarz inequality* (Problem 30, Chapter 1):

$$\int_{-\infty}^{\infty}|P(\omega)|^2\,d\omega\int_{-\infty}^{\infty}|Q(\omega)|^2\,d\omega \geq \left|\int_{-\infty}^{\infty}\frac{P^*(\omega)Q(\omega) + P(\omega)Q^*(\omega)}{2}\,d\omega\right|^2$$

which follows from the (discriminant) condition that the quadratic expression in x (independent of ω),

$$\int_{-\infty}^{\infty}|P(\omega) + xQ(\omega)|^2\,d\omega = ax^2 + bx + c$$

is negative for no value of x or equivalently possesses no real roots in x unless they coincide. The discriminant

$$b^2 - 4ac = \left|\int(P^*Q + PQ^*)\,d\omega\right|^2 - 4\int|P|^2\,d\omega\int|Q|^2\,d\omega$$

must then be nonpositive, whence the required inequality follows. For

$$P(\omega) = F'(\omega) \qquad \text{and} \qquad Q(\omega) = (\omega - \langle\omega\rangle)F(\omega)$$

the inequality becomes (noting the definitions of σ_t and σ_ω)

$$4\pi^2\sigma_t^2\sigma_w^2 \geq \left|\int_{-\infty}^{\infty}(\omega - \langle\omega\rangle)\left(\frac{F^*F' + FF'^*}{2}\right)d\omega\right|^2$$

or

$$2\pi\sigma_t\sigma_\omega \geq \tfrac{1}{2}\left|\int_{-\infty}^{\infty}(\omega - \langle\omega\rangle)\,d(|F|^2)\right|$$

and since $|F|^2$ is integrable $(-\infty,\infty)$, it must approach zero at $\pm\infty$:

$$\langle\omega\rangle\int_{-\infty}^{\infty}d\,|F|^2 = 0$$

while

$$\int_{-\infty}^{\infty}\omega d\,|F|^2 = -\int_{-\infty}^{\infty}|F|^2\,d\omega = -2\pi$$

so that

$$2\sigma_t\sigma_\omega \geq 1$$

as was to be proved. In terms of the statistical bandwidth $\sigma_\nu = \sigma_\omega/2\pi$ this becomes

$$4\pi\sigma_t\sigma_\nu \geq 1$$

while if the Planck hypothesis $E = h\nu = \hbar\omega$ of quantum mechanics is invoked, one has $\sigma_E = \hbar\sigma_\omega = h\sigma_\nu$ and

$$\sigma_t\sigma_E \geq \hbar/2 \tag{9.123}$$

the *uncertainty principle* for time t and energy E. Alternatively the de Broglie hypothesis $p = \hbar k(k = 2\pi/\lambda)$, applied to the bandwidth–duration principle for the Fourier analysis on position x and propagation number k,

$$2\sigma_x\sigma_k \geq 1$$

yields (with $\sigma_p = \hbar\sigma_k$, $h = 2\pi\hbar$ = Planck constant)

$$\sigma_x\sigma_p \geq \hbar/2 \tag{9.124}$$

the *uncertainty principle* for position x and momentum p.

Consequences of the bandwidth–duration principle are that an instantaneous signal (pulse of zero duration) has an infinite bandwidth [e.g., $\delta(t - a)$ and $e^{-i\omega a}$] and a signal of zero bandwidth has an infinite duration [e.g., $e^{i\omega_0 t}$ and $2\pi\delta(\omega - \omega_0)$]. The duration and bandwidth of a signal cannot be simultaneously reduced to arbitrarily small values.

Similarly, in quantum mechanics the uncertainties (standard deviations) in two canonically conjugate* (noncommuting) variables (e.g., energy and time, position and momentum) cannot be simultaneously reduced to arbitrarily small values.

The bandwidth–duration principle explains something of the difficulty in generating or detecting signals of extremely short duration or extremely narrow bandwidth.

9.13 LINEAR SYSTEMS AND STABILITY

The most common applications of Laplace and Fourier transforms are to linear differential equations with constant coefficients of the form

$$Ly(t) = \sum_{k=0}^{n} a_k d_t^k y(t) = x(t) \tag{9.125}$$

where the inhomogeneity $x(t)$ is often called the input or forcing term and $y(t)$ is often called the output or response to $x(t)$. It is known that for

* Canonically conjugate pairs of variables are those satisfying relations analogous to the relations between momentum and position in Hamilton's equations.

constant values of a_k and $x(t)$ equal to zero (the homogeneous case) the solutions of the equation ($Ly = 0$) are generally given by a linear combination of exponential functions with polynomial coefficients, which latter become constants in the case of distinct roots of the characteristic equation

$$P(\lambda) = \sum_{k=0}^{n} a_k \lambda^k = 0 \tag{9.126}$$

In the latter case the response to zero input is

$$y_H(t) = \sum_{k=1}^{n} c_k e^{\lambda_k t}$$

where $P(\lambda_k) = 0 \neq P'(\lambda_k)$ $(k = 1,2, \ldots ,n)$. Generally the roots λ_k are complex, but their real parts cannot be positive without *instability* in which terms in $y_H(t)$ would become indefinitely large in modulus as $t \to \infty$. Such a response could not be maintained in a physical system without an infinite source of energy for the system. Since the latter does not exist, terms with positive real parts for the characteristic roots λ_k cannot be of more than temporary interest (for the t variable as time) and are excluded as unstable solutions. If the system were delicately and precisely balanced, imaginary values of λ_k might be considered possible, in which case the system hypothetically would oscillate forever. Generally, damping (or resistance) will be present and the *stable* solutions will correspond to λ_k with *negative real parts only*.

All this refers to the case of zero input or more realistically to the case of a system capable of appreciable response to a negligible input. In the case of an undamped (resistanceless) system $\lambda_k = i\omega_k$ and the system hypothetically can oscillate forever (*resonance*) at the *natural frequencies* ω_k determined by the zeros of $P(\lambda)$.

The commonest kinds of nonzero input $x(t)$ are of exponential form

$$x(t) = x_0(s)e^{st}$$

where s is a complex constant (with Re $s \leq 0$). The response to this kind of input is also of exponential form

$$y(t) = y_0(s)e^{st}$$

since differentiation of $Ly - x = 0$ yields

$$\begin{aligned} d_t Ly - d_t x &= L(d_t y) - sx = L(d_t y) - sLy \\ &= L(d_t y - sy) = 0 \end{aligned}$$

which implies $d_t y - sy = 0$ corresponding to the nontrivial solution of $Ly = x$.

For s real and negative one has an exponentially *transient* input (e.g., starting at $t = 0$) producing a similar *transient* response (at some later time).

For s purely imaginary one has a persistent oscillatory input producing a persistent (or steady-state ac) output. Carrying out the above substitution, one has explicitly

$$Ly = P(d_t)y(t) = y_0(s)P(d_t)e^{st} = x_0(e)e^{st}$$

or

$$y_0(s)P(s) = x_0(s)$$

so that

$$y(t) = \frac{x_0(s)e^{st}}{P(s)} \tag{9.127}$$

whence again the tremendous response of the system is evident for inputs at ($s = i\omega$) the natural frequencies [zeros of $P(s)$] of the system.

Responses to a more general class of inputs may be generated by *linear superposition*. Thus to an input of the form

$$x(t) = \frac{1}{2\pi i}\int_{\sigma-i\infty}^{\sigma+i\infty} X(s)e^{st}\,ds \tag{9.128}$$

one should obtain a response of the form

$$y(t) = \frac{1}{2\pi i}\int_{\sigma-i\infty}^{\sigma+i\infty} Y(s)e^{st}\,ds \tag{9.129}$$

Actually $Y(s)$ can be obtained in terms of $X(s)$ by Laplace transforming the original differential equation. Thus,

$$P(s)Y(s) - \sum_{k=0}^{n} a_k \sum_{m=0}^{k-1} y^{(m)}(0)s^{k-m-1} = X(s)$$

so that

$$Y(s) = \frac{X(s)}{P(s)} + \frac{Q(s)}{P(s)}$$

with

$$Q(s) = \sum_{k=0}^{n} a_k \sum_{m=0}^{k-1} y^{(m)}(0)s^{k-m-1}$$

If the initial response of the system is zero, $Q(s) \equiv 0$ and

$$Y(s) = G(s)X(s) \tag{9.130}$$

where $G(s)P(s) \equiv 1$. The evaluation of $y(t)$ is generally carried out by the residue theorem, and each simple zero s_k of $P(s)$ contributes a term

$$\frac{X(s_k)}{P'(s_k)}\,e^{s_k t}$$

to the solution $y(t)$. Again *stability* requires that Re s_k should *avoid being positive*. If Re $s_k = 0$, the system is sometimes called *conditionally stable*, while *strict stability* requires Re $s_k < 0$ for all roots of $P(s)$.

9.14 FILTERS

The function $G(s)$ which characterizes the relationship between input $X(s)$ and output $Y(s)$ in the transformed s domain is called the *transfer function* of the system associated with the original differential equation. Generally such a system with a well-defined output for a well-defined input is called a *filter*.

Instead of determining the response of the system (filter) to an input of the form of an inverse Laplace transform, one could have supposed the input to be of the form (superposition of ac signals):

$$x(t) = \frac{1}{2\pi} \int_{-\infty}^{\infty} e^{i\omega t} X_1(\omega)\, d\omega \tag{9.131}$$

with response of the form

$$y(t) = \frac{1}{2\pi} \int_{-\infty}^{\infty} e^{i\omega t} Y_1(\omega)\, d\omega \tag{9.132}$$

This time $Y_1(\omega)$ can be obtained by Fourier transforming the original differential equation. If the response at $t = \pm\infty$ is zero, one has

$$P(i\omega) Y_1(\omega) = X_1(\omega)$$

so that

$$Y_1(\omega) = \frac{X_1(\omega)}{P(i\omega)} = G_1(\omega) X_1(\omega) \tag{9.133}$$

where $G_1(\omega)P(i\omega) \equiv 1$. The function $G_1(\omega)$ is often called the *system function* or *frequency response* of the filter.

A filter for which $G_1(\omega)$ is zero *outside* a finite interval of frequencies is called a strict *band-pass filter*. This of course is an idealization, since a filter with a strictly zero response for a finite interval of frequencies cannot be constructed. The (experimental) proof that such a construction had been accomplished would require a device to detect arbitrarily small signal strengths, and because all equipment has intrinsic noise levels this is impossible. Nevertheless, the ideal concept of a band-pass filter turns out to be of some conceptual use.

Similarly, an ideal filter for which $G_1(\omega)$ is zero *within* a finite interval of frequencies is called a strict *band-suppression filter*.

A *statistical band-pass filter* is one for which

$$\int_{-\infty}^{\infty} |\omega - \langle\omega\rangle|^2 \, |G_1(\omega)|^2 \, d\omega$$

is finite. Here $|G_1(\omega)|^2 \, d\omega$ is proportional to the probability that frequencies between ω and $\omega + d\omega$ are passed and

$$\langle\omega\rangle = \frac{\displaystyle\int_{-\infty}^{\infty} \omega \, |G_1(\omega)|^2 \, d\omega}{\displaystyle\int_{-\infty}^{\infty} |G_1(\omega)|^2 \, d\omega}$$

is the "center" frequency of the filter, while σ_ω is the statistical bandwidth

$$\sigma_\omega^2 = \frac{\displaystyle\int_{-\infty}^{\infty} |\omega - \langle\omega\rangle|^2 \, |G_1(\omega)|^2 \, d\omega}{\displaystyle\int_{-\infty}^{\infty} |G_1(\omega)|^2 \, d\omega} \tag{9.134}$$

With this definition, essentially all constructible filters are statistical band-pass filters. If $\langle\omega\rangle = 0$, the filter becomes a *statistical low-pass* filter. A strict low-pass filter has $G_1(\omega) = 0$ for $|\omega| > \omega_c$, while a strict high-pass filter has $G_1(\omega) = 0$ for $|\omega| < \omega_c$.

A signal is said to be strictly bandwidth-limited if its spectrum (Fourier transform) is zero outside a limited range of frequencies. In this case it will be analytic and cannot be strictly duration-limited without being zero for all t. Actually functions of physical interest are always of *statistically* limited bandwidth and duration subject to the bandwidth–duration principle.

A filter is said to be *amplitude-distortionless* if $|G_1(\omega)|$ is constant. It is said to be *phase-distortionless* if $d_\omega \arg G_1(\omega)$ is constant.

Until now two important restrictions on a physically realizable filter have been ignored. First, if the input $x(t)$ and output $y(t)$ of the filter are to be real functions, then

$$X_1(\omega) = \int_{-\infty}^{\infty} e^{-i\omega t} x(t) \, dt = X_1^*(-\omega^*)$$

$$Y_1(\omega) = \int_{-\infty}^{\infty} e^{-i\omega t} y(t) \, dt = Y_1^*(-\omega^*)$$

or, with real frequencies $\omega = \omega^*$,

$$X_1(\omega) = X_1^*(-\omega)$$

$$Y_1(\omega) = Y_1^*(-\omega)$$

Consequently, real input and output imply (for real $\omega = \omega^*$)

$$G_1(\omega) = G_1^*(-\omega^*) = G_1^*(-\omega) \tag{9.135}$$

which implies that $|G_1(\omega)|$ is an *even* function of frequency and $\arg G_1(\omega)$ is an *odd* function of frequency.

Second, the physical filter must not respond to a signal before it enters the input. This may seem like an obvious requirement, but it has not been implicitly imposed in the Fourier analysis of the input–output relationship. For instance, if $x(t) = \delta(t-2)$ and $G_1(\omega) = e^{i\omega}$,

$$X_1(\omega) = e^{-2i\omega}$$

and

$$Y_1(\omega) = G_1(\omega)X_1(\omega) = e^{-i\omega}$$

so that $y(t) = \delta(t-1)$, and the output (response) occurs at $t = 1$ *before* the input at $t = 2$ in spite of the fact that

$$G_1(\omega) = G_1^*(-\omega)$$

Such an unphysical filter (where output precedes input) is called *acausal* or *noncausal* and is a simple example of how physically unreasonable a perfectly simple mathematical construct can be.

It is not difficult to adduce a condition on the filter in order that it be *causal* (that its input precedes its output). One simply recalls that

$$Y_1(\omega) = G_1(\omega)X_1(\omega)$$

corresponds to

$$y(t) = \int_{-\infty}^{\infty} g_1(t-\tau)x(\tau)\,d\tau$$

with

$$g_1(t) = \frac{1}{2\pi}\int_{-\infty}^{\infty} e^{i\omega t}G_1(\omega)\,d\omega$$

Then, for a causal filter, $x(t) = \delta(t-a)$ must imply that

$$y(t) = g_1(t-a)$$

is zero for $t < a$ or equivalently that

$$g_1(t) = 0 \qquad \text{for} \quad t < 0 \tag{9.136}$$

The function $g_1(t)$ is generally called the *impulse response* of the filter, since it describes the output due to a delta function (impulse) input.

The two methods of Laplace transformation and Fourier transformation of the original problem are seen to be intimately related. In fact for complex values of s and ω they are essentially equivalent, while for imaginary values of s and real ω the Laplace transform* becomes the Fourier transform [take $c = 0$ in (9.95)] for $s = i\omega$. Thus the relationship between the transfer function $G(s)$ and the system function $G_1(\omega)$ is simply

$$G(i\omega) = G_1(\omega) \tag{9.137}$$

so that there is no need to carry out a Fourier analysis and Laplace analysis separately. The values of the transfer function on the imaginary axis in the s plane are the values of the system function for real frequencies.

For a strictly stable system the response to a delta function input $\delta(t - \varepsilon)$ has a Laplace transform

$$Y_\varepsilon(s) = G(s)e^{-\varepsilon s}$$

so that

$$Y(s) = \lim_{\varepsilon \to 0} Y_\varepsilon(s) = G(s)$$

and

$$y(t) = g(t) = \frac{1}{2\pi i} \int_{-i\infty}^{i\infty} e^{st}G(s)\, ds$$

since the poles of $G(s)$ must be in the left half-plane $\text{Re}\, s < 0$; otherwise there would be positive exponential factors approaching infinity as $t \to \infty$ arising from the evaluation by the residue theorem. The conclusion that the poles of $G(s)$ must be in $\text{Re}\, s < 0$ means that the abscissa of convergence cannot be to the right of the imaginary axis in the s plane; hence c can be taken to be zero in the inversion integral. With $s = i\omega$, one then has

$$g(t) = \frac{1}{2\pi} \int_{-\infty}^{\infty} e^{i\omega t}G(i\omega)\, d\omega$$

Hence,

$$g(t) = g_1(t)$$

and the inverse Laplace transform of the transfer function is identical with the (inverse) Fourier transform of the system function.

If the existence of the limit

$$\lim_{|s| \to \infty} |s^\varepsilon G(s)| \qquad \text{for} \qquad |\arg s| < \frac{\pi}{2} \qquad 0 < \varepsilon < 1$$

is assumed, the restriction of the poles of $G(s)$ to $\text{Re}\, s < 0$ suffices to ensure that $g(t)$ is causal [i.e., $g(t) = 0$ for $t < 0$] since then

$$g(t) = \frac{1}{2\pi i} \oint e^{st}G(s)\, ds = 0 \qquad t < 0$$

* More generally the bilateral Laplace transform.

with the contour completed by an infinite semicircular arc to the right. The integral along the infinite arc is zero.

A similar argument could have been made for the contour integration evaluation of the Fourier transform

$$g_1(t) = \frac{1}{2\pi}\int_{-\infty}^{\infty} e^{i\omega t}G_1(\omega)\, d\omega$$

Here for $t < 0$ the contour would have to be completed (to secure vanishing of the arc part) by an infinite semicircular arc in the lower half (complex) ω plane so that $G_1(\omega)$ would have to be free of poles in Im $\omega < 0$. This corresponds to $G(s)$ free of poles in Re $s > 0$ since ($s = i\omega$)

$$\omega = -is$$

indicates that the complex ω plane is obtained from the complex s plane by a *clockwise* rotation of $\pi/2$. This simple relationship should be recalled whenever statements about the s plane are compared with the corresponding statements about the ω plane.

The characteristics of a linear system or filter are specified by the functions $G(s)$ or $G_1(\omega)$. Naturally the output will depend not only on these functions but also on the particular inputs, and it is often customary to specify the system with respect to certain standard inputs such as $\delta(t - a)$, $u(t - a)$, or $k(t - a)u(t - a)$, which are called, respectively, impulse, step, and ramp inputs.

The corresponding outputs then have Laplace transforms $G(s)e^{-sa}$, $G(s)e^{-sa}/s$, and $kG(s)e^{-sa}/s^2$, so that on the basis of the convolution theorem for Laplace transforms the corresponding outputs are (for $t \geq a$)

$$g(t - a) \qquad \int_a^t g(t - \tau)\, d\tau \qquad k\int_a^t (\tau - a)g(t - \tau)\, d\tau$$

the impulse response, step response, and ramp response, respectively.

15 SERVOMECHANISMS AND FEEDBACK AMPLIFIERS

Instead of a system in which the output is determined directly by the input passing through a filter, one can consider a system in which part of the output is fed back to the input so that the output is specified by

$$Y(s) = G(s)[X(s) - Y(s)]$$

Then

$$Y(s) = \left[\frac{G(s)}{1 + G(s)}\right]X(s) \tag{9.138}$$

and $G(s)$ is called the *open-loop transfer* function, while

$$G_c(s) = \frac{G(s)}{1 + G(s)} \tag{9.139}$$

is called the *closed-loop transfer* function.

In such a system the response can be automatically "controlled" via the feedback loop. Thus,

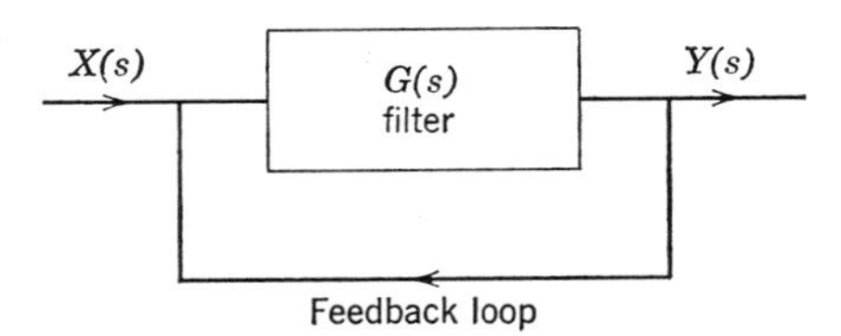

This is called a *servomechanism* or *feedback amplifier*. More usually one imagines the filter to have a variable gain control which can change its transfer function by a real factor K so that the system becomes

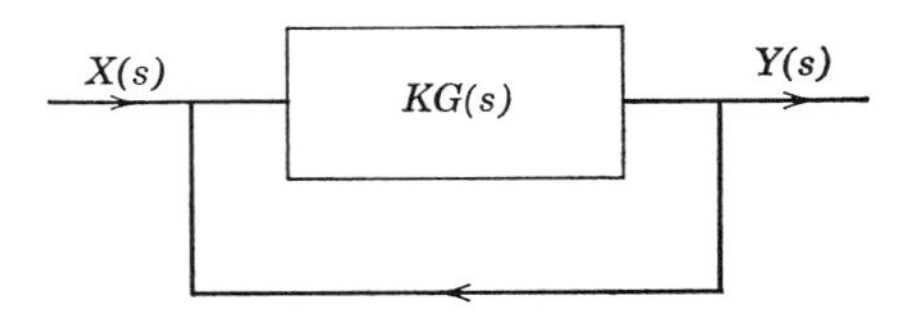

with closed-loop transfer function

$$G_c(s) = \frac{KG(s)}{1 + KG(s)} \tag{9.140}$$

The output of such a closed-loop system could be calculated by using the Laplace inversion formula

$$y(t) = \frac{1}{2\pi i} \int_{\sigma-i\infty}^{\sigma+i\infty} \left\{\frac{KG(s)X(s)}{1 + KG(s)}\right\} e^{st}\, ds \tag{9.141}$$

and would depend on the location of the poles of the integrand. For the case of an impulsive (delta function) input,

$$X(s) = \int_0^\infty e^{-st}\delta(t - a)\, dt = e^{-sa}$$

and for $a \to 0$, $X(s) \to 1$ so that

$$y(t) = \frac{1}{2\pi i} \int_{\sigma-i\infty}^{\sigma+i\infty} G_c(s) e^{st}\, ds \tag{9.142}$$

and the output depends on the poles of $G_c(s)$ alone. These occur at the locations of the zeros of $1 + KG(s) = 0$. By using the residue theorem, one could calculate the contribution to $y(t)$ at each pole of the integrand. However, if the real part of s satisfying

$$1 + KG(s) = 0 \tag{9.143}$$

were *positive*, there would be a positive exponential factor in the result. Thus, if $G(s) = -1/s$ with K positive,

$$G_c(s) = \frac{K}{K - s}$$

and

$$y(t) = -Ke^{Kt}$$

which is clearly *unstable*, since $|y|$ would increase indefinitely with increasing time. The restriction to *stable* solutions would generally require that no roots of

$$1 + KG(s) = 0$$

should occur in Re $s > 0$, the right half s plane. The question of whether a given closed-loop system is stable or unstable can be decided by considering the region of the $KG(s)$ plane, which is the image of the right half s plane. This is called a *Nyquist diagram*, and if it includes the point -1 [in the $KG(s)$ plane], the system will be unstable, since then there will exist a root of $1 + KG(s) = 0$ [and therefore a pole of $G_c(s)$] for which Re $s > 0$, producing an increasing exponential term in the output due to an impulsive input.

Usually the Nyquist diagram is conceived as a limiting case of the image of a region of the right half s plane between two concentric (at $s = 0$) semicircular arcs (with bounding diameters on the imaginary axis) as the radii approach zero and infinity, respectively. One imagines this contour in the s plane to be traversed in the direction of increasing (frequencies) imaginary part of s. The corresponding bounding trajectory in the $KG(s)$ plane encloses the area to the *right* of the image path [in $KG(s)$ plane] and the *Nyquist criterion for stability* is that this enclosed area should *not* include the point $KG(s) = -1$.

9.16 ROOT LOCUS METHOD

Another useful method in studying or designing servomechanisms is the *root locus* method of plotting the loci of the closed-loop system poles of $G_c(s)$ in the s plane as a function of the open-loop gain parameter K. In the usual case where the open-loop transfer function $G(s)$ is a rational function of s (ratio of polynomials) one has

$$G(s) = \frac{N(s)}{D(s)} \tag{9.144}$$

and

$$G_c(s) = \frac{KG}{1 + KG} = \frac{KN}{D + KN} \tag{9.145}$$

so that the poles of $G_c(s)$ occur at the zeros of $[D(s) + KN(s)]$, the zeros of $G(s)$ occur at the zeros of $N(s)$, and the poles of $G(s)$ occur at the zeros of $D(s)$. Thus it may be said that *the root loci start* ($K = 0$) *at the open-loop poles of* $G(s)$ *and end* ($K = \infty$) *at the open-loop zeros of* $G(s)$:

$$\lim_{K \to 0} (D + KN) = D$$

$$\lim_{K \to \infty} \left(\frac{D + KN}{K} \right) = N$$

Since the loci start at poles of $G(s)$ and end at zeros of $G(s)$, there are as many branches as there are such poles (or zeros). The places where *the loci cross to the right half s plane* correspond to values of K above which *instability* results. For the usual case where $D(s)$ and $N(s)$ have real coefficients the roots of $D(s) + KN(s) = 0$ must occur in conjugate pairs so the loci will be *symmetrical* to the real s axis. For those values of K for which the loci lie on the negative real s axis there will be solutions (outputs) exponentially decaying with time, but when the loci extend away from the real s axis there will be oscillatory outputs with an exponentially decaying "envelope." Exponentially increasing solutions would correspond to loci passing into the right half s plane, as seen above.

The Regge trajectories of the poles of the partial wave-scattering amplitudes in the complex angular momentum plane are analogous to the root loci above. These are of use in the theory of scattering of particles of given energy E which plays the role of the gain factor K above.

● EXAMPLES 9

1. Find the abscissa of convergence of the Laplace transform of $t^5 \sinh 3t$.
   Ans. For large t, $t^5 \sinh 3t \approx \frac{1}{2} t^5 e^{3t}$, so that $e^{-st} t^5 e^{3t} = e^{-(s-3)t} t^5$ would be integrable $(0, \infty)$ to a finite result provided $\operatorname{Re} s > 3$. Hence the abscissa of convergence is 3.

2. Find the convergence strip of the bilateral Laplace transform of $t^3e^{-5|t|}\sinh 3t$.

Ans. For large positive t the integrand will approximate $\frac{1}{2}t^3e^{-(s+2)t}$, so that Re $s > -2$ is required for convergence. For large negative t the integrand will approximate $\frac{1}{2}t^3e^{-(s-2)t}$, so that Re $s < 2$ for convergence. Hence the convergence strip is $|\text{Re } s| < 2$.

3. Find the function $f(t)$ with Laplace transform $F(s) = e^{-3s/(s+2)}$.

Ans. $$f(t) = \frac{1}{2\pi i}\int \frac{e^{s(t-3)}}{s+2}\,dS = \begin{cases} e^{-2(t-3)} & \text{for } t > 3 \\ 0 & \text{for } t < 3 \end{cases}$$

4. Find the function $f(t)$ with Laplace transform $F(s) = 1/s(e^{2s} - 1)$.

Ans. $$F(s) = \frac{e^{-2s}}{s(1 - e^{-2s})} = \frac{1}{s}\sum_{k=1}^{\infty} e^{-2ks} = \sum_{k=1}^{\infty} F_k(s)$$

The inverse transform of $F_k(s)$ is

$$f_k(t) = \frac{1}{2\pi i}\int_{\varepsilon-i\infty}^{\varepsilon+i\infty} \frac{e^{s(t-2k)}\,ds}{s} = \begin{cases} 1 & \text{for } t > 2k \\ 0 & \text{for } t < 2k \end{cases}$$

so that

$$f(t) = \begin{cases} k & \text{for } 2k + 2 > t > 2k \\ 0 & \text{for } 2 > t \end{cases}$$

5. Find the Laplace transform of

$$f(t) = \begin{cases} t^2 & \text{for } 0 < t < 3 \\ t^2 + 1 & \text{for } t > 3 \\ 0 & \text{for } t < 0 \end{cases}$$

Ans. $$F(s) = \int_0^{\infty} t^2e^{-st}\,dt + \int_3^{\infty} e^{-st}\,dt = \frac{2}{s^3} + \frac{e^{-3s}}{s}$$

6. Find the Laplace–Stieltjes transform of f:

$$df = \begin{cases} 2t\,dt & \text{for } 3 \neq t \geq 0 \\ 1 & \text{for } t = 3 \\ 0 & \text{for } t < 0 \end{cases}$$

Ans. $$\int e^{-st}\,df = 2\int_0^{\infty} e^{-st}t\,dt + e^{-3s} = \frac{2}{s^2} + e^{-3s}$$

7. Find an asymptotic approximation to the integral

$$2\pi i f(t,a) = \int_{\varepsilon-i\infty}^{\varepsilon+i\infty} e^{-a\sqrt{s}+st}\,ds \qquad \text{for} \quad t > 0$$

by the method of steepest descents.

Ans. Let

$$\phi(s) = st - a\sqrt{s}$$

$$\phi'(s) = t - \frac{a}{2\sqrt{s}} = 0 \qquad \text{for} \quad s = \left(\frac{a}{2t}\right)^2 = b$$

$$\phi(b) = -\frac{a^2}{4t}$$

$$\phi''(b) = \frac{2t^3}{a^2}$$

so that

$$2\pi i f(t,a) \sim e^{-a^2/4t} \int e^{(s-b)^2(t^3/a^2)}\,ds = \frac{ia}{t}\sqrt{\frac{\pi}{t}}\,e^{-a^2/4t}$$

In this case (but not in general) the asymptotic approximation turns out to be *exact* so that

$$f(t;a) = \frac{ae^{-a^2/4t}}{2t\sqrt{\pi t}}$$

is precisely the function with Laplace transform

$$F(s;a) = e^{-a\sqrt{s}}$$

8. Verify that

$$I = \int_{\varepsilon-i\infty}^{\varepsilon+i\infty} e^{-a\sqrt{s}+st}\,ds$$

satisfies the partial differential equation

$$\partial_a^2 I = \partial_t I$$

Ans.

$$I = I_0 \frac{e^{-a^2/4t}a}{t\sqrt{t}} \qquad \text{by Example 7}$$

$$\log I = \log I_0 - \frac{a^2}{4t} + \log a - \tfrac{3}{2}\log t$$

$$\frac{\partial_t I}{I} = \frac{a^2}{4t^2} - \frac{3}{2t}$$

$$\frac{\partial_a I}{I} = -\frac{a}{2t} + \frac{1}{a}$$

$$\partial_a I = \left(\frac{1}{a} - \frac{a}{2t}\right) I$$

$$\partial_a^2 I = -\left(\frac{1}{2t} + \frac{1}{a^2}\right) I + \left(-\frac{a}{2t} + \frac{1}{a}\right) \partial_a I$$

$$\partial_a^2 I = \left[-\left(\frac{1}{2t} + \frac{1}{a^2}\right) + \left(-\frac{a}{2t} + \frac{1}{a}\right)^2\right] I$$

$$\partial_a^2 I = \left[-\frac{1}{2t} - \frac{1}{a^2} + \frac{a^2}{4t^2} + \frac{1}{a^2} - \frac{1}{t}\right] I$$

$$\partial_a^2 I = \partial_t I$$

9. Find the function $h(t)$ with Laplace transform $H(s) = 1/s^3(s^2 + \omega^2)$.

Ans. $3!\, H(s) = \left(\frac{3!}{s^4}\right)\left(\frac{s}{s^2 + \omega^2}\right)$

so, by the convolution theorem,

$$3!\, h(t) = \int_0^t (t - \tau)^3 \cos(\omega\tau)\, d\tau$$

Alternatively,

$$2!\, \omega H(s) = \frac{2!}{s^3}\left(\frac{\omega}{s^2 + \omega^2}\right)$$

so

$$2!\, h(t) = \int_0^t (t - \tau)^2 \sin(\omega\tau)\, d\tau$$

10. If the temperature T in a semi-infinite rod satisfies the heat-conduction equation (with unit diffusion coefficient)

$$\partial_t T = \partial_x^2 T$$

and $T(x,0) = 0$, $T(0,t) = f(t)$. Find $T(x,t)$ for $t > 0$.

Ans. Laplace transforming the equations with

$$\mathscr{T}(x,s) = \int_0^\infty e^{-st} T(x,t)\, dt$$

$$F(s) = \int_0^\infty e^{-st} f(t)\, dt$$

one has [with $T(x,0) = 0$]

$$s\mathscr{T}(x,s) = d_x^2 \mathscr{T}(x,s)$$

$$\mathscr{T}(0,s) = F(s)$$

so that $\mathscr{T}(x,s) = A(s)e^{-x\sqrt{s}} + B(s)e^{x\sqrt{s}}$, and since $\mathscr{T}(x,\infty)$ should be zero, one takes $B(s) = 0$. Then $A(s) = F(s)$ and

$$\mathscr{T}(x,s) = F(s)e^{-x\sqrt{s}}$$

and, by the convolution theorem and Example 7, one has

$$T(x,t) = \frac{x}{2\sqrt{\pi}} \int_0^t \frac{f(t-\tau)e^{-x^2/4\tau}\,d\tau}{\tau\sqrt{\tau}}$$

11. Solve by Laplace transforms for the displacement $x(t)$ of a *forced damped classical harmonic oscillator* satisfying the equation of motion

$$m\ddot{x} + \gamma m\dot{x} + m\omega_0^2 x = ma(t) = f(t) \qquad \gamma > 0$$

where

γ = damping coefficient = damping frequency
ω_0 = natural frequency of undamped unforced oscillator
$f(t)$ = external force
$a(t)$ = external force/mass of oscillator.

Ans. Let

$$X(s) = \int_0^\infty e^{-st}x(t)\,dt \qquad x(t) = \frac{1}{2\pi i}\int e^{st}X(s)\,ds$$

$$A(s) = \int_0^\infty e^{-st}a(t)\,dt \qquad a(t) = \frac{1}{2\pi i}\int e^{st}A(s)\,ds$$

Then upon Laplace transforming each side of the equation, one has

$$s^2X(s) - \dot{x}(0) - sx(0) + \gamma sX(s) - \gamma x(0) + \omega_0^2X(s) = A(s)$$

or

$$X(s) = \frac{\dot{x}(0) + (\gamma + s)x(0) + A(s)}{s^2 + \gamma s + \omega_0^2}$$

The roots of $s^2 + \gamma s + \omega_0^2 = 0$ are

$$s = -\frac{\gamma}{2} \pm \sqrt{\left(\frac{\gamma}{2}\right)^2 - \omega_0^2}$$

or

$$s = s_+ \qquad \text{and} \qquad s = s_-$$

The inverse Laplace transform of $1/(s^2 + \gamma s + \omega_0^2)$ is given by

$$\frac{1}{2\pi i}\int \frac{e^{st}\,ds}{s^2 + \gamma s + \omega_0^2} = \frac{e^{s_+t}}{2s_+ + \gamma} + \frac{e^{s_-t}}{2s_- + \gamma} \qquad \text{for } t > 0$$

and since $2s_+ + \gamma = -(2s_- + \gamma) = \sqrt{\gamma^2 - 4\omega_0^2}$, the inverse transform of $A(s)/(s^2 + \gamma s + \omega_0^2)$ is (using the convolution theorem)

$$\frac{1}{\sqrt{\gamma^2 - 4\omega_0^2}} \int_0^t (e^{s_+\tau} - e^{s_-\tau})a(t-\tau)\, d\tau$$

so that the solution $x(t)$ for $t > 0$ is

$$x(t)\sqrt{\gamma^2 - 4\omega_0^2} = [x_0(\gamma + s_+) + \dot{x}_0]e^{s_+t} - [x_0(\gamma + s_-) + \dot{x}_0]e^{s_-t} + \int_0^t (e^{s_+\tau} - e^{s_-\tau})a(t-\tau)\, d\tau$$

For the case of no forcing $a \equiv 0$ and three cases are distinguished:

$\gamma^2 > 4\omega_0^2 > 0$	overdamping	$s_+ < 0 > s_-$
$\gamma^2 = 4\omega_0^2 > 0$	critical damping	$s_+ = s_- = -\gamma/2 < 0$
$0 < \gamma^2 < 4\omega_0^2$	underdamping	$s_+ = s_-^*$

For all three cases Re s_+ and Re s_- are negative. For the overdamping case, $x(t)$ decays exponentially with increasing time. For the underdamped case, $x(t)$ decays sinusoidally with an exponentially decreasing envelope. For the critical damping case

$$\frac{1}{2\pi i}\int \frac{e^{st}\, ds}{[s + (\gamma/2)]^2} = te^{-(\gamma/2)t} \qquad \text{for} \quad t > 0$$

so

$$x(t) = x_0 e^{-(\gamma/2)t} + \left(\dot{x}_0 + \frac{\gamma}{2}x_0\right)te^{-(\gamma/2)t}$$

12. For the case $x(0) = \dot{x}(0) = 0$ the transform of the equation of motion in Example 11 becomes $(\omega_0 > 0)$

$$m(s^2 + \gamma s + \omega_0^2)X(s) = mA(s) = F(s)$$

The ratio of the transforms of $f(t)$ and $\dot{x}(t)$ is called the (driving point) *impedance* $Z(s)$ of the system. Thus,

$$F(s) = Z(s)sX(s)$$

and

$$Z(s) = m\left(s + \gamma + \frac{\omega_0^2}{s}\right)$$

Explain why the zeros of $Z(s)$ are associated with *resonances* of the system (for the case of no damping).

Ans. For finite $F(s)$ at the zeros of $Z(s)$, $X(s)$ becomes infinite. If the system were sinusoidally forced, $f(t) = \varepsilon \sin \omega t$ at frequency ω

$$X(s) = \frac{\varepsilon\omega}{m(s^2 + \omega_0^2)(s^2 + \omega^2)}$$

$$x(t) = \frac{\varepsilon}{m\omega_0}\int_0^t \sin \omega_0\tau \sin \omega(t - \tau)\, d\tau$$

$$x(t) = \frac{\varepsilon}{2m\omega_0}\left[\frac{\sin \omega_0 t + \sin \omega t}{\omega_0 + \omega} - \frac{\sin \omega_0 t - \sin \omega t}{\omega_0 - \omega}\right] \quad \text{for} \quad \omega \neq \omega_0$$

$$x(t) = \frac{\varepsilon}{2m\omega_0}\left[\frac{\sin \omega_0 t}{\omega_0} - t \cos \omega_0 t\right] \quad \text{for} \quad \omega_0 = \omega$$

Thus the displacement would become infinitely oscillatory as $t \to \infty$ for $\omega^2 = \omega_0^2$ if the system did not rupture first.

13. For a passive network of electric circuits the ratio of the Laplace transforms of voltage $E(s)$ and current $I(s)$ with respect to time is called the (driving point) *impedance* $Z(s)$ of the network. Thus,

$$E(s) = I(s)Z(s)$$

analogously to Ohm's law. Generally, $Z(s)$ will turn out to be a rational function (ratio of polynomials) of s. Again the zeros of $Z(s)$ will correspond to frequencies which in the absence of damping would yield infinite currents for a finite input voltage (resonance), while the poles of $Z(s)$ would yield infinite voltages for a finite input current (antiresonances). Explain why the poles and zeros of $Z(s)$ cannot lie in the right half s plane.

Ans. Any zero of $Z(s)$ in the right half s plane would lead to a term with a positive exponential time-dependence for the current. Thus if $Z(a) = 0$ with $a = \alpha + i\beta$; $\alpha > 0$, there would be such a term in the expression for $i(t)$ from the inversion of

$$I(s) = \frac{E(s)}{Z(s)}$$

$$i(t) = \frac{1}{2\pi i}\int \frac{e^{st}E(s)}{Z(s)}\, ds$$

This term would be the residue at the pole $s = a$. Thus, for a pole of order m,

$$\lim_{s\to a} \frac{1}{(m-1)!}\, d_s^{m-1}\left[\frac{e^{st}E(s)(s-a)^m}{Z(s)}\right]$$

is the residue and would still contain a factor $e^{\alpha t}e^{i\beta t}$ which would become infinite as $t \to \infty$ for $\alpha > 0$. For a stable network without internal energy sources such behavior is impossible, so $\alpha > 0$ must be excluded.

The location of the poles of $Z(s)$ in the right half s plane would have a similar effect on

$$e(t) = \frac{1}{2\pi i} \int e^{st} I(s) Z(s)\, ds$$

14. Solve $\ddot{x} + \gamma\dot{x} + \omega_0^2 x = a(t)$ by Fourier transforms.

Ans. Let

$$X(\omega) = \int_{-\infty}^{\infty} e^{-i\omega t} x(t)\, dt$$

$$2\pi x(t) = \int_{-\infty}^{\infty} e^{i\omega t} X(\omega)\, d\omega$$

$$A(\omega) = \int_{-\infty}^{\infty} e^{-i\omega t} a(t)\, dt$$

$$2\pi a(t) = \int_{-\infty}^{\infty} e^{i\omega t} A(\omega)\, d\omega$$

Then Fourier transforming both sides of the equation, one has

$$-\omega^2 X(\omega) + i\gamma\omega X(\omega) + \omega_0^2 X(\omega) = A(\omega)$$

or

$$X(\omega) = \frac{A(\omega)}{(\omega_0^2 - \omega^2) + i\gamma\omega}$$

so that

$$2\pi x(t) = \int_{-\infty}^{\infty} \frac{e^{i\omega t} A(\omega)\, d\omega}{(\omega_0^2 - \omega^2) + i\gamma\omega}$$

There are two roots of the denominator of the integrand:

$$\omega_+ = i\frac{\gamma}{2} + \sqrt{\omega_0^2 - \left(\frac{\gamma}{2}\right)^2}$$

$$\omega_- = i\frac{\gamma}{2} - \sqrt{\omega_0^2 - \left(\frac{\gamma}{2}\right)^2}$$

For $\omega_0^2 > 0$, Im $\omega_+ > 0 <$ Im ω_-. There are three cases:

(I)	$(2\omega_0)^2 > \gamma^2$	underdamping	Im $\omega_+ =$ Im ω_-
(II)	$(2\omega_0)^2 = \gamma^2$	critical damping	$\omega_+ = \omega_-$
(III)	$(2\omega_0)^2 < \gamma^2$	overdamping	Re $\omega_+ =$ Re $\omega_- = 0$

For $t > 0$ the integral for $x(t)$ may be made part of a semicircular contour integral in the upper half (complex) ω plane. The limit of the integral along the circular arc will be zero, so that the integral along the real axis may be expressed in terms of the residues of the integrand. For cases I and III the residues are

$$\frac{e^{i\omega_+ t}A(\omega_+)}{i\gamma - 2\omega_+} \quad \text{and} \quad \frac{e^{i\omega_- t}A(\omega_-)}{i\gamma - 2\omega_-}$$

so that

$$x(t) = i\left\{\frac{e^{i\omega_+ t}A(\omega_+)}{i\gamma - 2\omega_+} + \frac{e^{i\omega_- t}A(\omega_-)}{i\gamma - 2\omega_-}\right\}$$

With $\mu = \sqrt{|\omega_0^2 - (\gamma^2/2)|}$ case I yields $\omega_+ = i(\gamma/2) + \mu$ and $\omega_- = i(\gamma/2) - \mu$, so that the underdamped case gives

$$x(t) = \frac{ie^{-(\gamma/2)t}}{2}\left[A(\omega_-)e^{-i\mu t} - A(\omega_+)e^{i\mu t}\right]$$

while case III yields

$$\omega_+ = i\left(\frac{\gamma}{2} + \mu\right) \quad \text{and} \quad \omega_- = i\left(\frac{\gamma}{2} - \mu\right)$$

so that the overdamped case gives

$$x(t) = \frac{e^{-(\gamma/2)t}}{2}\left[A(\omega_-)e^{\mu t} - A(\omega_+)e^{-\mu t}\right]$$

Case II corresponds to a double pole, so that

$$x(t) = i\left[A'\left(\frac{i\gamma}{2}\right) - \frac{\gamma}{2}A\left(\frac{i\gamma}{2}\right)\right]e^{-(\gamma/2)t}$$

For $t < 0$; $x(t) = 0$.

15. Suppose that in

$$\sum_{k=0}^{n} a_k d_t^k y = 0$$

$$\sum_{k=0}^{n-1} |d_t^k y| \neq 0 \qquad \text{for} \quad t = 0$$

and that there does *not* exist a λ for which $P(\lambda) = P'(\lambda) = 0$ with

$$P(\lambda) = \sum_{k=0}^{n} a_k \lambda^k$$

What is the difference between solving the above differential equation by exponential substitution and solving it by Laplace transform?

Ans. Exponential substitution of $y \sim e^{\lambda t}$ yields n distinct roots λ_k of $P(\lambda) = 0$, so that the general solution is

$$y = \sum_{k=1}^{n} c_k e^{\lambda_k t}$$

Laplace transform of $y(t)$ is

$$Y(s) = \frac{\sum_{k=0}^{n} a_k \sum_{m=0}^{k-1} s^m [d^{(k-1)-m} y]_{t=0}}{\sum_{k=0}^{n} a_k s^k}$$

which has n distinct poles at $s = \lambda_q$, so that by the Laplace inversion formula

$$y(t) = \sum_{q=1}^{n} c_q e^{\lambda_q t}$$

where

$$c_q = \frac{1}{P'(\lambda_q)} \sum_{k=0}^{n} a_k \lambda_q^m [d^{(k-1)-m} y]_{t=0}$$

16. In

$$\sum_{k=0}^{n} a_k d_t^k y = f(t)$$

with

$$\sum_{k=0}^{n-1} |d_t y| = 0 \qquad \text{for} \quad t = 0$$

how can the response $y(t)$ to oscillatory forcing $f(t) = f_0 \cos \omega t$ be found from the Laplace transform?

Ans. The Laplace transform of the equation with $f(t) = f_0 e^{i\omega t}$ is

$$\left[\sum_{k=0}^{n} a_k s^k\right] Y(s) = F(s)$$

$$Y(s) = \frac{F(s)}{\sum_{k=0}^{n} a_k s^k} = \frac{f_0}{(s - i\omega)(\sum_{k=0}^{n} a_k s^k)}$$

If the polynomial

$$P(s) = \sum_{k=0}^{n} a_k s^k = a_n \prod_{q=1}^{n} (s - s_q)$$

with no two roots equal,

$$Y(s) = f_0 \left\{ \frac{1}{P(i\omega)(s - i\omega)} + \sum_{k=1}^{n} \frac{1}{(s_q - i\omega) P'(s_q)(s - s_q)} \right\}$$

and by the Laplace inversion formula

$$y(t) = f_0\left\{\frac{e^{i\omega t}}{P(i\omega)} + \sum_{k=1}^{n} \frac{e^{s_q t}}{(s_q - i\omega)P'(s_q)}\right\}$$

Provided $\operatorname{Re} s_q < 0$ for all q (otherwise the system will be *unstable*, growing exponentially large with time), the summation will approach zero as $t \to \infty$, i.e., it represents a *transient* response while the first term represents the *steady-state* oscillatory response to the forcing. Actually, for $f = f_0 \cos \omega t$, one obtains

$$y(t) = f_0 \operatorname{Re}\left\{\frac{e^{i\omega t}}{P(i\omega)} + \sum_{k=1}^{n} \frac{e^{s_q t}}{(s_q - i\omega)P'(s_q)}\right\}$$

rather than the previous result. Hence the simple result that the steady-state response for this case is found by taking the real part of $f_0 e^{i\omega t}$ divided by $P(i\omega)$.

17. The Laplace transform of a function $y(t)$ is given by

$$Y(s) = 2\pi i \sum_{k=1}^{n} c_k \delta(s - \lambda_k)$$

Find the lowest-order differential equation satisfied by $y(t)$.

Ans. $y(t) = \dfrac{1}{2\pi i} \displaystyle\int Y(s) e^{st}\, ds$

$$y(t) = \sum_{k=1}^{n} c_k e^{\lambda_k t}$$

so the differential equation is

$$\sum_{q=0}^{n} a_q d_t^q y = 0$$

where λ_k are the roots of $\sum_{q=0}^{n} a_q \lambda^q = 0$. Naturally the $\delta(s - \lambda_k)$ have only a formal significance related to the Cauchy principal values associated with portions of the inversion integral (see Chapter 11).

18. Calculate about $2\,|z + 3| = 1$

$$i \oint e^{z^2} \csc\left(\frac{1}{\Gamma(z)}\right) dz$$

Ans. The function $\Gamma(z)$ has a simple pole of residue $-\frac{1}{3}!$ at $z = -3$, so its reciprocal has a simple zero there, and for z sufficiently close to -3 the integrand will be approximately

$$\frac{ie^{3^2}}{-3!\,(z+3)}$$

with an "error" which is analytic and free of singularities within $2\,|z+3| = 1$. Thus, by the theorem of residues, the result is

$$\frac{\pi e^9}{3}$$

19. Find the function $h(t)$ of which the Fourier transform $H(\omega)$ is the square modulus $|F(\omega)|^2$ of the Fourier transform $F(\omega)$ of a given function $f(t)$.

Ans. $H(\omega) = |F(\omega)|^2$ which can be regarded as the convolution theorem (9.116) by taking $G(\omega) = F^*(\omega)$, in which case

$$g(t) = \frac{1}{2\pi}\int_{-\infty}^{\infty} e^{i\omega t} G(\omega)\, d\omega = \frac{1}{2\pi}\int_{-\infty}^{\infty} e^{i\omega t} F^*(\omega)\, d\omega$$

so that

$$g(t) = f^*(-t)$$

and the corresponding convolution becomes

$$h(t) = \int_{-\infty}^{\infty} f(\tau) f^*(\tau - t)\, d\tau$$

With $u = \tau - t$ this becomes

$$h(t) = \int_{-\infty}^{\infty} f^*(u) f(u + t)\, du$$

This operation on the function $f(t)$ is called the *autocorrelation* of $f(t)$. We note that

$$h^*(t) = h(-t)$$

so that $|h(t)|^2 = h(t)h(-t)$ is an even function of t.

20. Evaluate $\int_0^\infty e^{i\omega t^2}\, dt$ by Laplace transform.

Ans. Let $i\omega = -s$ and $t^2 = \tau$ so that $dt = d\tau/2\sqrt{\tau}$. Then the integral becomes

$$\int_0^\infty e^{-s\tau} \frac{d\tau}{2\sqrt{\tau}} = \frac{1}{2}\sqrt{\frac{\pi}{s}} = \frac{(1+i)}{2\sqrt{2}}\sqrt{\frac{\pi}{\omega}}$$

21. Show that if

$$I_T(\omega) = \int_T^\infty e^{i\omega t^2}\, dt$$

then

$$2\omega T\, |I_T(\omega)| \leq 1$$

Ans. $I_T(\omega) = I_T^*(-\omega)$, so with $v = t - T$, one has

$$I_T(\omega) = I_T^*(-\omega) = \frac{ie^{i\omega T^2}}{\sqrt{\omega}} \int_0^\infty e^{-iv^2} e^{-2vT\sqrt{\omega}}\, dv$$

so that

$$|I_T(\omega)| \leq \frac{1}{\sqrt{\omega}} \int_0^\infty e^{-2vT\sqrt{\omega}}\, dv = \frac{1}{2T\omega}$$

and the error in approximating $\int_0^T e^{i\omega t^2}\, dt$ by $\int_0^\infty e^{i\omega t^2}\, dt$ for fixed T approaches zero as ω becomes large. Thus the approximation improves as $\omega \to \infty$.

22. Approximate

$$I(\omega) = \int_1^8 e^{i\omega(t^4+4t^3-18t^2-108t)} \cosh t\, dt$$

by the method of stationary phase.

Ans. $\phi(t) = t^4 + 4t^3 - 18t^2 - 108t$

$\phi'(t) = 4(t^3 + 3t^2 - 9t - 27) = 4(t - 3)(t + 3)^2$

$\phi''(t) = 12(t^2 + 2t - 3)$

Hence there are points of stationary phase at $t = \pm 3$, but only $t = 3$ is in the range of integration, so

$$\phi(2) = -297$$

$$\phi''(3) = 144$$

$$I(\omega) \approx \cosh 3 e^{-297i\omega} \frac{(1 + i)}{12} \sqrt{\frac{\pi}{\omega}}$$

23. The function

$$\phi(t) = \sum_{n=-\infty}^{\infty} f(t + nT)$$

constructed from an $f(t)$ for which the series is convergent, is clearly periodic: $\phi(t + T) = \phi(t)$. Find the Fourier coefficients of $\phi(t)$ in terms of the Fourier transform of $f(t)$. Supposing $\phi(t)$ to be of bounded variation (i.e., that the integral $\int_0^T |d\phi|$ exists), the Fourier series may be equated to $\phi(t)$. Write out this result, which is called the *Cauchy–Poisson sum formula*.

Ans. The Fourier series for $\phi(t)$ may be written

$$\phi(t) = \sum_{k=-\infty}^{\infty} \phi_k e^{ik\omega t}$$

where $\omega = 2\pi/T$, and by the Fourier coefficient formula

$$\phi_k = \frac{1}{T}\int_0^T \phi(t)e^{-ik\omega t}\,dt$$

Substituting the series for $\phi(t)$, in terms of $f(t + nT)$, one has

$$\phi_k = \frac{1}{T}\sum_{n=-\infty}^{\infty}\int_0^T f(t + nT)e^{-ik\omega t}\,dt$$

With $\tau = t + nT$, one has $\omega t = \omega\tau - 2\pi n$, so that

$$\phi_k = \frac{1}{T}\sum_{-\infty}^{\infty}\int_{nT}^{(n+1)T} f(\tau)e^{-ik\omega\tau}\,d\tau$$

or

$$\phi_k = \frac{1}{T}\int_{-\infty}^{\infty} f(\tau)e^{-ik\omega\tau}\,d\tau$$

Here it is slightly more convenient [unlike (9.97 and (9.96)] to define the Fourier transform of $f(t)$ by

$$F(\omega) = \frac{1}{2\pi}\int_{-\infty}^{\infty} f(t)e^{-i\omega t}\,dt$$

so that

$$f(t) = \int_{-\infty}^{\infty} F(\omega)e^{i\omega t}\,d\omega$$

Then

$$\phi_k = \frac{2\pi}{T}F(k\omega) = \omega F(k\omega)$$

and one has the Cauchy–Poisson formula

$$\sum_{-\infty}^{\infty} f(t + nT) = \omega\sum_{-\infty}^{\infty} F(k\omega)e^{ik\omega t}$$

with the corollary for $t = 0$

$$\sum_{-\infty}^{\infty} f(nT) = \omega\sum_{-\infty}^{\infty} F(k\omega)$$

24. Establish the *sampling theorem:* The values of a function $f(t)$ for all values of t can be expressed in terms of its values at the set $\{t = n\pi/\omega_c\}$ provided the Fourier transform $F(\omega)$ of $f(t)$ is zero for $|\omega| > \omega_c$.

Ans. $$2\pi f(t) = \int_{-\omega_c}^{\omega} e^{i\omega t}F(\omega)\,d\omega$$

Hence,

$$2\pi f\left(\frac{n\pi}{\omega_c}\right) = \int_{-\omega_c}^{\omega} e^{in\pi(\omega/\omega_c)}F(\omega)\,d\omega$$

A periodic function of period $2\omega_c$, taking the same values as $F(\omega)$ for $|\omega| \le \omega_c$, is given by the Fourier series

$$F(\omega) = \sum_{-\infty}^{\infty} F_n e^{-2\pi i(n\omega/2\omega_c)} \qquad |\omega| \le \omega_c$$

Then

$$2\omega_c F_n = \int_{-\omega_c}^{\omega_c} e^{in\pi(\omega/\omega_c)} F(\omega)\, d\omega = 2\pi f(n\pi/\omega_c)$$

If

$$\Psi(\omega) = \begin{cases} 1 & \text{for} \quad |\omega| \le \omega_c \\ 0 & \text{for} \quad |\omega| > \omega_c \end{cases}$$

$$F(\omega) = \sum_{-\infty}^{\infty} F_n \Psi(\omega) e^{-in\pi(\omega/\omega_c)}$$

Since the transform of $e^{-in\pi(\omega/\omega_c)}$ is $\delta[t - (n\pi/\omega_c)]$ and the transform of $\Psi(\omega)$ is

$$\frac{1}{2\pi}\int_{-\omega_c}^{\omega} e^{i\omega t}\, d\omega = \frac{\sin(\omega_c t)}{\pi}$$

inverting the transforms

$$f(t) = \sum_{-\infty}^{\infty} \frac{F_n \omega_c}{\pi} \int_{-\infty}^{\infty} \delta\left(\tau - \frac{n\pi}{\omega_c}\right) \frac{\sin[\omega_c(t-\tau)]}{\omega_c(t-\tau)}\, d\tau$$

one has

$$f(t) = \sum_{-\infty}^{\infty} f\left(\frac{n\pi}{\omega_c}\right) \frac{\sin(\omega_c t - n\pi)}{\omega_c t - n\pi}$$

the *sampling theorem* expressing $f(t)$ in terms of the "samples" $f(n\pi/\omega_c)$.

25. Establish the inversion formula for the bilateral Laplace transform

$$F(s) = \int_{-\infty}^{\infty} e^{-st} f(t)\, dt = F_+(s) + F_-(s)$$

if

$$F_+(s) = \int_0^{\infty} e^{-st} f(t)\, dt$$

$$F_-(s) = \int_0^{\infty} e^{st} f(-t)\, dt$$

where $F_+(s)$ converges for $\operatorname{Re} s > \sigma_1$, $F_-(s)$ converges for $\operatorname{Re} s < \sigma_2$ and there exist positive numbers ε_1 and ε_2 for which the limits $|s^{\varepsilon_1} F_+(s)|$ and $|s^{\varepsilon_2} F_-(s)|$ exist as $|s| \to \infty$ $(\sigma_2 > \sigma_1)$.

Ans. Integrating clockwise for $F_+(s)$ and counterclockwise for $F_-(s)$,

$$F(s) = F_+(s) + F_-(s) = \frac{1}{2\pi i}\left[\oint \frac{F_+(z)\,dz}{s - z} + \oint \frac{F_-(z)\,dz}{z - s}\right]$$

$$F(s) = \frac{1}{2\pi i}\left[\int_{\sigma_1 - i\infty}^{\sigma_1 + i\infty} \frac{F_+(z)\,dz}{s - z} + \int_{\sigma_2 - i\infty}^{\sigma_2 + i\infty} \frac{F_-(z)\,dz}{z - s}\right]$$

The integrals along the arcs vanish because of the limit conditions. In the first integral, $\operatorname{Re} s > \operatorname{Re} z$ inside the region of convergence, hence

$$\frac{1}{s - z} = \int_0^\infty e^{-(s-z)u}\,du$$

Also, in the second integral, $\operatorname{Re} s < \operatorname{Re} z$ inside the region of convergence, so

$$\frac{1}{z - s} = \int_{-\infty}^0 e^{-(s-z)u}\,du$$

Furthermore, since $F_+(s)$ has no singularities to the right of σ_1 and $F_-(s)$ has no singularities to the left of σ_2, the abscissae of the integration lines may *both* be replaced by *any* σ for which $\sigma_1 < \sigma < \sigma_2$. Thus, interchanging the order of integration, one has

$$2\pi i F(s) = \int_0^\infty e^{-su} \int_{\sigma - i\infty}^{\sigma + i\infty} F_+(z) e^{zu}\,dz\,du + \int_{-\infty}^0 e^{-su} \int_{\sigma - i\infty}^{\sigma + i\infty} F_-(z) e^{zu}\,dz\,du$$

or

$$2\pi i F(s) = \int_{-\infty}^\infty e^{-su}\left[\int_{\sigma - i\infty}^{\sigma + i\infty} F_+(z) e^{zu}\,dz\right] du + \int_{-\infty}^\infty e^{-su}\left[\int_{\sigma - i\infty}^{\sigma + i\infty} F_-(z) e^{zu}\,dz\right] du$$

since the result of the inversion integral for $F_+(z)$ is zero for $u < 0$, while the result of the inversion integral for $F_-(z)$ is zero for $u > 0$. Finally, since $F(z) = F_+(z) + F_-(z)$,

$$F(s) = \int_{-\infty}^\infty e^{-su} \frac{1}{2\pi i} \int_{\sigma - i\infty}^{\sigma + i\infty} F(z) e^{zu}\,dz\,du$$

so that replacing z by s and u by t,

$$f(t) = \frac{1}{2\pi i} \int_{\sigma - i\infty}^{\sigma + i\infty} F(s) e^{st}\,ds$$

Thus, one has an inversion formula formally identical with that for the unilateral Laplace transform.

26. Use Laplace transforms to find the current $i(t)$ in a series RLC circuit subjected to a step voltage $Vu(t)$ commencing at $t = 0$ ($i \neq \sqrt{-1}$ here).

Ans. The voltage $e(t) = Vu(t)$ is the sum of inductive, resistive, and capacitative voltage drops across the three elements;

$$Ld_t i + Ri + \frac{1}{C}\int_0^t i\,dt = Vu(t)$$

The Laplace transform of this equation is

$$LsI(s) + RI(s) + \frac{I(s)}{Cs} = \frac{V}{s}$$

where $I(s)$ is the Laplace transform of $i(t)$. Thus,

$$I(s) = \frac{V}{Ls^2 + Rs + (1/C)} = \frac{V}{LCs^2 + RCs + 1}$$

corresponding to an impedance (for $s = i\omega$)

$$Z(s) = Ls + R + \frac{1}{Cs} = i\omega L + R + \frac{1}{i\omega C}$$

For $R^2C^2 \neq 4LC$, $I(s)$ has simple poles at

$$s_+ = -\lambda + \sqrt{\lambda^2 - \mu^2} \qquad \text{and} \qquad s_- = -\lambda - \sqrt{\lambda^2 - \mu^2}$$

with $2L\lambda = R$ and $LC\mu^2 = 1$. Thus,

$$I(s) = \frac{V\mu^2}{(s + \lambda)^2 + (\mu^2 - \lambda^2)}$$

and for $\mu > \lambda$ (underdamped)

$$i(t) = \frac{1}{2\pi i}\int e^{st} I(s)\,ds = \frac{V\mu^2 e^{-\lambda t}}{\sqrt{\mu^2 - \lambda^2}} \sin(t\sqrt{\mu^2 - \lambda^2})u(t)$$

while for $\mu < \lambda$ (overdamped)

$$i(t) = \frac{V\mu^2 e^{-\lambda t}}{\sqrt{\lambda^2 - \mu^2}} \sinh(t\sqrt{\lambda^2 - \mu^2})u(t)$$

and for $\mu = \lambda$ (critically damped) $I(s)$ has a double pole at $s = -\lambda$ so that

$$I(s) = \frac{V\lambda^2}{(s + \lambda)^2} \qquad \text{and} \qquad i(t) = V\lambda^2 t e^{-\lambda t} u(t)$$

27. By using a Fourier method, find the values of a harmonic function $u(x,y)$ for $y > 0$ if $u(x, 0)$ is known.

Ans. $\Delta u = (\partial_x^2 + \partial_y^2)u = 0$

If u is assumed to be the product of a function of x alone and a function of y alone (Bernoulli separation argument), $u(x,y) = X(x)\,Y(y)$, then $X''Y + XY'' = 0$ or $X''/X = -Y''/Y$, and the only way this can be true is if both these functions are the same constant. (Otherwise a function of x alone would have to be equal to a function of y alone, which is absurd.) Thus if the constant is $-\omega^2$,

$$X'' + \omega^2 X = 0 = Y'' - \omega^2 Y$$

and

$$X = e^{\pm i\omega x} \qquad Y = e^{\pm \omega y}$$

are solutions. Hence, $e^{i\omega x}e^{\pm\omega y}$ is a solution of $\Delta u = 0$ and, by linear superposition,

$$u(x,y) = \int_{-\infty}^{\infty} e^{i\omega x}e^{-|\omega|y}U(\omega)\,d\omega$$

is also a solution of $\Delta u = 0$. For $y = 0$

$$u(x,0) = \int_{-\infty}^{\infty} e^{i\omega x}U(\omega)\,d\omega$$

so

$$U(\omega) = \frac{1}{2\pi}\int_{-\infty}^{\infty} e^{-i\omega x}u(x,0)\,dx$$

Hence, for $y > 0$,

$$u(x,y) = \frac{1}{2\pi}\int_{-\infty}^{\infty} e^{i\omega x}e^{-|\omega|y}\left[\int_{-\infty}^{\infty} e^{-i\omega s}u(s,0)\,ds\right]d\omega$$

and reversing the order of integration, one has

$$2\pi u(x,y) = \int_{-\infty}^{\infty} u(s,0)\left[\int_{-\infty}^{\infty} e^{i\omega x}e^{-|\omega|y}e^{-i\omega s}\,d\omega\right]ds$$

or

$$u(x,y) = \frac{y}{\pi}\int_{-\infty}^{\infty}\frac{u(s,0)\,ds}{y^2 + (x-s)^2}$$

This is called the *Poisson integral for the half-plane*, and it is readily recognized as being harmonic for $y > 0$, since (s = real)

$$\frac{y}{y^2 + (x-s)^2} = -\tfrac{1}{2}\,\mathrm{Im}\left(\frac{1}{z-s}\right)$$

is the imaginary part of an analytic function. Also, for $u(s,0) \equiv 1$,

$$u(x,y) = \frac{y}{\pi} \int_{-\infty}^{\infty} \frac{ds}{y^2 + (x - s)^2} = 1$$

and it remains so as $y \to 0$. Thus the curve of $y/\pi[y^2 + (x - s)^2]$ has unit area below it, and since it becomes arbitrarily narrowly concentrated about $x = s$ as $y \to 0$, the limit of the Poisson integral agrees with $u(x,0)$, as it should.

28. A plane metal plate consists of two half-planes insulated from each other with one grounded and the other at voltage V. Find the potential of the configuration in the (charge-free) half-space bounded by the plate.

Ans. $u(s,0) = Vu(s)$

$$u(x,y) = \frac{y}{\pi} \int_{-\infty}^{\infty} \frac{u(s,0)\, ds}{y^2 + (x - s)^2} = \frac{Vy}{\pi} \int_{0}^{\infty} \frac{ds}{y^2 + (x - s)^2}$$

$$u(x,y) = \frac{V}{\pi}\left[\frac{\pi}{2} + \arctan\left(\frac{x}{y}\right)\right]$$

(Compare Example 16, Chapter 8.)

29. Find the Laplace transform of the Bessel function of zero order, $J_0(kr)$, by a physical argument.

Ans. The potential of a unit charge at the origin is $1/\sqrt{r^2 + z^2}$ in terms of cylindrical coordinates. By a separation argument it can be verified that $e^{-kz}J_0(kr)$ is a solution of the Laplace equation in these coordinates (for $\partial_\theta \phi = 0$)

$$\frac{1}{r}\, \partial_r(r\partial_r\phi) + \partial_z^2\phi = 0$$

Since the potential of the unit charge vanishes at infinity, it must be expressible as the linear superposition

$$\frac{1}{\sqrt{r^2 + z^2}} = \int_0^{\infty} \psi(k) e^{-kz} J_0(kr)\, dk$$

where $\psi(k)$ is yet to be determined. For $z > 0 = r$, this becomes

$$\frac{1}{z} = \int_0^{\infty} \psi(k) e^{-kz}\, dk$$

because $J_0(0) = 1$. Hence, $\psi(k) = 1$. Thus,

$$\frac{1}{\sqrt{r^2 + z^2}} = \int_0^\infty e^{-kz} J_0(kr)\, dk \qquad z \geq 0$$

or

$$\frac{1}{\sqrt{s^2 + \sigma^2}} = \int_0^\infty e^{-st} J_0(\sigma t)\, dt \qquad s \geq 0$$

30. Prove that the perimeter L and area S of the closed curve

$$z(s) = \sum_{-\infty}^{\infty} z_n e^{in\omega s}$$

with $\omega L = 2\pi$, are related by

$$4\pi S \leq L^2$$

the *isoperimetric inequality*.

Ans. $$2S = \mathrm{Im} \int_0^L z^*\, dz$$

$$L = \int_0^L |d_s z|\, ds = \int_0^L |d_s z|^2\, ds$$

since $|d_s z| = 1$. Also,

$$d_s z = \sum_{-\infty}^{\infty} in\omega z_n e^{in\omega s}$$

$$d_s z^* = -\sum_{-\infty}^{\infty} im\omega z_m^* e^{-im\omega s}$$

so that

$$L = \int_0^L |d_s z|^2\, ds = \sum_{m,n} mn\omega^2 z_m^* z_n \int_0^L e^{i(n-m)\omega s}\, ds$$

Hence,

$$L = \sum_{m,n} mn\omega^2 z_m^* z_n L \delta_{mn} = L \sum_{-\infty}^{\infty} n^2 |z_n|^2 \omega^2$$

Consequently,

$$L^2 = 4\pi^2 \sum_{-\infty}^{\infty} n^2 |z_n|^2$$

Also, by a similar argument,

$$2S = \sum_{-\infty}^{\infty} n\omega |z_n|^2 L$$

so

$$S = \pi \sum_{-\infty}^{\infty} n |z_n|^2$$

and

$$L^2 - 4\pi S = 4\pi^2 \sum_{-\infty}^{\infty} (n^2 - n)\, |z_n|^2 \geq 0$$

This will actually be equal to zero for $|z_n| = 0$ for $n(n-1) \neq 0$. For this case

$$z(s) = z_0 + z_1 e^{i\omega s}$$

so the curve for which $L^2 = 4\pi S$,

$$|z - z_0| = |z_1|$$

is a circle. The method is due to Hurwitz.

31. The *transmission line equations*

$$\partial_x e + Rj + L\partial_t j = 0$$
$$\partial_x j + Ge + C\partial_t e = 0$$

govern the transmission of signals in a long parallel pair of wires (Fig. 9.2) connecting a source (of voltage or current) with a load impedance Z_L. The voltage between corresponding points (same x) on the two wires is called $e(x,t)$, while the current in one wire is $j(x,t)$ The equations simply state that there is a loss in voltage in traveling a distance dx which is due to the resistance $R\,dx$ and the inductance $L\,dx$ of this infinitesimal portion dx of the line, and that there is a loss of current for distance dx due to leakage (imperfect insulation) between the two lines of conductance $G\,dx$ and due to displacement current passing between the lines of capacitance $C\,dx$. Find $e(x,t)$ and $j(x,t)$ given $e(0,t)$ and $j(0,t)$. If $e(0,t) = e_0 e^{i\omega_0 t}$ and $j(0,t) = j_0 e^{i(\omega_0 t + \phi)}$, what does the solution become? Assume Z_L is infinitely remote from the source.

Ans. The transmission line equations are clearly satisfied by the *Ansatz*

$$e(x,t) = Ae^{i(\omega t - kx)}$$
$$j(x,t) = Be^{i(\omega t - kx)}$$

provided (by substitution)

$$-ikA + (R + i\omega L)B = 0$$
$$(G + i\omega C)A - ikB = 0$$

which possess nontrivial (i.e., nonvanishing) solutions for A and B only if

$$\begin{vmatrix} -ik & (R + i\omega L) \\ (G + i\omega C) & -ik \end{vmatrix} = 0$$

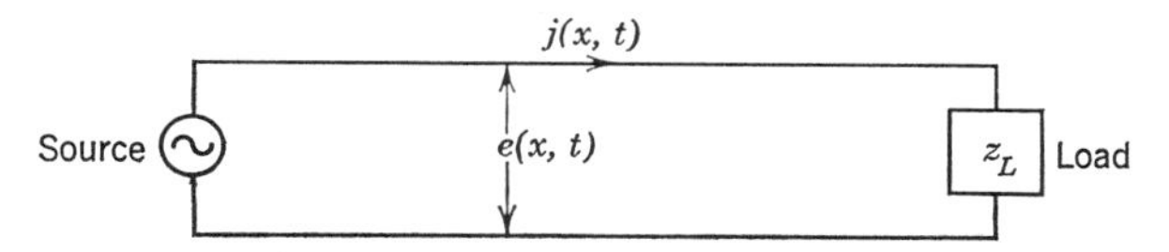

Figure 9.2. Current $j(x,t)$, voltage $e(x,t)$ and load impedance z_L in transmission line.

or

$$k^2 + (R + i\omega L)(G + i\omega C) = 0$$

so that the propagation "constant" becomes

$$k = -i\sqrt{ZY} = k(\omega)$$

where $Z = R + i\omega L =$ series impedance and $Y = G + i\omega C =$ shunt admittance. Since the transmission line equations are linear, not only does the *Ansatz* above yield a solution for this choice of $k(\omega)$ but also the linear superpositions

$$e(x,t) = \int_{-\infty}^{\infty} e^{i[\omega t - k(\omega)x]} A(\omega)\, d\omega$$

$$j(x,t) = \int_{-\infty}^{\infty} e^{i[\omega t - k(\omega)x]} B(\omega)\, d\omega$$

are solutions. At the source end of the line

$$e(0,t) = \int_{-\infty}^{\infty} e^{i\omega t} A(\omega)\, d\omega$$

$$j(0,t) = \int_{-\infty}^{\infty} e^{i\omega t} B(\omega)\, d\omega$$

so that, by the Fourier inversion formula,

$$A(\omega) = \frac{1}{2\pi} \int_{-\infty}^{\infty} e^{-i\omega t} e(0,t)\, dt$$

$$B(\omega) = \frac{1}{2\pi} \int_{-\infty}^{\infty} e^{-i\omega t} j(0,t)\, dt$$

which may then be substituted into the integrals for $e(x,t)$ and $j(x,t)$ to yield voltage and current for any x and t values. Clearly, for real $e(0,t)$ and $j(0,t)$ one has $A^*(-\omega) = A(\omega)$ and $B^*(-\omega) = B(\omega)$, while $k^*(-\omega) = -k(\omega)$ in any case.

For $e(0,t) = e_0 e^{i\omega_0 t}$ and $j(0,t) = j_0 e^{i(\omega_0 t+\phi)}$,

$$A(\omega) = e_0 \delta(\omega - \omega_0)$$
$$B(\omega) = j_0 e^{i\phi}\delta(\omega - \omega_0)$$

so that

$$e(x,t) = e_0 e^{i[\omega_0 t - k(\omega_0)x]}$$
$$j(x,t) = j_0 e^{i[\omega_0 t - k(\omega_0)x+\phi]}$$

and from the first transmission line equation one has

$$ik(\omega_0)e(x,t) = (R + i\omega_0 L)j(x,t)$$

and from the second one

$$ik(\omega_0)j(x,t) = (G + i\omega_0 C)e(x,t)$$

or equivalently

$$ike(x,t) = Zj(x,t)$$
$$ikj(x,t) = Ye(x,t)$$

so that the *characteristic impedance* of the line is given by

$$\frac{e(x,t)}{j(x,t)} = \frac{Z}{ik} = \frac{ik}{Y} = \sqrt{\frac{Z}{Y}} = Z_K$$

32. Determine the form of the transmission line equations at high frequency (radio equations) and the corresponding general solution.

Ans. For large ω, $Z \approx i\omega L$, $Y \approx i\omega C$, $k \approx \omega\sqrt{LC}$ so that the equations become

$$\partial_x e + L\partial_t j = 0$$
$$\partial_x j + C\partial_t e = 0$$

and thus e and j satisfy wave equations

$$\partial_x^2 e = LC\partial_t^2 e$$
$$\partial_x^2 j = LC\partial_t^2 j$$

so that the general solutions are

$$e(x,t) = f_1(t - x\sqrt{LC}) + g_1(t + x\sqrt{LC})$$
$$j(x,t) = f_2(t - x\sqrt{LC}) + g_2(t + x\sqrt{LC})$$

where f_1, g_1, f_2, g_2 are arbitrary functions.

33. Find the condition under which a voltage signal will be transmitted along a transmission line with exponential decay with time but no change in "shape," i.e., distortionlessly.

Ans. Elimination of $j(x,t)$ between the two equations yields

$$-\partial_x^2 e + LC\partial_t^2 e + (RC + LG)\partial_t e + RGe = 0$$

For a distortionless transmission in the above sense

$$e(x,t) = e^{-\lambda t}[f(t - \sqrt{LC}\,x) + g(t + \sqrt{LC}\,x)]$$

Hence, $e^{\lambda t}e(x,t)$ is a wave (solution of wave equation)

$$\partial_x^2[e^{\lambda t}e(x,t)] = \partial_{ct}^2[e^{\lambda t}e(x,t)]$$

$$c^2 e^{\lambda t}\partial_x^2 e = \lambda^2 e^{\lambda t}e + e^{\lambda t}\partial_t^2 e + 2e^{\lambda t}\lambda\partial_t e$$

or

$$c^2\partial_x^2 e - \partial_t^2 e - 2\lambda\partial_t e - \lambda^2 e = 0$$

whence comparison with the first equation above yields

$$LCc^2 = 1 \qquad 2\lambda = (RC + LG)c^2 \qquad \lambda^2 = RGc^2$$

or

$$2\lambda = \frac{R}{L} + \frac{G}{C} \qquad \lambda^2 = \frac{R}{L}\frac{G}{C} \qquad \text{whence } \frac{R}{L} = \frac{G}{C} = \lambda$$

34. Separable particular complex solutions of the transmission line equations are given by

$$e(x,t) = M(x)e^{i\omega t}$$
$$j(x,t) = N(x)e^{i\omega t}$$

Find the differential equations connecting $M(x)$ and $N(x)$ and solve these to find $e(x,t)$ and $j(x,t)$.

Ans. $M' + RN + i\omega LN = 0$

$N' + GM + i\omega CM = 0$

or

$$M' = -ZN$$
$$N' = -YM$$

so that

$$M'' - ZYM = 0 = M'' + k^2 M$$
$$N'' - ZYN = 0 = N'' + k^2 N$$

and

$$M(x) = M_+ e^{-ikx} + M_- e^{ikx}$$
$$N(x) = N_+ e^{-ikx} + N_- e^{ikx}$$
$$M(0) = M_+ + M_- = e_s = \text{source voltage amplitude}$$
$$N(0) = N_+ + N_- = j_s = \text{source current amplitude}$$

Also, by the first-order equations,

$$-ikM_+ + ikM_- = -Z(N_+ + N_-) = -Zj_s$$
$$-ikN_+ + ikN_- = -Y(M_+ + M_-) = -Ye_s$$

so that

$$2M_+ = e_s + Z_K j_s$$
$$2M_- = e_s - Z_K j_s$$
$$2N_+ = j_s + \frac{e_s}{Z_K}$$
$$2N_- = j_s - \frac{e_s}{Z_K}$$

and therefore

$$M(x) = e_s \cos kx - iZ_K j_s \sin kx$$
$$N(x) = j_s \cos kx - i\frac{e_s}{Z_K} \sin kx$$

35. In the preceding example, how do the voltage and current depend upon terminal impedance Z_L and length l?

Ans. $M(l) = Z_L N(l)$
Hence,

$$Z_K j_s = \left(\frac{Z_K \cos kl + iZ_L \sin kl}{Z_L \cos kl + iZ_K \sin kl}\right) e_s = \Lambda e_s$$

so, with $Z_s = Z_K/\Lambda$ = source impedance,

$$M(x) = e_s(\cos kx - i\Lambda \sin kx)$$
$$N(x) = \frac{e_s}{Z_K}(\Lambda \cos kx - i \sin kx) = \frac{e_s}{Z_s}\left(\cos kx - \frac{i}{\Lambda} \sin kx\right)$$

For $Z_K = Z_L$ one has $\Lambda = 1$ so that

$$M(x) = e_s e^{-ikx}$$
$$N(x) = \frac{e_s}{Z_K} e^{-ikx}$$

corresponding to

$$e(x,t) = e_s e^{i(\omega t - kx)}$$
$$j(x,t) = \frac{e_s}{Z_K} e^{i(\omega t - kx)}$$

representing a wave traveling to the right *without any reflections.*

Thus one concludes that if the line is terminated by a load equal to its characteristic impedance, it behaves like an infinite line in having *no* reflections. If the line is not terminated by Z_K but by $Z_L \neq Z_K$, another impedance Z can be inserted (shunting Z_L) so that the impedance of Z and Z_L in parallel is Z_K. Thus,

$$\frac{1}{Z} + \frac{1}{Z_L} = \frac{1}{Z_K}$$

and

$$Z = \frac{Z_K Z_L}{Z_L - Z_K}$$

is the correct *matching impedance* to insert so that the line is "matched" to Z_L.

36. Show that if the line in Example 35 is half a wavelength, the source impedance is equal to the load impedance. Show that if the line is a quarter wavelength, the source impedance is Z_K^2/Z_L.

Ans. For $\lambda = 2l$, $k = 2\pi/\lambda = \pi/l$ so $kl = \pi$ and $\Lambda Z_L = Z_K$. Hence,

$$N(0) = \frac{e_s}{Z_K}\Lambda = \frac{e_s}{Z_L} \qquad \text{and} \qquad Z_s = Z_L$$

For $\lambda = 4l$, $kl = \pi/2$ so

$$\Lambda Z_K = Z_L$$

and

$$N(0) = \frac{e_s Z_L}{Z_K^2} \qquad \text{so} \qquad Z_s = \frac{Z_K^2}{Z_L}$$

Thus $Z_L = 0$ implies $Z_s = \infty$ (opened source) and $Z_L = \infty$ implies $Z_s = 0$ (shorted source).

37. Determine the reflection coefficient (ratio of power reflected to power incident) at a location on a transmission line where the line impedance $Z = R + i\omega L$ and admittance $Y = G + i\omega C$ change discontinuously.

Ans. Take the point of discontinuity to be the origin $x = 0$ supposing the source and terminal load to be at $x = -\infty$ and $x = +\infty$. Then supposing a transmitted wave e_T traveling to the right for $x > 0$, an incident wave e_i traveling right, and a reflected wave e_R traveling left for $x < 0$, one has

$$\begin{aligned} e_T(x,t) &= A_T e^{i(\omega t - k_T x)} = Z_K j_T(x,t) \\ e_R(x,t) &= A_R e^{i(\omega t + k_R x)} = Z_K j_R(x,t) \\ e_i(x,t) &= e^{i(\omega t - k_R x)} = Z_K j_i(x,t) \end{aligned}$$

For $x = 0$,

$$e_T(0,t) = e_i(0,t) + e_R(0,t)$$

$$\partial_x e_T(0,t) = \partial_x e_i(0,t) + \partial_x e_R(0,t)$$

Hence,

$$A_T = 1 + A_R$$

$$-ik_T A_T = -ik_R + ik_R A_R$$

so that

$$A_R = \frac{k_R - k_T}{k_R + k_T}$$

$$A_T = \frac{2k_R}{k_R + k_T}$$

and the reflection coefficient is

$$|A_R|^2 = \left| \frac{k_R - k_T}{k_R + k_T} \right|^2 = \left| \frac{\sqrt{Z_R Y_R} - \sqrt{Z_T Y_T}}{\sqrt{Z_R Y_R} - \sqrt{Z_T Y_T}} \right|^2$$

38. Find an asymptotic expansion for (valid for large z) $I(z) = \int_z^\infty e^{-t}\, dt/t$.

Ans. $I(z) = e^{-z} \displaystyle\int_0^\infty \frac{e^{-u}}{u + z}\, du \sim \frac{e^{-z}}{z} \int_0^\infty e^{-u} \sum_{k=0}^{\infty} \frac{(-1)^k u^k}{z^k}\, du$

$$I(z) \sim \frac{e^{-z}}{z} \sum_{k=0}^{\infty} \frac{(-1)^k k!}{z^k}$$

To show it to be asymptotic it is better to integrate by parts and apply (9.82) to the remainder.

39. Find the currents $j_1(t)$ and $j_2(t)$ in the *capacitatively coupled* circuits shown in Figure 9.3 given the applied voltage $e(t)$, the resistances R_1, R_2, the inductances L_1, L_2, the capacitances C_1, C_2, C_{12} and assuming $e(0) = 0 = j_1(0) = j_2(0)$ and the derivatives $d_t j_1(t)$ and $d_t j_2(t)$ are zero at $t = 0$. Assume nonresonant circuits.

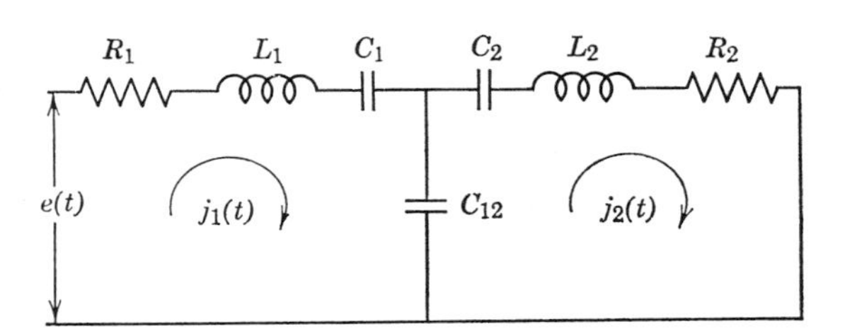

Figure 9.3. Capacitatively coupled circuits.

Ans. The (differentiated) sum of voltage changes around each mesh yields an equation:

$$\left[L_1 d_t + R_1 d_t + \frac{1}{C_1}\right] j_1(t) + \frac{1}{C_{12}} [j_1(t) - j_2(t)] = d_t e(t)$$

$$\left[L_2 d_t^2 + R_2 d_t + \frac{1}{C_2}\right] j_2(t) + \frac{1}{C_{12}} [j_2(t) - j_1(t)] = 0$$

Laplace transforming, with

$$J_1(s) = \int_0^\infty e^{-st} j_1(t)\, dt$$

$$J_2(s) = \int_0^\infty e^{-st} j_2(t)\, dt$$

$$E(s) = \int_0^\infty e^{-st} e(t)\, dt$$

one has

$$\left[L_1 s^2 + R_1 s + \frac{1}{C_1}\right] J_1(s) + \frac{1}{C_{12}} [J_1(s) - J_2(s)] = sE(s)$$

$$\left[L_2 s^2 + R_2 s + \frac{1}{C_2}\right] J_2(s) + \frac{1}{C_{12}} [J_2(s) - J_1(s)] = 0$$

or with

$$L_1 s^2 + R_1 s + \frac{1}{C_1} + \frac{1}{C_{12}} = sZ_1(s)$$

$$L_2 s^2 + R_2 s + \frac{1}{C_2} + \frac{1}{C_{12}} = sZ_2(s)$$

one has

$$sZ_1(s) J_1(s) - \frac{J_2(s)}{C_{12}} = sE(s)$$

$$-\frac{J_1(s)}{C_{12}} + sZ_2(s) J_2(s) = 0$$

which may be solved for the transforms of the currents provided

$$\begin{vmatrix} sZ_1(s) & -\dfrac{1}{C_{12}} \\ -\dfrac{1}{C_{12}} & sZ_i(s) \end{vmatrix} \neq 0$$

If this is fulfilled (nonresonance),

$$J_1(s) = \frac{s^2E(s)Z_2(s)}{s^2Z_1(s)Z_2(s) - (1/C_{12}^2)} = \frac{s^2E(s)Z_2(s)C_{12}^2}{C_{12}^2s^2Z_1(s)Z_2(s) - 1}$$

$$J_2(s) = \frac{sE(s)C_{12}}{C_{12}^2s^2Z_1(s)Z_2(s) - 1}$$

and the currents are then found by the Laplace inversion formula (9.11)

40. Analyze the coupled circuit of the preceding example by a Fourier transform technique.

Ans. Take

$$J_1(\omega) = \int_{-\infty}^{\infty} e^{-i\omega t} j_1(t)\, dt$$

$$J_2(\omega) = \int_{-\infty}^{\infty} e^{-i\omega t} j_2(t)\, dt$$

$$E(\omega) = \int_{-\infty}^{\infty} e^{-i\omega t} e(t)\, dt$$

$$\left[-\omega^2L_1 + i\omega R_1 + \frac{1}{C_1} + \frac{1}{C_{12}}\right]J_1(\omega) - \frac{J_2(\omega)}{C_{12}} = +i\omega E(\omega)$$

$$-\frac{J_1(\omega)}{C_{12}} + \left[-\omega^2L_2 + i\omega R_2 + \frac{1}{C_2} + \frac{1}{C_{12}}\right]J_2(\omega) = 0$$

Substitution of $J_2(\omega)$ from the second equation yields

$$i\omega\frac{E(\omega)}{J_1(\omega)} = \left[-\omega^2L_1 + i\omega R_1 + \frac{1}{C_1} + \frac{1}{C_{12}}\right] - \frac{1}{C_{12}^2}\left[\frac{1}{-\omega^2L_2 + i\omega R_2 + (1/C_2) + (1/C_{12})}\right]$$

It is convenient here to introduce certain abbreviations ($k = 1,2$)

$$Z_k = \sqrt{\frac{L_k(C_k + C_{12})}{C_kC_{12}}} = \text{characteristic impedance of } k\text{th mesh}$$

$$\alpha_k = \frac{R_k}{Z_k} = \text{normalized resistance of } k\text{th mesh}$$

$$\omega_k = \frac{Z_k}{L_k} = \sqrt{\frac{C_k + C_{12}}{L_kC_kC_{12}}}$$

$$= \text{natural (resonant) frequency of } k\text{th mesh}$$

$$\delta_k = \frac{\omega}{\omega_k} - \frac{\omega_k}{\omega} = \text{normalized reactance for } k\text{th mesh}$$

$$Z_{12} = \frac{1}{i\omega C_{12}} = \text{coupling impedance}$$

In terms of these, the (input) impedance is

$$\frac{E(\omega)}{J_1(\omega)} = Z_1(\alpha_1 + i\delta_1) - \frac{Z_{12}^2}{Z_2(\alpha_2 + i\delta_2)}$$

$$= Z_1\left[(\alpha_1 + i\delta_1) + \frac{\omega_1\omega_2 k^2}{\omega^2(\alpha_2 + i\delta_2)}\right]$$

where k defined by

$$k^2 = -\frac{Z_{12}^2\omega^2}{\omega_1 Z_1 \omega_2 Z_2} = \frac{C_1 C_2}{(C_1 + C_{12})(C_2 + C_{12})}$$

is called the *coupling factor*. Thus,

$$J_1(\omega) = \frac{\omega^2 E(\omega)[\alpha_2 + i\delta_2(\omega)]}{Z_1[(\alpha_1 + i\delta_1(\omega))(\alpha_2 + i\delta_2(\omega))\omega^2 + k^2\omega_1\omega_2]}$$

$$J_2(\omega) = \frac{-i\omega C_{12} E(\omega)}{Z_1 Z_2[(\alpha_1 + i\delta_1(\omega))(\alpha_2 + i\delta_2(\omega))\omega^2 + k^2\omega_1\omega_2]}$$

and $j_1(t)$ and $j_2(t)$ are then found by the Fourier inversion theorem. The "monochromatic case" where $E(\omega)$, $J_1(\omega)$, $J_2(\omega)$ are delta functions then corresponds to solution of the differential equations by exponential substitution.

41. Find the natural (resonant) frequencies of the coupled circuit of the preceding example for the case where both meshes are identical (except for the applied voltage).

Ans. These correspond to excitation of oscillations under vanishingly small voltages. For the case of identical meshes the input impedance found in the preceding example becomes

$$Z_1\left[(\alpha_1 + i\delta_1) + \frac{\omega_1^2 k^2}{\omega^2(\alpha_1 + i\delta_1)}\right]$$

and setting this equal to zero, one has

$$\omega^2(\alpha_1 + i\delta_1)^2 = -\omega_1^2 k^2$$

or

$$\omega(\alpha_1 + i\delta_1) = \pm i\omega_1 k$$

Substituting for δ_1, one then has, for the (positive) resonant frequency,

$$\omega = \omega_1 \left[\frac{\sqrt{\alpha_1^2 + 4(1 \pm k)} - \alpha_1}{2} \right]$$

For the undamped case ($R_1 = R_2 = 0 = \alpha_1 = \alpha_2$)

$$\omega = \omega_1 \sqrt{1 \pm k}$$

which exhibits explicitly the effect of the coupling which *splits* the original resonant frequency ω_1 of either mesh into a pair (doublet) of frequencies *above* and *below* the original ω_1. This is sometimes expressed by saying that the two identical meshes possess a single *degenerate* resonant frequency. The coupling then removes the degeneracy and leads to the new pair of frequencies for the coupled circuit. The circuit then has a *double resonance*. This extremely important argument is the classical analog of the *covalent bond* in the quantum theory of solids and of the *nuclear bond*. In the quantum version, frequencies are replaced by energies via the Planck hypothesis $E = h\nu = \hbar\omega$ (h = Planck's constant) and the two oscillators (meshes) become two *identical* particles, each with a characteristic energy (eigenvalue) $E_1 = \hbar\omega_1$ considered as separate noninteracting (uncoupled) systems. When the two particles interact (are coupled) the original energy (level) E_1 can then split into *two* new energies (levels) $E = E_1\sqrt{1 \pm k}$ above and *below* the original level. Since there now exists a *lower* energy (level) available for the coupled particles than for the separate particles, it becomes energetically favorable for the two identical particles to come together and form such a coupled system. This can equivalently be interpreted in terms of an *attractive force* between such particles. In this way particles of like charge which repel one another electrostatically (e.g., protons in the nucleus of atoms) also attract one another via the coupling (covalent) effect.

42. Find the solution of the differential equation

$$d_t x = x$$

by Laplace transforms.

Ans. With

$$X(s) = \int_0^\infty e^{-st} x(t)\, dt$$

the differential equation becomes

$$sX(s) - x(0) = X(s)$$

so

$$X(s) = \frac{x(0)}{s - 1}$$

which is recognized as the transform of

$$x(t) = x(0)e^t$$

43. With regard to Example 42, what is wrong with the following argument? For $s = 1$ the transformed equation becomes

$$X(1) - x(0) = X(1)$$

which implies $x(0) = 0$ upon cancellation.

Ans. This equation only implies $x(0) = 0$ if $X(1)$ is finite. Here in fact $X(1)$ is infinite, so one cannot conclude $x(0) = 0$.

44. For the *Mellin transform*, defined by

$$M(s) = \int_0^\infty m(t)t^{s-1}\,dt$$

and convergent in a strip $\sigma_2 > \operatorname{Re} s > \sigma_1$, show that an inversion formula is

$$m(t) = \frac{1}{2\pi i}\int_{\sigma - i\infty}^{\sigma + i\infty} M(s)t^{-s}\,ds$$

where $\sigma_2 > \sigma > \sigma_1$.

Ans. The result follows immediately upon taking $-t = \log \tau$ in the bilateral Laplace inversion formula.

45. Find the solution for the system of differential equations ($u(t) \neq$ unit step here)

$$d_t u = 2v$$
$$d_t v = 2w$$
$$d_t w = -16u - 24v - 12w$$

Ans. Laplace transforming with

$$U(s) = \int_0^\infty e^{-st}u(t)\,dt$$

$$V(s) = \int_0^\infty e^{-st}v(t)\,dt$$

$$W(s) = \int_0^\infty e^{-st}w(t)\,dt$$

yields

$$sU(s) - u(0) = 2V(s)$$
$$sV(s) - v(0) = 2W(s)$$
$$sW(s) - w(0) = -16U(s) - 24V(s) - 12W(s)$$

which may be solved to yield

$$(s+4)^3U(s) = (s^2 + 12s + 48)u(0) + 2(s + 12)v(0) + 4w(0)$$
$$(s+4)^3V(s) = s(s + 12)v(0) + 2sw(0) - 32u(0)$$
$$(s+4)^3W(s) = s^2w(0) - 8(3s + 4)v(0) - 16su(0)$$

which may also be written in the form (calling $u(0) = u_0$, $v(0) = v_0$, $w(0) = w_0$)

$$U(s) = \frac{u_0}{(s+4)} + \frac{2(v_0 + 2u_0)}{(s+4)^2} + \frac{4(w_0 + 4v_0 + 4u_0)}{(s+4)^3}$$
$$V(s) = \frac{v_0}{(s+4)} + \frac{2(2v_0 + w_0)}{(s+4)^2} - \frac{8(w_0 + 4v_0 + 4u_0)}{(s+4)^3}$$
$$W(s) = \frac{w_0}{(s+4)} - \frac{8(w_0 + 3v_0 + 2u_0)}{(s+4)^2} + \frac{16(w_0 + 4v_0 + 4u_0)}{(s+4)^3}$$

The functions $u(t)$, $v(t)$, $w(t)$ with these transforms are then recognized (or calculated by the Laplace inversion formula 9.11) to be

$$u(t) = [u_0 + 2(v_0 + 2u_0)t + 2(w_0 + 4v_0 + 4u_0)t^2]e^{-4t}$$
$$v(t) = [v_0 + 2(2v_0 + w_0)t - 4(w_0 + 4v_0 + 4u_0)t^2]e^{-4t}$$
$$w(t) = [w_0 - 8(w_0 + 3v_0 + 2u_0)t + 8(w_0 + 4v_0 + 4u_0)t^2]e^{-4t}$$

46. Plot the Nyquist diagram for

$$G_c(s) = \frac{G_o(s)}{1 + G_o(s)}$$

with

$$G_o(s) = \frac{K}{s(s+a)(s+b)} \qquad K, a, b > 0$$

Ans. The Nyquist diagram is the limit of the map of the interior of the region $\varepsilon \le |s| \le R$; $|\arg s| \le \pi/2$ as $\varepsilon \to 0$ and $R \to \infty$ on the $G_o(s)$ plane. Since $G_o^*(s) = G_o(s^*)$, the diagram is symmetrical about the real $G_o(s)$ axis; hence it suffices to construct the map for $0 \leq \arg s \leq \pi/2$. Call the portions of the boundary in the s plane

$$\begin{aligned} J_1&: \quad |s| = \varepsilon; \ 0 \leq \arg s \leq \pi/2 \\ \mathrm{I}&: \quad \varepsilon < \operatorname{Im} s < R; \ \operatorname{Re} s = 0 \\ J_2&: \quad |s| = R; \ 0 \leq \arg s \leq \pi/2 \end{aligned}$$

and their images in the $G_o(s)$ plane J_1', I', J_2', respectively. For large R, one has on J_2' approximately

$$G_o(s) = G_o(Re^{i\theta}) \approx \frac{Ke^{-3i\theta}}{R^3}$$

so that the change in argument is

$$\Delta_{J_2'} \arg G_o(s) = -3\Delta_{J_2} \arg s = -3\Delta\theta$$

and since $\theta = \arg s$ decreases by $\pi/2$ upon traversal of J_2 (in a clockwise sense), $\arg G_o(s)$ must increase on J_2' (which is close to $G_o = 0$) by $3\pi/2$. For small ε, one has on J_1' approximately

$$G_o(s) = G_o(\varepsilon e^{i\theta}) \approx \frac{Ke^{-i\theta}}{\varepsilon ab}$$

so that

$$\Delta_{J_1'} \arg G_o(s) = -\Delta_{J_1} \arg s = -\Delta\theta$$

and since $\theta = \arg s$ increases by $\pi/2$ upon traversal of J_1, $\arg G_o(s)$ must decrease by $\pi/2$. On I

$$G_o(s) = G_o(iy) = \frac{K}{iy(iy + a)(iy + b)}$$

or

$$G_o(iy) = u + iv$$

with

$$u = \frac{-K(a + b)}{(a^2 + y^2)(b^2 + y^2)} \leq 0$$

$$v = \frac{K(y^2 - ab)}{y(a^2 + y^2)(b^2 + y^2)}$$

so that as y increases from ε to R, u and v have approximately the values indicated in Table 9.1. It is noticed that u increases toward zero steadily and that there is a vertical asymptote at $u = -K(a + b)/a^2b^2$. Also u approaches zero faster than v as $R \to \infty$. From Figures 9.4a–c it is readily discerned that if K becomes larger than $ab(a + b)$, the point $G_o = -1$ will be included in the Nyquist diagram and instability will result.

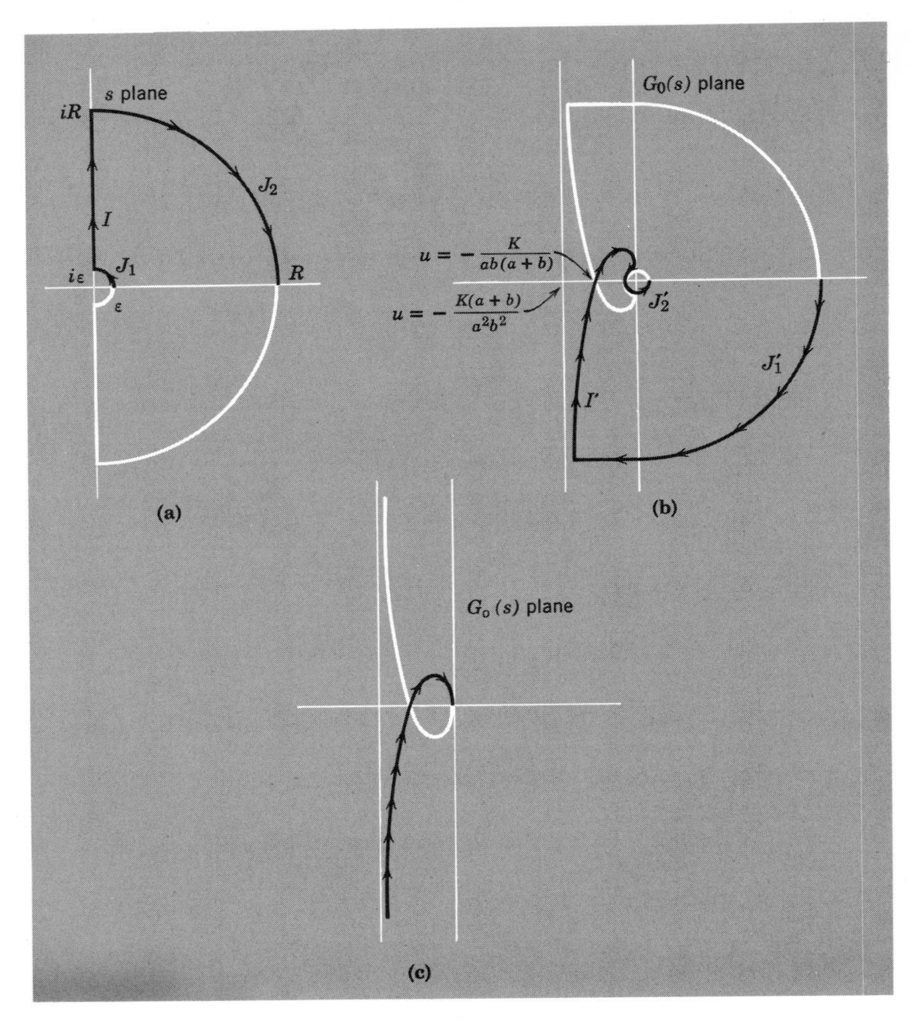

Figure 9.4. (a) The s plane region to be imaged into prelimit Nyquist diagram; (b) the prelimit Nyquist diagram (before $R \rightarrow \infty$, $\varepsilon \rightarrow 0$) $G_o(s) = u + iv$; (c) the postlimit Nyquist diagram (after $R \rightarrow \infty$, $\varepsilon \rightarrow 0$) $G_o(s) = u + iv$.

Table 9.1

y	ε		$\sqrt{ab}$		R
u	$-\dfrac{K(a+b)}{a^2b^2}$	$-$	$-\dfrac{K}{ab(a+b)}$	$-$	$-\dfrac{K(a+b)}{R^4}$
v	$-\dfrac{K}{ab\varepsilon}$	$-$	0	$+$	$\dfrac{K}{R^3}$

47. Analyze the stability behavior of the feedback system of the preceding problem by the root locus method. (Take $b > a > 0$.)

Ans. The "trajectories" in the s plane of the poles of $G_c(s)$ as K increases from 0 to ∞ must start at the poles of $G_0(s)$ (for $K = 0$) and end on the zeros of $G_0(s)$ (for $K = \infty$). The part of the locus on the real axis can only exist to the left of an *odd number* of poles and zeros (on the real axis) of $G_0(s)$. This follows from the locus equation

$$D(s) + KN(s) = 0$$

which may be written

$$\frac{N}{D} = -\frac{1}{K}$$

Whence

$$\arg N - \arg D = \arg(-1) = \pi$$

For

$$N = N_0 \prod_k (s - z_k) \qquad D = D_0 \prod_k (s - p_k)$$

one has

$$\sum_k \arg(s - z_k) - \sum_k \arg(s - p_k) = \pi$$

For any point s on the real axis, only poles and zeros to the right of it contribute to the total, and if an even number of poles and zeros are to the right of s, it is impossible for the total to be π. Since $G_0(s^*) = G_0^*(s)$, the locus is symmetric to the real axis. For

$$G_0(s) = \frac{1}{s(s+a)(s+b)}$$

the loci start at $s = 0, -a, -b$. The locus equation is

$$s(s+a)(s+b) + K = 0$$

or

$$s^3 + (a + b)s^2 + abs + K = 0$$

and

$$[3s^2 + 2(a + b)s + ab]\,ds + dK = 0$$

Since $dK > 0$, one has

$$ds = \frac{dK}{a(b - a)} > 0 \qquad \text{for} \quad s = -a$$

$$ds = -\frac{dK}{b(b - a)} < 0 \qquad \text{for} \quad s = -b$$

$$ds = -\frac{dK}{ab} < 0 \qquad \text{for} \quad s = 0$$

Hence the loci proceed initially along the real axis in the directions indicated above. They will continue along the real axis until two roots meet, since

$$ds = -\frac{dK}{[3s^2 + 2(a + b)s + ab]}$$

will remain real and finite as long as the denominator is not zero. The roots of the denominator are

$$s = -\tfrac{1}{3}[(a + b) \mp \sqrt{a^2 - ab + b^2}]$$

One root is between $-a$ and $-b$ outside the real axis portion of the loci:

$$3b > (a + b) + \sqrt{(b - a)^2 + ab} > 3a$$

The other root occurs at the breakaway point s_o at which the locus leaves the real axis:

$$s_o = -\tfrac{1}{3}[(a + b) - \sqrt{(b - a)^2 + ab}]$$

One has

$$s_o^3 + (a + b)s_o^2 + abs_o + K_o = 0$$

which gives the breakaway gain K_o. Also,

$$0 = (s_o + ds)^3 + (a + b)(s_o + ds)^2 + ab(s_o + ds) + K_o + dK$$

Subtraction then yields

$$[3s_o^2 + 2(a + b)s_o + ab]\,ds + [3s_o + (a + b)]\,ds^2 + ds^3 + dK = 0$$

Since at s_o the root is double, the coefficient of ds is zero, which shows that a first-order argument will *not* suffice to determine ds.

Then neglecting $ds^3 \ll ds^2$, one has

$$ds^2 = \frac{-dK}{3s_o + (a + b)}$$

so that

$$ds = \pm i\sqrt{\frac{dK}{3s_o + (a + b}}$$

which shows that the loci depart from the real axis orthogonally. For $K \to \infty$, $|s| \to \infty$ so that the locus equation becomes approximately

$$s^3 + (a + b)s^2 + abs + K \approx \left(s + \frac{a + b}{3}\right)^3 + K = 0$$

and thus for large $|s|$ the loci extend in the directions [from $-(a + b)/3$] of the cube roots of -1. To find the critical gain

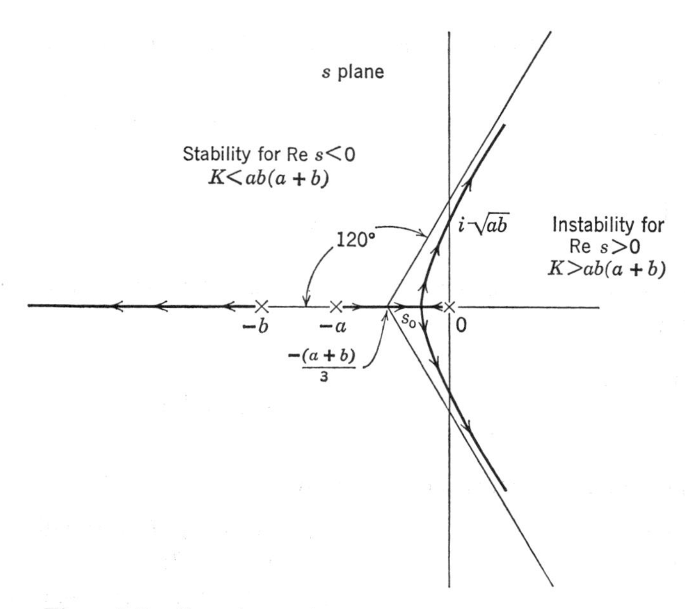

Figure 9.5. Root locus plot in s plane of trajectories of poles of $G_c(s) = KG_o(s)/[1 + KG_o(s)]$ given by roots of $s(s + a)(s + b) + K = 0$ as K increases from zero to infinity; s_o is the location of the breakaway point at which the locus leaves the real axis. Arrows indicate directions of increasing K.

value at which the system becomes unstable one may set $s = iy$ in the locus equation

$$-iy^3 - (a + b)y^2 + iyab + K = 0$$

so that

$$iy(ab - y^2) + K - (a + b)y^2 = 0$$

so

$$y^2 = ab$$

and

$$K = ab(a + b)$$

as was also found in Example 46. The root locus plot is shown in Figure 9.5.

● PROBLEMS 9

Find the abscissae of convergence of the (unilateral) Laplace transforms of the following:

1. $t^5 \sin t$
2. $\dfrac{1}{2e^t - 1}$
3. $1 - u(t - 1)$ with $u(t)$ the unit step
4. e^{t^2}

Find the strip of convergence of the bilateral Laplace transforms of the following:

5. $e^{-3|t|} \tanh t$
6. e^{t^3}
7. $\dfrac{1}{1 + \sinh t}$

Find the Laplace transforms of the following:

8. $t \sinh kt$
9. $t^n e^{kt}$
10. $e^{-kt}/\sqrt{t}$
11. $\sin kt/t$
12. $u(t - a)u(b - t)$ with $b > a$
13. $t^3 u(t - 2)$
14. $[t]$ = the integer part of t

Find the bilateral Laplace transform of the following:

15. $u(t - a)u(t + a)$
16. $e^{-2|t|} \sin kt$
17. $e^{-3|t-5|} \cos kt$
18. $e^{-\alpha t^2}$

Find the function with the following unilateral Laplace transform:

19. e^{-ks}/s

20. $\log(s + k)/(s - k)$

21. $1/s^2(s^2 - 1)$

22. $1/(s - a)(s - b)(s^2 + 1)$

23. $1/\sqrt{s^2 + 1}$

24. $1/s(1 + e^{-2s})$

25. $(e^{-s} + s - 1)/s^2(1 - e^{-s})$

26. $(s + 1 - e^s)/s^2(1 - e^s)$

By use of the Laplace transform solve the following differential equations:

27. $d_t^2 f + f = 0$ with $f(0) = 1$ and $f'(0) = 0$

28. $(d_t^2 + \omega^2)f = ce^{-\alpha t}$ with $f(0) = f'(0) = 0$ and $\omega \neq 0 \neq c$

29. $(d_t - 1)g = e^{-t}$ $\qquad f(0) = 1$ and $f'(0) = 2$

$(d_t - 2)f = g$

30. $(d_t^2 - 4d_t + 4)f = 4t + 4e^{2t}$ with $f(0) = f'(0) = 0$

31. If a plane metal plate consists of parallel strips insulated from each other and maintained either at voltage π or 0, show that the potential in the half-space bounded by the configuration is the sum of the angles subtended by the nonzero voltage strips at any line parallel to the strips.

32. Show that an axially symmetric potential (independent of angle) can be expressed in terms of cylindrical coordinates

$$\phi(r,z) = \frac{1}{2\pi i}\int_{\sigma-i\infty}^{\sigma+i\infty} \frac{\phi(0,s)\,ds}{\sqrt{(z - s)^2 + r^2}}$$

where $\phi(0,s)$ is the value of the potential on the axis $r = 0$.

33. Find the potential of a uniformly charged circular disk.

34. Show that the argument of the Fourier transform $F(\omega)$ of a real function $f(t)$ is an odd function of ω.

35. Show that an (ideal) low-pass filter free of phase distortion with amplitude $A(\omega) = A_0 u(\omega_c - |\omega|)$ is *not* strictly casual.

36. Find the response (output) $g(t)$ of an (ideal) low-pass filter with cutoff frequency ω_c to an arbitrary input $f(t)$.

37. Find the response $g(t)$ of a *cosine filter* which is phase-distortionless with amplitude function $A(\omega) = [a + b\cos(n\pi\omega/\omega_c)]u(\omega_c - |\omega|)$ to an arbitrary input $f(t)$. Show that there is an echo and a pre-echo.

38. Show that the Laplace transform of the product of two functions $f(t)$ and $g(t)$ in terms of $F(s)$ and $G(s)$, the Laplace transforms, is

$$\frac{1}{2\pi i}\int_{\sigma-i\infty}^{\sigma+i\infty} F(z)G(s - z)\,dz$$

with $\sigma_F < \sigma < \operatorname{Re} s - \sigma_G$ given. The abscissae of convergence of $F(s)$ and $G(s)$ are σ_F and σ_G, respectively.

39. Establish the *generalized Parseval relationship* for Laplace transforms

$$2\pi \int_0^\infty e^{-(\sigma_F+\sigma_G)t} f(t) g^*(t)\, dt = \int_{-\infty}^{\infty} F(\sigma_F + iy) G^*(\sigma_G + iy)\, dy$$

40. Plot a Nyquist diagram for the open-loop transfer function $G(s) = 1/s(s + 1)$. When is the system unstable? Construct the corresponding closed-loop transfer function $G_c(s)$ and the associated root locus plot.
41. Repeat the procedures of Problem 40 for the open-loop transfer function $G(s) = 1/s(s + 2)(s + 4)$.
42. Find the root locus for the open-loop transfer function $G(s) = K(s + 2)/(s + 1)^2$.
43. Repeat the procedures of Problem 42 for $G(s) = K/(s + 2)^3$.
44. If the open-loop transfer function of a system is $G(s) = K\, N(s)/D(s)$, with $N(s)$ and $D(s)$ polynomials, show that the condition for a breakaway point s (where root locus leaves real axis) is

$$\begin{vmatrix} D(s) & N(s) \\ D'(s) & N'(s) \end{vmatrix} = 0$$

How can the corresponding value of K be found?

45. Illustrate the results of Problem 44 on the transfer functions of Problems 42 and 43.
46. Evaluate $\oint \tanh z \csc[1/\Gamma(z)]\, dz$ about $2\,|z + 5| = 1$.
47. Evaluate $\int_{-\infty}^{\infty} e^{-tx^2}\, dx$ by Laplace transforming with respect to t.
48. Show that the asymptotic approximation for large s

$$\int_0^\infty e^{-s(\sinh z - z)}\, dx \sim \frac{2}{6^{2/3}} \frac{\Gamma(\frac{1}{3})}{s^{1/3}}$$

may be obtained by the method of steepest descents.

49. Suppose $f(z)$ is meromorphic with no poles on the negative real axis and let all its poles lie in the left half-plane, $\operatorname{Re} z < 0$ being simple with residue c_k at $z = z_k$. Evaluate

$$\int_L f(z)\Gamma(z)\, dz$$

for the contour of (9.44) supposing that $|z^2 f(z)\Gamma(z)| \to 0$ as $|z| \to \infty$ for $\operatorname{Re} z \leq 0$.

50. Evaluate [with $f(z)$ as in Problem 49]

$$\int_{\varepsilon-i\infty}^{\varepsilon+i\infty} f(z)\Gamma(z)\, dz$$

51. The specific heat of a solid is given by

$$C(\alpha) = Nk\int_0^\infty \frac{(\alpha\nu)^2 e^{\alpha\nu} D(\nu)\, d\nu}{(e^{\alpha\nu} - 1)^2}$$

where $\alpha kT = \hbar$. If the relationship between $D(\nu)$ and $C(\alpha)$ which holds at low temperatures were generally true, find a formula for $D(\nu)$ in terms of $C(\alpha)$.

52. Evaluate $\int_{-\infty}^{\infty} e^{itx^2}\, dx$ by Laplace transforming the integral.

53. For a real signal

$$x(t) = \int_{-\infty}^{\infty} |A(\nu)|\, e^{i[\omega t + \phi(\nu)]}\, d\nu$$

find the time of greatest amplitude by the stationary phase method.

54. Show that a filter which only produces a phase shift proportional to frequency will delay a pulse at a time $-d\phi/d\omega$ but does not affect its form.

55. Use the method of stationary phase to estimate the maximal frequencies in the (spectrum of the) signal $x(t) = u(t - |t|)\cos \omega_0 t$

56. Use the method of stationary phase to show that an asymptotic approximation to

$$\int_{-\infty}^{\infty} e^{-[u^2+(u-t)^2+(u^4/2)]} \cos[au^2 + (2\omega - at)u - \omega t + (at^2/2)]\, du$$

is

$$e^{-[v^2+(v-t)^2+(v^4/2)]}\sqrt{\frac{\pi}{2a}}$$

with $v = (t/2) - (\omega/a)$ and a large.

57. For what points of the complex z plane is $\Gamma(z) = 0$?

58. Describe the singularities of $\Gamma(z^2)$.

59. Is $\Gamma(|z|)$ ever infinite?

60. In a series RC circuit the initial charge on the capacitor is q_0. If a voltage $e(t)$ is applied across the terminals, one has

$$Rd_t q + \frac{q}{C} = e(t)$$

which Laplace transforms to

$$\left(Rs + \frac{1}{C}\right)Q(s) - Rq_0 = E(s)$$

with $Q(s)$ and $E(s)$ the transforms of charge and voltage. If ($u(t)$ = unit step at $t = 0$) a voltage pulse of width T is applied at $t = \tau$,

$e(t) = Vu(t - \tau)u(\tau + T - t)$, find $Q(s)$ and show that the charge on the capacitor is

$$q(t) = \begin{cases} q_0 e^{-t/RC} & \text{for} \quad t < \tau \\ q_0 e^{-t/RC} + CV[1 - e^{-(t-\tau)/RC}] & \text{for} \quad \tau < t < \tau + T \\ q_0 e^{-t/RC} + CVe^{-(t-\tau)/RC}[e^{T/RC} - 1] & \text{for} \quad t > \tau + T \end{cases}$$

61. Find the charge in Problem 60 for $VT \to k$ as $V \to \infty$ and $T \to 0$ both from the previous solution and by a delta function argument.
62. By the method of steepest descents find the asymptotic approximation for large s:

$$\int_0^3 z^2 e^{sze^{-3z}}\, dz \sim \frac{e^{s/3e}}{9}\sqrt{\frac{2\pi e}{3s}}$$

63. Why is the ratio of transmitted to incident power in Example 37 *not* equal to

$$|A_T|^2 = \frac{4\,|k_R^2|}{|k_R + k_T|^2}$$

Find the Fourier series for the following:

64. $\arctan\left(\dfrac{r \sin\theta}{1 - r\cos\theta}\right)$
65. $\dfrac{\cos(\theta/2) - \sin(\theta/2)}{2(\sin\theta)^{3/2}}$
66. $e^{r\cos\theta}\cos(r\sin\theta)$
67. $[(\theta^2/2) - (\pi^2/12)]$ for $|\theta| < \pi$ and periodic $(T = 2\pi)$
68. $\log(1 - 2r\cos\theta + r^2)$
69. $\log\tan(\theta/2)$

Find the Fourier transforms for the following:

70. $u(t - a)u(b - t)$ for $b > a$
71. $(t - a)^\nu u(t - a)$ for $-1 < \mathrm{Re}\,\nu < 0$
72. $(a + it)^{-\nu}$ for $\mathrm{Re}\,\nu > 0$
73. $(1 + i)u(t)u(a - t)$
74. $e^{-at}u(t)$
75. $[\sin(at)/t]u(t)$
76. $[\sin(at)\sin(bt)/t]u(t)$
77. $\mathrm{sech}(at)u(t)$

By using the Parseval relation, or otherwise, show that the following results hold:

78. $\displaystyle\int_0^\infty \frac{dt}{(t^2 + a^2)(t^2 + b^2)} = \frac{\pi}{2(a + b)}$

79. $$\frac{2}{\pi}\int_0^\infty \frac{\sin at \sin bt}{t^2}\,dt = \min(a,b)$$

80. $$\frac{1}{\pi^2}\int_0^\infty \log\left|\frac{a+t}{a-t}\right| \log\left|\frac{b+t}{b-t}\right| dt = \min(a,b)$$

81. Solve the Dirichlet problem for a harmonic function for a semi-infinite strip with zero boundary values along two adjacent edges.
82. Solve the harmonic Dirichlet problem in the first quadrant.
83. Show that an asymptotic expansion for

$$\Gamma(z) - \gamma(\nu,z) = \int_z^\infty e^{-t} t^{\nu-1}\,dt$$

is given by

$$\Gamma(z) - \gamma(\nu,z) \sim z^\nu e^{-z}\left[\frac{1}{z} + \frac{\nu-1}{z^2} + \frac{(\nu-1)(\nu-2)}{z^3} + \cdots\right]$$

84. Find an asymptotic expansion for

$$\frac{\sqrt{\pi}}{2} - \text{Erf}\, z = \int_z^\infty e^{-t^2}\,dt \sim e^{-z^2}\left[\frac{1}{2z} - \frac{1}{4z^3} + \frac{3}{8z^5} - \cdots\right]$$

85. Show that

$$\int_z^\infty \cos(t-z)\frac{dt}{t} \sim \frac{1}{z^2} - \frac{3!}{z^4} + \frac{5!}{z^6}$$

CHAPTER 10:

Infinite Product and Rational Fraction Expansions

10.1 WEIERSTRASS FACTORIZATION (PRODUCT) THEOREM

If $P_n(x)$ is a polynomial of degree n and of the form (with $a_m \neq 0$)

$$P_n(z) = \sum_{k=m}^{n} a_k z^k = a_m z^m + a_{m+1} z^{m+1} + \cdots + a_n z^n \tag{10.1}$$

it is readily verified that $P_n(z)$ may be expressed in the form

$$P_n(z) = a_m z^m \prod_{k=1}^{n-m} \left(1 - \frac{z}{z_k}\right) \tag{10.2}$$

[where $m!a_m = P_n^{(m))}(0) = m$th derivative of P_n at $z = 0$], and multiple roots are repeated in the product.

The series (10.1) and the product (10.2) each contain a finite number of terms and therefore trivially converge for all finite z. Indeed $P_n(z)$ is an *entire function* with no singularities in the finite z plane. It is natural to inquire whether there is a property similar to (10.2) for the general class of entire functions. Such functions might be regarded as "polynomials of infinite degrees" since they possess power series expansions

$$f(z) = \sum_{k=m}^{\infty} a_k z^k = a_m z^m + a_{m+1} a^{m+1} + \cdots \tag{10.3}$$

with *infinite radii of convergence*, but it is not customary to use such terminology.

One is thus led to speculate whether there might be an expansion analogous to (10.2) expressing a given entire function $f(z)$ as a product of factors vanishing precisely at the *zeros* of $f(z)$ (i.e., the values of z for which $f(z) = 0$). Apparently there are three cases for the finite z plane. Either $f(z)$ has *no* zeros, or it has a *finite positive number* of zeros, or it has an *infinite number* of zeros. In the first two cases if $Q_n(z)$ is a polynomial with roots identical with the zeros of $f(z)$, it is clear that the function

$$g(z) = \frac{f(z)}{Q_n(z)} \tag{10.4}$$

is also entire, since the zeros of $Q_n(z)$ are matched by those of $f(z)$ and the convergence properties of the power series for $g(z)$ and $f(z)$ are the same. In these cases the product representation for $f(z)$,

$$f(z) = Q_n(z)g(z) \tag{10.5}$$

involves only the (finite) product for $Q_n(z)$, and aside from the factor $g(z)$ (nonvanishing in the finite z plane) nothing is added to the expansion (10.2).

In case $f(z)$ has an infinite number of zeros, one might expect analogously to (10.2) to have

$$f(z) = A_m z^m \prod_{k=1}^{\infty} \left(1 - \frac{z}{z_k}\right) \tag{10.6}$$

[where $m!A_m = f^{(m)}(0)$], which would certainly appear to vanish for $z = z_k$. Unfortunately (10.6) contains an infinite number of factors in the product and therefore does not generally converge, as may be illustrated by the following simple example: Let

$$h_n(z) = \prod_{k=1}^{n} \left(1 - \frac{z}{k}\right) \tag{10.7}$$

Then

$$\log h_n(z) = \sum_{k=1}^{n} \log\left(1 - \frac{z}{k}\right)$$

But for $|z| < k$

$$-\log\left(1 - \frac{z}{k}\right) = \frac{z}{k} + \frac{1}{2}\left(\frac{z}{k}\right)^2 + \cdots = \sum_{q=1}^{\infty} \frac{1}{q}\left(\frac{z}{k}\right)^q$$

and

$$-\sum_{k=1}^{n} \log\left(1 - \frac{z}{k}\right) = z\sum_{k=1}^{n} \frac{1}{k} + \frac{z^2}{2}\sum_{k=1}^{n} \frac{1}{k^2} + \cdots \tag{10.8}$$

so that the limit of (10.8) will not exist as $n \to \infty$ because the first term on the right side of (10.8) contains the partial sum of the (divergent) harmonic series of reciprocals of the integers. The restriction to values of z for which $|z| < k$ is of no consequence, since for any finite z values of k sufficiently large so that $|z| < k$ will be encountered in (10.7), and only the terms further than this (i.e., for still larger k) will determine the convergence properties. Thus clearly, the factors $[1 - (z/z_k)]$ must be augmented by other factors if an infinite product of them is to converge.

In the case of (10.8) the convergence properties can be improved by transferring the harmonic series to the left side. In the limit as $n \to \infty$ one has then,

$$-\sum_{k=1}^{\infty}\left[\frac{z}{k} + \log\left(1 - \frac{z}{k}\right)\right] = \sum_{k=1}^{\infty}\sum_{q=2}^{\infty}\frac{1}{q}\left(\frac{z}{k}\right)^q \tag{10.9}$$

which is absolutely convergent for $|z| \neq k$ by the following argument: The modulus of the right side of (10.9) does not exceed

$$\sum_{k=1}^{\infty}\left|\frac{z}{k}\right|^2\sum_{p=0}^{\infty}\frac{1}{(p+2)}\left|\frac{z}{k}\right|^p \leq \frac{1}{2}\sum_{k=1}^{\infty}\left|\frac{z}{k}\right|^2\sum_{p=0}^{\infty}\left|\frac{z}{k}\right|^p \tag{10.10}$$

The convergence properties will be determined only by the summation over values of k larger than $|z|$ since only a *finite* number of terms occur for which $|z| > k$. Thus (10.10) may be replaced by

$$\sum_{k=m}^{\infty}\left|\frac{z}{k}\right|^2\sum_{p=0}^{\infty}\frac{1}{(p+2)}\left|\frac{z}{k}\right|^p \leq \frac{1}{2}\sum_{k=m}^{\infty}\left|\frac{z}{k}\right|^2\sum_{p=0}^{\infty}\left|\frac{z}{k}\right|^p \tag{10.11}$$

where m is the smallest value of k larger than $|z|$. But (10.11) does not exceed

$$\frac{1}{2}\sum_{k=m}^{\infty}\left|\frac{z}{k}\right|^2\sum_{p=0}^{\infty}\left|\frac{z}{m}\right|^p = \frac{|z|^2}{2(1 - |z/m|)}\sum_{k=m}^{\infty}\frac{1}{k^2}$$

which clearly converges.

This argument indicates that although the infinite product

$$\prod_{k=1}^{\infty}\left(1 - \frac{z}{k}\right) \tag{10.12}$$

does *not* converge, the related infinite product

$$g(z) = \prod_{k=1}^{\infty}\left(1 - \frac{z}{k}\right)e^{z/k} \tag{10.13}$$

of which the logarithm is

$$\log g(z) = \sum_{k=1}^{\infty}\left[\frac{z}{k} + \log\left(1 - \frac{z}{k}\right)\right] \tag{10.14}$$

does converge. The exponential factors (which do not affect the fact that $g(k) = 0$) make the difference between (10.12) and (10.13). From the product representation for the reciprocal gamma function (9.64) one can conclude that (10.13) is

$$g(z) = \frac{e^{\gamma z}}{\Gamma(1 - z)} \tag{10.15}$$

where γ is the Euler–Mascheroni constant ($\gamma \approx 0.5772$).

The exponential convergence factors required for an infinite product vanishing simply at the positive integers have been found in (10.13). What convergence factors would be necessary for an infinite product vanishing at points z_k with $|z_k| \to \infty$ as $k \to \infty$? The same general procedure could be used as before, but now it might be necessary to take more terms in the expansion of the logarithm to secure convergence. Thus, let

$$f(z) = \prod_{k=1}^{\infty} \left(1 - \frac{z}{z_k}\right) \tag{10.16}$$

Then

$$\log f(z) = \sum_{k=1}^{\infty} \log\left(1 - \frac{z}{z_k}\right)$$

and for convergence it suffices to consider terms for which $|z_k| > |z|$ which will hold for all sufficiently large k since $|z_k| \to \infty$ as $k \to \infty$. Then

$$-\log\left(1 - \frac{z}{z_k}\right) = \left(\frac{z}{z_k}\right) + \frac{1}{2}\left(\frac{z}{z_k}\right)^2 + \cdots = \sum_{s=1}^{\infty} \frac{1}{s}\left(\frac{z}{z_k}\right)^s \tag{10.17}$$

If $|z_k|$ does not become infinite sufficiently rapidly as k increases, the summation of the right side of (10.17) on k may diverge. Thus for $|z_k| \sim \sqrt{k}$ the terms for $s = 1$ and $s = 2$ on the right side of (10.17) will diverge when summed on k. This sort of behavior can be avoided by transferring a sufficient number m_k of terms from the right side of (10.17) so that

$$\sum_k \left|\frac{z}{z_k}\right|^{m_k+1}$$

converges. Then one has

$$\left|\log\left(1 - \frac{z}{z_k}\right) + \sum_{s=1}^{m_k} \frac{1}{s}\left(\frac{z}{z_k}\right)^s\right| = \sum_{s=m_k+1}^{\infty} \frac{1}{s}\left|\frac{z}{z_k}\right|^s \leq \sum_{s=m_k+1}^{\infty} \left|\frac{z}{z_k}\right|^s \tag{10.18}$$

For $|z| < q\,|z_k|$ with $0 < q < 1$, the right side of (10.18) cannot exceed

$$\left|\frac{z}{z_k}\right|^{m_k+1} \sum_{s=0}^{\infty} q^s = \frac{1}{(1 - q)}\left|\frac{z}{z_k}\right|^{m_k+1} \tag{10.19}$$

so that (10.18) will converge when summed on k provided that

$$\sum_{k=1}^{\infty} \left| \frac{z}{z_k} \right|^{m_k+1} \tag{10.20}$$

converges. Since $|z_k| \to \infty$, this must hold for all finite $|z|$. Thus, although (10.16) does *not* generally converge, (10.18) shows that

$$f(z) = \prod_{k=1}^{\infty} \left[\left(1 - \frac{z}{z_k}\right) \exp \sum_{s=1}^{m_k} \frac{1}{s} \left(\frac{z}{z_k}\right)^s \right] \tag{10.21}$$

does converge to an entire function $f(z)$ with zeros at $z = z_k$ provided the m_k are so chosen that

$$\sum_{k=1}^{\infty} \left| \frac{z}{z_k} \right|^{m_k+1}$$

converge. This is the *Weierstrass factorization theorem* for the case of simple zeros. If the zeros at z_k have multiplicities q_k and there is a pth-order zero at $z = 0$, (10.21) becomes

$$f(z) = z^p \prod_{k=1}^{\infty} \left[\left(1 - \frac{z}{z_k}\right) \exp \sum_{s=1}^{m_k} \frac{1}{s} \left(\frac{z}{z_k}\right)^s \right]^{q_k} \tag{10.22}$$

while the convergence criterion becomes

$$\sum_{k=1}^{\infty} q_k \left| \frac{z}{z_k} \right|^{m_k+1}$$

instead of (10.20).

Naturally the representation (10.22) is not unique, since it may be multiplied by the exponential of any entire function without affecting the distribution of the zeros, as such an exponential does not vanish in the finite plane.

Since a *meromorphic function* is the ratio of two entire functions it may be represented as the ratio of two products like (10.22) which then explicitly shows the location of its poles and zeros.

The Laurent expansion of a meromorphic function about one of its poles contains only a finite number of negative powers. Otherwise the point would be an essential singularity rather than a pole and the function would not be meromorphic unless the point were at infinity. The portion of the Laurent expansion consisting of the series of negative powers is called the *principal part* of the function.

10.2 MITTAG–LEFFLER RATIONAL (PARTIAL) FRACTION EXPANSION

There is an expansion theorem due to Mittag–Leffler for meromorphic functions (possessing only poles in the finite plane) in terms of partial fractions which is quite as important as the Weierstrass theorem giving product expansions for entire functions. In the simplest case it can be viewed as a kind of logarithmic derivative of the Weierstrass product expansion. An explicit statement of the Mittag–Leffler theorem follows.

For any set of points $\{a_k\}$ (finite in number in any finite region) for which the functions (partial fractions)

$$h_k(z) = \sum_{q=1}^{m} \frac{A_{kq}}{(z - a_k)^q}$$

have been assigned there exists a meromorphic function $f(z)$ free of singularities in the finite z plane except at points of the set $\{a_k\}$, and in a neighborhood of each such point $z = a_k$ the function $(f(z) - h_k(z))$ is analytic and free of singularities. The meromorphic function $f(z)$ is *not* uniquely defined by this process but is only determined to within an arbitrary additive entire function $E(z)$.

For the case where $\{a_k\}$ consists of a finite number N of points the result is trivial as in the case of the product representation of Weierstrass.

In fact,

$$f(z) = M(z) + E(z)$$

where

$$M(z) = \sum_{k=1}^{N} h_k(z)$$

and

$$f(z) - h_p(z) = E(z) + \sum_{k \neq p} h_k(z)$$

is clearly analytic about $z = a_p$.

If, however, N became infinite, the series for $M(z)$ might fail to converge. The procedure must then be modified by subtracting, from each $h_k(z)$ in the series, a polynomial $g_k(z)$ designed to produce convergence of the series for $M(z)$ without introducing any new singularities. One then obtains

$$f(z) = M(z) + E(z)$$

where

$$M(z) = \sum_{k=1}^{\infty} [h_k(z) - g_k(z)]$$

so that

$$f(z) - h_p(z) = E(z) - g_p(z) + \sum_{k \neq p} [h_k(z) - g_k(z)]$$

and it suffices to choose $g_k(z)$ of sufficiently high degree so that

$$|h_k(z) - g_k(z)| < \frac{1}{2^k}$$

and the series for $M(z)$ will be uniformly convergent. This is the basic idea of the Mittag–Leffler argument. For the detailed proof, one may consult K. Knopp, *Theory of Functions*, Vol. 2, New York: Dover, 1947, p. 39.

The Mittag–Leffler theorem is somewhat more general than the Weier strass product theorem in that the special case where the $h_k(z)$ have only simple poles leads to a meromorphic function with a partial fraction expansion that can be regarded as a logarithmic derivative of a Weierstrass product expansion.· However, the case of higher-order poles goes beyond the Weierstrass result.

The importance of both theorems in applications stems from the decomposition of functions into simpler factors or partial fractions. In the applications to circuits these terms correspond to simple impedances which can then be combined to produce more complicated impedances.

● EXAMPLES 10

1. By considering

$$I(z) = \oint \frac{z}{w(w - z)} \frac{J_{\nu+1}(w)}{J_\nu(w)} dw$$

establish the Weierstrass product for $J_\nu(z)$ in terms of the zeros $\{a_k\}$ of $J_\nu(a_k) = 0$. (Take $\nu > 0$; for a definition of J_ν see Chapter 13.)

Ans. Consider an increasing sequence of rectangular contours avoiding the zeros of $J_\nu(z)$. By letting them become infinite the integral $I(z)$ about the limiting contour approaches zero. On the other hand, if z is any point inside a rectangle of the sequence with $z \neq a_k$, the poles of the integrand occur at $w = z$ and $w = a_k$ with residues $J_{\nu+1}(z)/J_\nu(z)$ at the former and $-z/a_k(a_k - z)$ at the latter, since $J_{\nu+1}(z)/J_\nu(z)$ at the former and $-z/(a_k(a_k - z)$ at the latter, since $J_{\nu+1}(a_k)/J_\nu'(a_k) = -1$ (see recurrence relations for Bessel functions). Thus

$$-\frac{J_{\nu+1}(z)}{J_\nu(z)} = \sum_{k \neq 0} \left(\frac{1}{z - a_k} + \frac{1}{a_k}\right)$$

so that with

$$-\int_0^z \frac{J_{\nu+1}(z)\, dz}{J_\nu(z)} = \log[z^{-\nu} J_\nu(z)] + \log[2^\nu \Gamma(\nu + 1)]$$

from the recurrence relation,

$$zJ_{\nu+1} = \nu J_\nu - zJ_\nu'$$

one has

$$J_\nu(z) = \frac{z^\nu}{2^\nu \Gamma(\nu+1)} \prod_{k\neq 0} \left(1 - \frac{z}{a_k}\right) e^{z/a_k}$$

or pairing $\pm a_k$,

$$J_\nu(z) = \frac{z^\nu}{2^\nu \Gamma(\nu+1)} \prod_{k=1}^{\infty} \left(1 - \frac{z^2}{a_k^2}\right)$$

2. If $f(z)$ has simple poles at $z = a_k \neq 0$ where $\{a_k\}$ is a sequence approaching infinity, and there exists an infinite sequence of contours enclosing (but not passing through) these poles on which (contours) for a smallest nonnegative integer p

$$\lim_{|z|\to\infty} \left|\frac{f(z)}{z^{p+1}}\right| = 0$$

then

$$f(z) = \sum_{k=0}^{p} \frac{f^{(k)}(0) z^k}{k!} + \sum_{k=0}^{\infty} c_k \left[\frac{1}{z - a_k} + \sum_{s=0}^{p} \frac{z^s}{a_k^{s+1}}\right]$$

with $c_k = \lim_{z\to a_k} [(z - a_k) f(z)]$ the residue at $z = a_k$.

Ans. Consider

$$I(z) = \frac{1}{2\pi i} \oint \frac{f(w)\, dw}{w^{p+1}(w - z)}$$

As the contours become larger the limiting condition on $f(z)$ ensures that $I(z) \to 0$. On the other hand, the integrand has a simple pole at $w = z$ with residue $f(z)/z^{p+1}$, a $(p + 1)$th-order pole at the origin $w = 0$ with residue

$$-\frac{1}{z} \sum_{k=0}^{p} \frac{f^{(k)}(0)}{z^{p-k} k!}$$

[easily found by multiplying the Laurent series for $f(w)/w^{p+1}$ by the MacLaurin series for the reciprocal of $(w - z)$] and simple poles at $w = a_k$ with residues $c_k/a_k^{p+1}(a_k - z)$ [c_k is residue of $f(w)$ at $w = a_k$]. Thus the residue theorem yields

$$\frac{f(z)}{z^{p+1}} - \sum_{k=0}^{p} \frac{f^{(k)}(0)}{z^{p-k+1} k!} - \sum_{k=0}^{\infty} \frac{c_k}{a_k^{p+1}(a_k - z)} = 0$$

whence the result follows.

3. Find a partial fraction expansion for the function

$$f(z) = \frac{1}{e^z - 1} - \frac{1}{z}$$

Ans. By Example 2 one may here take $p = 0$. Then

$$f(0) = \lim_{z \to 0}\left\{\frac{1}{[z + (z^2/2)]} - \frac{1}{z}\right\} = -\tfrac{1}{2}$$

using the MacLaurin expansion for e^z. The function $f(z)$ has simple poles at $z = 2\pi ni$ with unit residue at each, hence,

$$f(z) = -\tfrac{1}{2} + \sum_{n \neq 0}^{\infty}\left(\frac{1}{z - 2\pi ni} + \frac{1}{2\pi ni}\right)$$

or pairing poles symmetrically located to the real axis

$$f(z) = -\tfrac{1}{2} + \sum_{n=1}^{\infty}\left(\frac{1}{z - 2\pi ni} + \frac{1}{z + 2\pi ni}\right)$$

so that

$$f(z) = -\tfrac{1}{2} + 2z\sum_{n=1}^{\infty}\frac{1}{z^2 + (2\pi n)^2}$$

4. If $g(z)$ is an entire function with simple zeros at $z = a_k \neq 0$ with $\{a_k\}$ as in Example 2 and if

$$\lim_{|z| \to \infty}\left|\frac{g'(z)}{zg(z)}\right| = 0$$

show that

$$g(z) = g(0)e^{zg'(0)/g(0)}\prod_{k=1}^{\infty}\left(1 - \frac{z}{a_k}\right)e^{z/a_k}$$

Ans. The logarithmic derivative of $g(z)$ has simple poles of residue unity at $z = a_k$ and satisfies the limit condition of Example 2 with $p = 0$. Hence,

$$d_z[\log g(z)] = \frac{g'(z)}{g(z)} = \frac{g'(0)}{g(0)} + \sum_{k=1}^{\infty}\left(\frac{1}{z - a_k} + \frac{1}{a_k}\right)$$

and integrating from 0 to z along a path avoiding the poles, one has

$$\log\left[\frac{g(z)}{g(0)}\right] = z\frac{g'(0)}{g(0)} + \sum_{k=1}^{\infty}\left\{\log\left(1 - \frac{z}{a_k}\right) + \frac{z}{a_k}\right\}$$

which gives the desired result by exponentiation.

5. Find an infinite product expansion for $\sin \pi z$.

Ans. $g(z) = \sin \pi z/\pi z$ is entire with simple zeros at the nonzero integers. Its logarithmic derivative $[\pi \cot \pi z - (1/z)]$ satisfies

$$\lim_{|z| \to \infty}\frac{1}{|z|}\left|\pi \cot \pi z - \frac{1}{z}\right| = 0$$

on a sequence of contours avoiding the poles. Thus Example 4 is applicable and $g(0) = 1$, $g'(0) = 0$, so

$$\frac{\sin \pi z}{\pi z} = \prod_{n \neq 0} \left(1 - \frac{z}{n}\right) e^{z/n}$$

or, pairing terms for $\pm n$,

$$\frac{\sin \pi z}{\pi z} = \prod_{n=1}^{\infty} \left(1 - \frac{z^2}{n^2}\right)$$

6. Construct a product representation for the transfer impedance of an acausal filter for pulse compression.

Ans. The desired output is a pulse $y(t) = u(T - 2|t|)$ with $u(t)$ the unit step. The input is $x(t) = u(NT - 2|t|)$. The ratio of their Fourier transforms is the system function $H(\omega)$:

$$H(\omega) = \frac{Y(\omega)}{X(\omega)} = \frac{\sin(\omega T/2)}{\sin(N\omega T/2)}$$

The original pulse width is N times the final pulse width. A product representation for $H(\omega)$ is obtained from products for the sines ($\omega = 2\pi\nu$):

$$H(\omega) = \prod_{k=1}^{\infty} \left(\frac{k^2 - (\nu T)^2}{k^2 - (N\nu T)^2}\right)$$

To the extent that a partial product of this is an approximant, a finite sequence of simpler functions may be multiplied together to yield $H(\omega)$ approximately. A finite product representing $H(\omega)$ exactly is given by

$$H(\omega) = 2^{-m} \prod_{k=0}^{m-1} \sec(2^{k-1}\omega T)$$

for $N = 2^m$. Extending to complex variables $s = \sigma + i\omega$, one has

$$H(s) = 2^{-m} \prod_{k=0}^{m-1} \operatorname{sech}(2^{k-1} sT)$$

The typical element is then $z_T = \operatorname{sech}(qs)$ with $q = 2^{k-1}T$, which may be constructed from a symmetrical lattice network (as shown in Fig. 10.1). Such a filter design is acausal because it ignores the requirement that output should follow input.

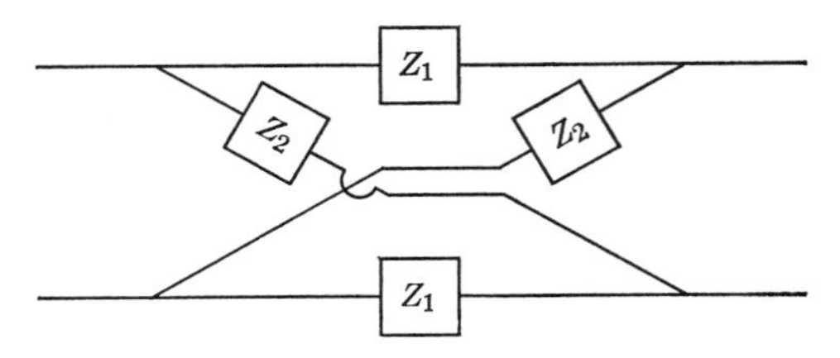

Figure 10.1. Symmetrical lattice network with $Z_1 = [\tanh(qs/2)]^2$ and $Z_2 = \coth(qs/2)$.

• PROBLEMS 10

1. Deduce the infinite product for the $\sin \pi z$ from that for $\Gamma(z)$ and the identity $\Gamma(z)\Gamma(1 - z) = \pi \csc \pi z$.

Find a partial fraction expansion for the following:

2. $\cot z$
3. $\csc z$
4. $\tan z$
5. the logarithmic derivative of $\Gamma(z)$
6. $\csc^2 z$

Find an infinite product representation for the following:

7. $\cos z^2$
8. $\tan \pi z$
9. $\sinh 2\pi z$
10. the beta function $B(z,w)$

CHAPTER 11:

Dispersion Relations

1.1 HILBERT TRANSFORMS

The Cauchy–Riemann equations (2.14) establish a connection between the real and imaginary parts of an analytic function $f(z) = u + iv$. Thus the partial derivatives of either u or v are determined by (2.14) as soon as the other function v or u is known. Furthermore, the value of u or v could be changed by a constant without affecting (2.14), so one cannot generally expect to determine the value of u from a knowledge of v except to within an additive constant. The same holds for v.

In the physical literature the formulae expressing the real part of an analytic function in terms of its imaginary part and vice versa are known as *dispersion relations*. As an illustration of the kind of problem which can arise, suppose the value of $f(z) = u + iv$ is given along the real axis and also that

$$\lim_{|z|\to\infty} |z^\delta f(z)| = 0 \tag{11.1}$$

for a small positive number δ and $\operatorname{Im} z \geq 0$. Then according to the Cauchy integral theorem

$$f(z) = \frac{1}{2\pi i} \oint \frac{f(w)\,dw}{w - z} \tag{11.2}$$

where $z = x + i\varepsilon (\varepsilon > 0)$ and the contour consists of the real axis and a (infinitely) large semicircle. The function $f(w)$ is supposed to be free of singularities in the upper half-plane including the real axis. The result (11.2) is unaltered if the original contour (Fig. 11.1a) is deformed to that of Figure 11.1b which includes a small semicircular deviation centered at $w = x$. If $\varepsilon \to 0$, the limiting location of z will be at $w = x$. If also the radius

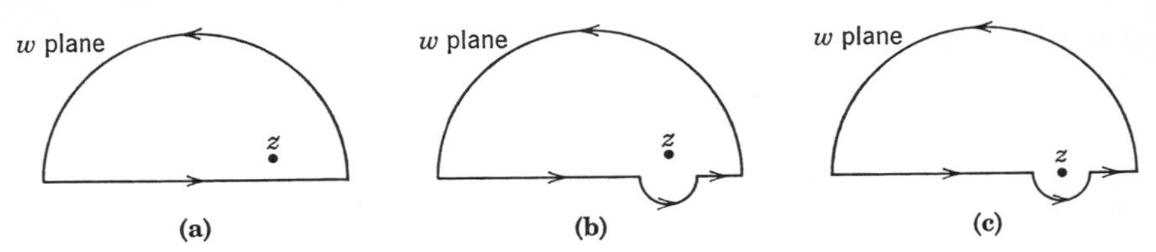

Figure 11.1. Relationship of Cauchy principal value to contour integral.

of the large semicircle becomes infinite, (11.1) ensures that the integral along the large arc will approach zero. Thus (11.2) becomes

$$f(x) = \frac{1}{2\pi i}\int_{-\infty}^{x-\rho} + \int_{x+\rho}^{\infty} \frac{f(w)\,dw}{w - x} + \frac{1}{2\pi i}\oint \frac{f(w)\,dw}{w - x} \tag{11.3}$$

where ρ is the radius of the small semicircular deviation. If the limit

$$\lim_{\rho\to 0}\left(\int_{-\infty}^{x-\rho} + \int_{x+\rho}^{\infty}\right) = P\int_{-\infty}^{\infty}$$

exists, it is called the *Cauchy principal value* (see also Section 6.4) and is often written as shown with a P before the integration sign. It can exist when the integral over an interval containing x might not exist, but whenever the latter does, it is equal to the Cauchy principal value which therefore represents a generalization of the notion of integral. The convention of writing a P before the integral will *not* be adhered to here and the principal value will be understood in all cases where the integral does not exist. The infinite limits here have nothing to do with the question; it is the neighborhood of the point x which is potentially dangerous. The range extirpated $(x - \rho, x + \rho)$ before the passage to the limit $\rho \to 0$ is always understood to be *symmetrical* about the point x. A simple example of an integral which does *not* exist but possesses a Cauchy principal value is (exceptionally we include the P)

$$P\int_{-1}^{2}\frac{dx}{x^3} = \lim_{\rho\to 0}\int_{-1}^{-\rho} + \int_{\rho}^{2}\frac{dx}{x^3} = \frac{3}{8}$$

Notice that if the range excluded were not symmetrical, say $(-\rho, 2\rho)$, the limit would *not* exist.

Returning now to (11.3), we may write as $\rho \to 0$

$$f(x) = \frac{1}{2\pi i}\int_{-\infty}^{\infty}\frac{f(w)\,dw}{w - x} + \frac{f(x)}{2}$$

where the "integral" is a Cauchy principal value (assumed to exist) and the last term is obtained from the limit of the last integral of (11.3) as $\rho \to 0$.

The result may be rewritten as

$$f(x) = \frac{1}{\pi i} \int_{-\infty}^{\infty} \frac{f(w)\, dw}{w - x} \tag{11.4}$$

or, decomposing $f(x)$ into real and imaginary parts,

$$u(x) = \frac{1}{\pi} \int_{-\infty}^{\infty} \frac{v(w)\, dw}{w - x} \tag{11.5}$$

$$v(x) = -\frac{1}{\pi} \int_{-\infty}^{\infty} \frac{u(w)\, dw}{w - x} \tag{11.6}$$

Two functions u and v satisfying (11.5) and (11.6) are called *Hilbert transforms* of each other. These relations then show how one can compute the real part of $f(z)$ from a knowledge of its imaginary part and vice versa. They should thus be viewed as an integrated version of the Cauchy–Riemann equations. Note also that v must be known almost everywhere on the real axis to find u at any particular point there and vice versa.

The particular form of the integral relation between u and v will depend on the curve along which values of one or the other are given. In this case it was the real axis. It might well have been a circle or other contour. A most important feature of such formulae as (11.5) and (11.6) is that a *local* knowledge of v for a portion of the integration range does *not* suffice to determine u and vice versa.

Generally it should be clearly understood that the possibility of determining the imaginary part of an analytic function to within an additive constant from a knowledge of its real part precludes the independent assignment of *both* parts on a closed contour, since if u and v were *both* arbitrary, they would generally contradict the Cauchy–Riemann equations.

11.2 PLEMELJ FORMULAE

The integral

$$F(z) = \frac{1}{2\pi i} \int_L \frac{f(t)\, dt}{t - z}$$

where L is a simple smooth non-self-intersecting (open) curve and $f(t)$ is bounded and integrable on L, is an analytic function of z free of singularities in the portion of the z plane exterior to L. Also $\lim_{|z|\to\infty} F(z) = 0$. However, this function $F(z)$ does not generally have the same limiting values as z approaches a point of L from *opposite* sides. For this reason $F(z)$ is said to be (sectionally analytic) sectionally holomorphic. To see that this is so one need only divide L into a portion L' exterior to a small circle of radius ε and center τ (on L) and a portion l (of L) interior to this circle. Then

$$F(z) = \frac{1}{2\pi i}\int_{L'} \frac{f(t)\,dt}{t - z} + \frac{1}{2\pi i}\int_{l} \frac{f(t) - f(\tau)\,dt}{t - z} + \frac{f(\tau)}{2\pi i}\int_{l} \frac{dt}{t - z}$$

If $f(t)$ is differentiable at $t = \tau$, the second integral will approach zero as $z \to \tau$ from either side of L and as $\varepsilon \to 0$ (subsequently). The last integral is the change in $\log(t - z)$ as t traverses l. Figure 11.2 shows that for z close to L the change in the log will approximate $\pm\pi i$ since $\log|t - \tau|$ will *not* change (being initially and finally $\log \varepsilon$) while the $\arg(t - z)$ will increase for z to the left of L and decrease for z to the right of L. Thus the limiting values of $F(z)$, which are designated by $F^+(\tau)$ from the left and $F^-(\tau)$ from the right, are

$$F^+(\tau) = \frac{1}{2\pi i}\int_L \frac{f(t)\,dt}{t - \tau} + \frac{f(\tau)}{2} \tag{11.7}$$

$$F^-(\tau) = \frac{1}{2\pi i}\int_L \frac{f(t)\,dt}{t - \tau} - \frac{f(\tau)}{2} \tag{11.8}$$

The integrals are understood to be Cauchy principal values. These are the *Plemelj formulae*, and it is understood that τ is an *interior* point of the curve

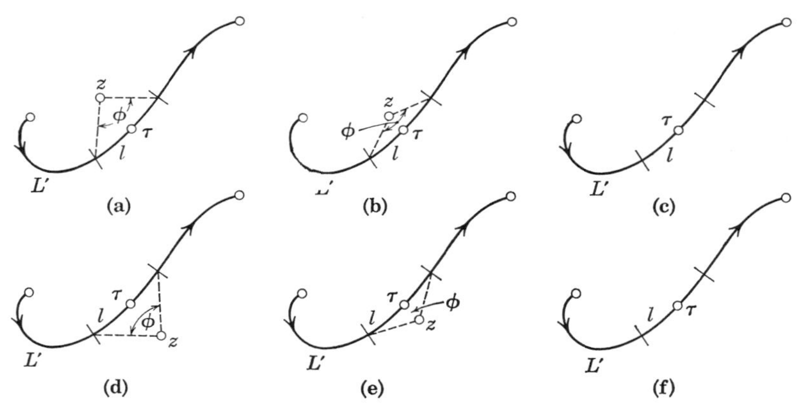

Figure 11.2. Discontinuity of $F(z)$ along arc L: ϕ = angle subtended by l at z, z = point not on arc L approaching τ, τ = point on L approached by z, l = portion of L included in circle of radius ε about τ, L' = portion of L remaining after l is removed from L. Parts (a), (b), and (c) show sequence of configurations as $z \to \tau$ from left; parts (d), (e), and (f) show sequence as $z \to \tau$ from right

$$\varepsilon \to 0 \begin{cases} \text{after (c)} & \phi \to \pi \\ \text{after (f)} & \phi \to -\pi \end{cases}$$

L. The condition that $f(t)$ be differentiable can be replaced by

$$|f(t) - f(\tau)| \leq K\,|t - \tau|^{\mu} \tag{11.9}$$

where $0 < \mu \leq 1$, which is called a *Hölder condition*. It is important in the limiting processes to let $\varepsilon \to 0$ *after* z has attained L.

Clearly the derivation applies also to the case where L is a closed contour C. Also, since for a closed contour

$$\frac{1}{2\pi i}\oint \frac{dt}{t-\tau} = \frac{1}{2\pi i}\oint d\log(t-\tau) = \tfrac{1}{2} \tag{11.10}$$

(where the integral is a Cauchy principal value), the Plemelj formulae become, for a closed contour,

$$F^{+}(\tau) = \frac{1}{2\pi i}\oint \left[\frac{f(t) - f(\tau)}{t-\tau}\right] dt + f(\tau) \tag{11.11}$$

$$F^{-}(\tau) = \frac{1}{2\pi i}\oint \left[\frac{f(t) - f(\tau)}{t-\tau}\right] dt \tag{11.12}$$

The difference and sum of the Plemelj formulae are

$$F^{+}(\tau) - F^{-}(\tau) = f(\tau) \tag{11.13}$$

$$F^{+}(\tau) + F^{-}(\tau) = \frac{1}{\pi i}\int_L \frac{f(t)\,dt}{t-\tau} \tag{11.14}$$

Examples of $F(z)$, $f(t)$ pairs are given in Table 11.1.

Table 11.1. Sectionally holomorphic $F(z)$ versus discontinuity $f(t)$

$f(t)$	$2\pi i F(z)$†
1	$\log\left(\dfrac{z-b}{z-a}\right)$
t	$(b-a) + z\log\left(\dfrac{z-a}{z-b}\right)$
t^k	$\displaystyle\sum_{\lambda+\mu=k-1}\left(\frac{b^{\lambda+1} - a^{\lambda+1}}{\lambda+1}\right)z^{\mu} + z^k \log\left(\frac{z-b}{z-a}\right)$
$1/t$	$\dfrac{1}{z}\log\left[\dfrac{a(z-b)}{b(z-a)}\right]$

† Here, a, b are end points of L.

Alternatively the properties of $F(z)$ may be derived from the Stieltjes integrals

$$2\pi i F(z) = \int_L f(t)\, d \log(t - z)$$

$$2\pi i F(z) = f(b)\log(b - z) - f(a)\log(a - z) - \int_L \log(t - z)\, df(t)$$

11.3 CROSSING RELATIONS AND SUM RULES

In case the Fourier transform of $f(x)$ is *real*, the condition

$$f(x) = f^*(-x) \tag{11.15}$$

for all real x must be satisfied. This case occurs very commonly and implies that

$$\begin{aligned} u(x) &= u(-x) \\ v(x) &= -v(-x) \end{aligned} \tag{11.16}$$

so that $\operatorname{Re} f(x)$ is even and $\operatorname{Im} f(x)$ is odd. In scattering problems, conditions (11.16) are called *crossing relations*. Using these in (11.5) and (11.6), the range of integration can be reduced to the positive real axis. Thus,

$$u(x) = \frac{1}{\pi}\int_0^\infty v(w)\left[\frac{1}{w - x} + \frac{1}{w + x}\right] dw$$

$$v(x) = \frac{1}{\pi}\int_0^\infty u(w)\left[\frac{1}{w + x} - \frac{1}{w - x}\right] dw$$

so that

$$u(x) = \frac{2}{\pi}\int_0^\infty \frac{wv(w)\, dw}{w^2 - x^2} \tag{11.17}$$

$$v(x) = \frac{-2x}{\pi}\int_0^\infty \frac{u(w)\, dw}{w^2 - x^2} \tag{11.18}$$

These imply certain asymptotic relations between $u(x)$ and $v(x)$ for large x. Thus,

$$\lim_{x\to\infty} x^2 u(x) = \frac{-2}{\pi}\int_0^\infty wv(w)\, dw$$

$$\lim_{x\to\infty} xv(x) = \frac{2}{\pi}\int_0^\infty u(w)\, dw$$

so that for large x

$$u(x) \sim \frac{-2}{\pi x^2} \int_0^\infty w v(w)\, dw \tag{11.19}$$

$$v(x) \sim \frac{2}{\pi x} \int_0^\infty u(w)\, dw \tag{11.20}$$

Also from (11.17) and (11.18) for $x = 0$,

$$u(0) = \frac{2}{\pi} \int_0^\infty \frac{v(w)}{w}\, dw \tag{11.21}$$

$$\lim_{x \to 0} \frac{v(x)}{x} = \frac{-2}{\pi} \int_0^\infty \frac{u(w)}{w^2}\, dw$$

so that for small x

$$v(x) \sim \frac{-2x}{\pi} \int_0^\infty \frac{u(w)}{w^2}\, dw \tag{11.22}$$

Relations such as (11.19) and (11.20) are often called *sum rules* in quantum mechanics.

11.4 SUBTRACTION PROCEDURE

The assumption that the Cauchy principal values (11.5) and (11.6) exist may not be fulfilled due to failure of the convergence for large values of w. In such cases it is convenient to consider

$$u(x) - u(0) = \frac{x}{\pi} \int_{-\infty}^{\infty} \frac{v(w)\, dw}{w(w - x)} \tag{11.23}$$

which for any fixed value of x has improved convergence properties, since w occurs quadratically in the denominator of the integrand. If in spite of this improvement the integral in (11.23) does not converge, one can consider

$$\frac{u(x) - u(0)}{x} - u'(0) = \frac{x}{\pi} \int_{-\infty}^{\infty} \frac{v(w)\, dw}{w^2(w - x)} \tag{11.24}$$

This process is called "*subtraction*" and may be expected to lead eventually to a "convergent" Cauchy principal value if it is repeated often enough. Formally one has

$$u(x) = \frac{1}{\pi} \int_{-\infty}^{\infty} \frac{v(w)}{w} \sum_{k=0}^{\infty} \left(\frac{x}{w}\right)^k dw = \sum_{k=0}^{\infty} \frac{u^{(k)}(0) x^k}{k!}$$

The MacLaurin expansion of $u(x)$ is

$$u(x) - \sum_{k=0}^{n-1} \frac{u^{(k)}(0) x^k}{k!} = \sum_{k=n}^{\infty} \frac{u^{(k)}(0) x^k}{k!} \tag{11.25}$$

But, from (11.5),

$$u^{(k)}(x) = \frac{k!}{\pi} \int_{-\infty}^{\infty} \frac{v(w)\, dw}{(w - x)^{k+1}}$$

and

$$u^{(k)}(0) = \frac{k!}{\pi} \int_{-\infty}^{\infty} \frac{v(w)\, dw}{w^{k+1}}$$

so that, substituting into (11.25),

$$u(x) - \sum_{k=0}^{n-1} \frac{u^{(k)}(0)x^k}{k!} = \frac{1}{\pi} \int_{-\infty}^{\infty} \frac{v(w)}{w} \sum_{k=n}^{\infty} \left(\frac{x}{w}\right)^k dw$$

and

$$\frac{u(x) - \sum_{k=0}^{n-1} [u^{(k)}(0)x^k/k!]}{x^n} = \frac{1}{\pi} \int_{-\infty}^{\infty} \frac{v(w)\, dw}{w^n(w - x)} \tag{11.26}$$

which is then the result of n subtractions and will converge if $v(w) \sim w^{n-\varepsilon}$ as $|w| \to \infty$.

11.5 KRAMERS–KRONIG DISPERSION FORMULAE

The *Kramers–Kronig dispersion relation* between electrical permittivity ε and conductivity σ (or alternatively between absorption and phase shift) for an electromagnetic wave in a medium will now be derived.

According to Maxwell's theory of electromagnetic wave propagation the electric intensity $\vec{E}$ satisfies the wave equation (obtained by eliminating the magnetic intensity between the Faraday induction law and the Ampere circuital law including displacement current). According to the theory of Fourier series a plane wave of electric intensity may be regarded as a superposition of terms

$$\vec{E} = \vec{E}_0 e^{-i\omega[t-(ns/c)]} \tag{11.27}$$

where

$$\begin{aligned}
c &= \text{phase velocity of wave in vacuum} \\
s &= \text{distance in direction of propagation} \\
t &= \text{time} \\
\nu &= \omega/2\pi = \text{frequency} \\
\vec{E}_0 &= \text{amplitude} \\
n &= \text{index of refraction} \\
c/n &= \text{phase velocity of wave in medium} \\
n^2 &= \varepsilon\mu \approx \varepsilon = \text{electric permittivity} \\
1 \approx \mu &= \text{magnetic permeability}
\end{aligned}$$

Under the assumptions

$$\vec{E} = \vec{E}_0 e^{-i\omega[t-(ns/c)]}$$
$$\vec{H} = \vec{H}_0 e^{-i\omega[t-(ns/c)]}$$
$$\vec{J} = \sigma\vec{E} \qquad \text{Ohm's law}$$

Maxwell's equations may be written

$$\nabla \times \vec{E} = i\omega\mu\vec{H}$$
$$\nabla \times \vec{H} = (\sigma - i\omega\varepsilon)\vec{E}$$
$$\varepsilon\nabla \cdot \vec{E} = \rho$$
$$\nabla \cdot \vec{H} = 0$$

where

ρ = charge density = charge/volume
σ = conductivity
$\vec{J}$ = current density = current/area
$\vec{E}$ = electric intensity
$\vec{H}$ = magnetic intensity

In the second Maxwell equation above σ and ε are supposed to be real quantities, but it is seen that one may combine them into one complex quantity $(\sigma - i\omega\varepsilon) = \sigma^+$ interpreted as a *complex conductivity* or determined by a *complex permittivity* $\varepsilon^+ = \varepsilon + (i\sigma/\omega)$. The use of such a complex quantity permits one to describe simultaneously the properties of convective and displacement currents. As long as the index of refraction n is real the exponential solutions represent *unattenuated* waves, but as soon as the permittivity ε becomes complex the index of refraction also becomes complex:

$$n^2 = \varepsilon + (i\sigma/\omega) = (u + iv)^2$$

so that

$$\operatorname{Re} n = \frac{\sqrt{\sqrt{\varepsilon^2 + (\sigma/\omega)^2} + \varepsilon}}{\sqrt{2}} = u$$
$$\operatorname{Im} n = \frac{\sqrt{\sqrt{\varepsilon^2 + (\sigma/\omega)^2} - \varepsilon}}{\sqrt{2}} = v \qquad (11.28)$$

and the wave experiences an *attenuation factor* in amplitude due to v:

$$e^{-\omega vs/c}$$

in addition to the phase shift due to u.

Although at low frequencies ε and σ can be treated as approximately independent of frequency, this is not true at high frequencies. Generally the medium can be supposed to consist of equal quantities of positive and negative charge. The effect of the externally imposed electric intensity is to increase the separation of unlike charges a vectorial measure of which is the polarization vector $\vec{P}$ (which may be defined as the discrepancy between the electric displacement in a vacuum and in the medium). If the frequency is high, the charge separation and consequently the polarization will not be able to change as rapidly as the currently applied field, and therefore it will depend upon *previous* values of the electric intensity. The simplest assumption that can be made to fit the observed facts is that the medium behaves like a filter, with input in the frequency domain as the Fourier transform of the electric intensity and output in the frequency domain as the Fourier transform of the polarization. If the system function or frequency response is $G(\omega)$, one has, upon transforming back to the time domain,

$$\vec{P}(t) = \varepsilon^{+}\vec{E}(t) - \vec{E}(t) = \int_0^{\infty} g(\tau)\vec{E}(t-\tau)\, d\tau \tag{11.29}$$

using the causality condition (9.136) so that the polarization does *not* precede the applied field $\vec{E}(t)$. Under the specific assumption that the time-dependence of $\vec{E}$ occurs via a factor of the form $e^{-i\omega t}$, (11.29) may be written in terms of ε^{+} to yield

$$\varepsilon^{+}(\omega) = 1 + \int_0^{\infty} g(\tau)e^{i\omega\tau}\, d\tau \tag{11.30}$$

the frequency-dependence of the permittivity in terms of the "impulse response of the medium" $g(\tau)$. Since $g(\tau) = g^{*}(\tau)$ is real, it is seen that although $\varepsilon^{+}(\omega)$ is complex, it satisfies

$$\varepsilon^{+*}(-\omega) = \varepsilon^{+}(\omega) \tag{11.31}$$

which implies the "crossing relations"

$$\varepsilon(-\omega) = \varepsilon(\omega) \tag{11.32}$$

$$\sigma(-\omega) = \sigma(\omega) \tag{11.33}$$

Naturally, at low frequencies one has

$$\varepsilon^{+} \approx \frac{i\sigma(0)}{\omega} \qquad \sigma(0) = \sigma_{DC}$$

for a metal (opaque) and

$$\varepsilon^{+} \approx \varepsilon(0) \qquad \varepsilon(0) = \varepsilon_{DC}$$

for a dielectric (transparent) medium.

If the consideration of (11.30) is extended to complex values of the frequency $\omega^{+} = x + iy$ and it is supposed that the Laplace transform of $|g(t)|$ exists, one concludes that $\varepsilon^{+}(\omega^{+})$ is analytic for $y > 0$. Also, $\varepsilon^{+}(\omega^{+})$

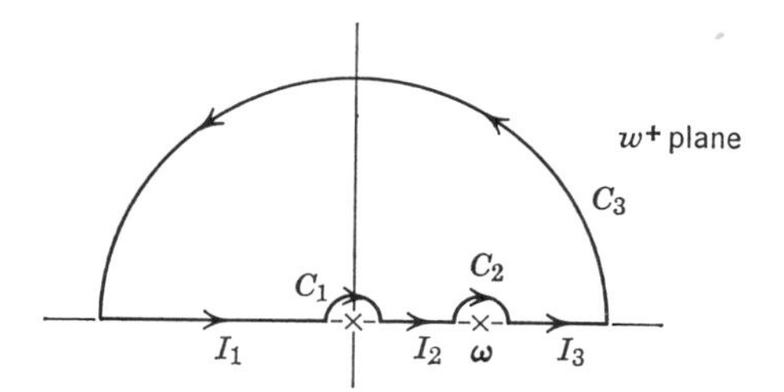

Figure 11.3. Contour of (11.34) for the case of a metal.

has no singularity on $y = 0$ except in the case of a metal when it has a simple pole of residue $i\sigma(0)$. Furthermore, $\varepsilon^{+*}(iy) = \varepsilon^{+}(iy)$ so that $\varepsilon^{+}(\omega^{+})$ is real on $x = 0$, as follows from $\varepsilon^{+*}(-\omega^{+*}) = \varepsilon^{+}(\omega^{+})$.

To establish the Kramers–Kronig relations one considers

$$\oint \left[\frac{\varepsilon^{+}(\omega^{+}) - 1}{\omega^{+} - \omega}\right] d\omega^{+} = 0 \tag{11.34}$$

about the contour shown in Figure 11.3 for the case of a metal. If there exists a positive number $\delta > 0$ such that $|\varepsilon^{+}(\omega^{+}) - 1| < |\omega^{+}|^{-\delta}$ as $|\omega^{+}| \to \infty$ for $\mathrm{Im}\, \omega^{+} \geq 0$, the integral on the large semicircle C_3 will approach zero. The sum of the three integrals along the real axis will approach the principal value of

$$\int_{-\infty}^{\infty} \left[\frac{\varepsilon^{+}(x) - 1}{x - \omega}\right] dx$$

while as the radii of the small semicircles approach zero the limit of the integral along C_1 is $-\pi\sigma(0)/\omega$ and the limit of the integral along C_2 is $-i\pi(\varepsilon^{+}(\omega) - 1)$. Hence,

$$\int_{-\infty}^{\infty} \left[\frac{\varepsilon^{+}(x) - 1}{x - \omega}\right] dx - \pi \frac{\sigma(0)}{\omega} - i\pi[\varepsilon^{+}(\omega) - 1] = 0$$

Since $\varepsilon^{+}(\omega^{+}) = \varepsilon(\omega^{+}) + [i\sigma(\omega^{+})/\omega^{+}]$, the real and imaginary parts of the above yield

$$\frac{\sigma(\omega)}{\omega} = -\frac{1}{\pi}\int_{-\infty}^{\infty} \left[\frac{\varepsilon(x) - 1}{x - \omega}\right] dx + \frac{\sigma(0)}{\omega} \tag{11.35}$$

$$\varepsilon(\omega) - 1 = \frac{1}{\pi}\int_{-\infty}^{\infty} \frac{\sigma(x)\, dx}{x(x - \omega)} \tag{11.36}$$

for the case of a metal.

For the case of a dielectric

$$\frac{\sigma(\omega)}{\omega} = -\frac{1}{\pi}\int_{-\infty}^{\infty}\left[\frac{\varepsilon(x)-1}{x-\omega}\right] dx \tag{11.37}$$

$$\varepsilon(\omega) - 1 = \frac{1}{\pi}\int_{-\infty}^{\infty}\left[\frac{\sigma(x)\,dx}{x(x-\omega)}\right] \tag{11.38}$$

which could also have been obtained directly from (11.5) and (11.6). Using the crossing relations (11.32) and (11.33), one has for a dielectric

$$\sigma(\omega) = \frac{2\omega^2}{\pi}\int_{0}^{\infty}\left[\frac{\varepsilon(x)-1}{\omega^2-x^2}\right] dx \tag{11.39}$$

$$\varepsilon(\omega) - 1 = \frac{2}{\pi}\int_{0}^{\infty}\frac{\sigma(x)\,dx}{x^2-\omega^2} \tag{11.40}$$

while for a metal $\sigma(0)$ must be subtracted from the left side of (11.39). These are the Kramers–Kronig relations.

Since $\varepsilon'(\omega)$ and $[\sigma(\omega)/\omega]'$ are also Hilbert transform pairs, it follows from the Parseval relation (see Example 4) that if $\varepsilon'(\omega) \neq 0$, for any interval of frequencies, $\sigma(\omega)$ cannot be zero for all frequencies. Thus one concludes that *there must be absorption in a dispersive medium.* Actually if $\sigma(\omega^+) = 0$, the Cauchy–Riemann conditions applied to $\varepsilon^+(\omega^+)$ result in the conclusion that $\varepsilon(\omega^+)$ is constant; hence the medium is dispersionless.

Equation (11.40) indicates that $\varepsilon(\omega) \to 1$ as $\omega \to \infty$ so that both dielectrics and metals are seen to become transparent at sufficiently high frequencies. Actually, many dielectrics are transparent at optical frequencies.

● EXAMPLES 11

1. Show that the Hilbert transform

$$u(x) = \frac{1}{\pi}\int_{-\infty}^{\infty}\frac{v(y)\,dy}{y-x}$$

can be considered as a repeated Laplace transform.

Ans. Consider the bilateral Laplace transform

$$V(s) = \int_{-\infty}^{\infty} e^{-st}v(-t)\,dt$$

$$V(s) = \int_{-\infty}^{\infty} e^{sy}v(y)\,dy$$

Then

$$\int_0^\infty e^{-sx}V(s)\,ds = \int_{-\infty}^\infty \frac{v(y)}{x-y}\,dy = -\pi u(x)$$

2. Find the Hilbert transform $v(x)$ if $u(x) = \delta(x)$ formally.

Ans. $v(x) = \dfrac{1}{\pi}\displaystyle\int_{-\infty}^\infty \frac{\delta(y)\,dy}{y-x} = -\frac{1}{\pi x}$

and since

$$u(x) = \frac{1}{\pi}\int_{-\infty}^\infty \frac{v(y)\,dy}{x-y}$$

one concludes (formally)

$$\delta(x) = \frac{1}{\pi^2}\int_{-\infty}^\infty \frac{dy}{y(y-x)}$$

3. Show that (formally)

$$\delta(a-b) = \frac{1}{\pi^2}\int_{-\infty}^\infty \frac{dt}{(t-a)(t-b)}$$

Ans. $\delta(a-b) = \dfrac{1}{\pi^2}\displaystyle\int_{-\infty}^\infty \frac{dy}{y(y-a+b)}$

and the result follows with $t = a - y$.

4. Show that

$$\int_{-\infty}^\infty |u(x)|^2\,dx = \int_{-\infty}^\infty |v(x)|^2\,dx$$

the Parseval relation, holds for Hilbert transforms.

Ans. $|u(x)|^2 = \dfrac{1}{\pi^2}\displaystyle\int_{-\infty}^\infty \frac{v(y)\,dy}{y-x}\int_{-\infty}^\infty \frac{v(s)\,ds}{s-x}$

$$\int_{-\infty}^\infty |u(x)|^2\,dx = \frac{1}{\pi^2}\int_{-\infty}^\infty\int_{-\infty}^\infty v(y)v(s)\int_{-\infty}^\infty \frac{dx}{(y-x)(s-x)}\,dy\,ds$$

$$\int_{-\infty}^\infty |u(x)|^2\,dx = \int_{-\infty}^\infty\int_{-\infty}^\infty v(y)v(s)\delta(y-s)\,dy\,ds = \int_{-\infty}^\infty |v(y)|^2\,dy$$

Strictly, the relation only has meaning if u and v are of integrable square modulus.

5. Find dispersion relations for the impedance of a parallel *RLC* circuit subjected to a voltage $e(t)$.

Ans. The impedance $Z(\omega)$ of the circuit is given by

$$\frac{1}{Z(\omega)} = \frac{1}{i\omega L} + \frac{1}{R} + i\omega C$$

$$J(\omega) = \frac{1}{2\pi}\int_{-\infty}^{\infty} e^{-i\omega t} j(t)\, dt$$

$$E(\omega) = \frac{1}{2\pi}\int_{-\infty}^{\infty} e^{-i\omega t} e(t)\, dt$$

are the Fourier transforms of the current $j(t)$ and voltage $e(t)$, $E(\omega) = Z(\omega)J(\omega)$

The impedance has two poles in the upper half complex ω plane at

$$\omega = i\left[\frac{1}{2RC} \pm \sqrt{\frac{1}{(2RC)^2} - \frac{1}{LC}}\right]$$

corresponding to positive resistance ($R > 0$). For $(2RC)^2 \leq LC$ they are located on the imaginary axis, and for $(2RC)^2 > LC$ they are located symmetrically on each side of the imaginary axis. For $(2RC)^2 = LC$ they coalesce to a double pole. Otherwise they are simple poles. Since the capacitative reactance predominates at high frequencies, $|\omega|\,|Z(\omega)|\,C \to 1$ as $|\omega| \to \infty$. Thus, invoking Cauchy's integral theorem

$$Z(\omega) = \frac{1}{2\pi i}\oint \frac{Z(w)\, dw}{w - \omega}$$

along a contour conjugate to that of Figure 11.1c (i.e., completed with semicircular arc in the *lower* half w plane), one has

$$-2\pi i Z(\omega) = \int_{-\infty}^{\infty} \frac{Z(w)\, dw}{w - \omega} - \oint \frac{Z(w)\, dw}{w - \omega}$$

$$-2\pi i Z(\omega) = \int_{-\infty}^{\infty} \frac{Z(w)\, dw}{w - \omega} - \pi i Z(\omega)$$

as in (11.3). Hence,

$$-Z(\omega) = \frac{1}{\pi i}\int_{-\infty}^{\infty} \frac{Z(w)\, dw}{w - \omega}$$

with the integral as usual a Cauchy principal value. For real ω, then, with

$$Z(\omega) = P(\omega) + iX(\omega)$$

decomposed into real part $P(\omega)$ (resistance) and imaginary part $X(\omega)$ (reactance), one has

$$-P(\omega) = \frac{1}{\pi}\int_{-\infty}^{\infty}\frac{X(w)\,dw}{w-\omega}$$

$$X(\omega) = \frac{1}{\pi}\int_{-\infty}^{\infty}\frac{P(w)\,dw}{w-\omega}$$

6. Since $e(t)$ and $j(t)$ are real, $E^*(-\omega) = E(\omega)$ and $J^*(-\omega) = J(\omega)$. From this derive the *crossing relations* and express the integral relationship between the resistance and reactance of the $\|RLC$ circuit in terms of integrations over *positive* frequencies *only*.

   Ans. The reality of $e(t)$ and $j(t)$ imply $Z^*(-\omega) = Z(\omega)$ so that

$$P(\omega) = P(-\omega) \qquad X(\omega) = -X(-\omega)$$

   are the crossing relations. Substitution into the dispersion relations then yields, as in (11.17) and (11.18),

$$X(\omega) = \frac{2\omega}{\pi}\int_0^{\infty}\frac{P(w)\,dw}{w^2-\omega^2}$$

$$P(\omega) = -\frac{2}{\pi}\int_0^{\infty}\frac{wX(w)\,dw}{w^2-\omega^2}$$

7. Deduce a sum rule for high-frequency reactance of the parallel RLC circuit in terms of resistance at all frequencies.

   Ans. Since $Z(\omega) = P(\omega) + iX(\omega)$, one has for large ω

$$P(\omega) = \frac{\omega^2RL^2}{R^2(1-\omega^2LC)^2+\omega^2L^2} \approx \frac{1}{\omega^2RC^2}$$

$$X(\omega) = \frac{\omega R^2L(1-\omega^2LC)}{R^2(1-\omega^2LC)^2+\omega^2L^2} \approx -\frac{1}{\omega C}$$

   and the sum rule becomes (for large ω in the expression for $X(\omega)$ in Example 6)

$$-\omega X(\omega) = \frac{1}{C} \approx \frac{2}{\pi}\int_0^{\infty}P(w)\,dw$$

● PROBLEMS 11

1. If $u(x)$ and $v(x)$ are Hilbert transform pairs, show that $u(x+k)$ and $v(x+k)$ and also $u(kx)$ and $v(kx)$ for $k > 0$ are transform pairs.

2. Show that if $u(x)$ and $v(x)$ are Hilbert transforms, so also are $u'(x)$ and $v'(x)$.

Show that the following are Hilbert transform pairs for $k > 0$:

3. $u(x) = k/(x^2 + k^2); \quad v(x) = x/(x^2 + k^2)$
4. $u(x) = ie^{ikx}; \quad v(x) = e^{ikx}$
5. $u(x) = \cos(kx); \quad v(x) = \sin(kx)$
6. $u(x) = J_1(kx)\cos(kx); \quad v(x) = J_1(kx)\sin(kx)$
7. Show that an alternative form of the Hilbert formulae is

$$u(x) = \frac{1}{\pi}\int_0^\infty \left[\frac{v(x+y) - v(x-y)}{y}\right] dy$$

$$v(x) = -\frac{1}{\pi}\int_0^\infty \left[\frac{u(x+y) - u(x-y)}{y}\right] dy$$

8. How is Example 2 related to the phenomenon of *anomalous dispersion*? Imagine $u(x)$ to be an even continuous nonnegative function with a single maximum and appreciable values only near $x = 0$.
9. Show that the differentials $du(x)$ and $dv(x)$ satisfy the same dispersion relations as $u(x)$ and $v(x)$. Do not assume that $df(z)$ is analytic.
10. By using the Cauchy integral theorem, establish the following formal rule ($\text{Im}\, a = 0$):

$$\lim_{\varepsilon\to 0} \frac{1}{x - (a + i\varepsilon)} = P\frac{1}{x - a} + i\pi\delta(x - a)$$

as a concise (but sloppy) representation of

$$\lim_{\varepsilon\to 0} \int_{-\infty}^{\infty} \frac{f(x)\,dx}{x - (a + i\varepsilon)} = P\int_{-\infty}^{\infty} \frac{f(x)\,dx}{x - a} + i\pi f(a)$$

Assume that $|f(z)| \to 0$ on the large semicircular completion arc.

11. Construct a sectionally holomorphic function $f(z)$ equal to unity inside a closed curve C and equal to zero outside C.
12. Construct a sectionally holomorphic function $F(z)$ with $F^-(\tau) = \tau^2$ and $F^+(\tau) = \tau^2 + 1$ on an arc L (ends a, b).
13. Construct a sectionally holomorphic function $F(z)$ with $F^-(\tau) = \tau^3$ and $F^+(\tau) = \tau^3 + (1/\tau)$ on an arc L (ends a, b).
14. Construct a sectionally holomorphic function $F(z)$ with $F^-(\tau) = \sin\tau$ and $F^+(\tau) = \sin\tau + \tau$.
15. How would one construct a sectionally holomorphic approximant to the nonanalytic function $e^{-|z|^2}$?
16. Derive the dispersion relations for a series *RLC* circuit.
17. Use the crossing relations for the series *RLC* circuit to express the integrations in terms of positive frequencies.
18. Find a sum rule expressing the inductance in terms of an integration over all frequencies.

CHAPTER 12:

Elliptic Functions and Integrals

.1 WEIERSTRASS ELLIPTIC FUNCTION $\wp(z)$

An analytic function $f(z)$ is called *periodic* if there exists a (complex) constant $a \neq 0$ such that

$$f(z + a) \equiv f(z) \tag{12.1}$$

for all values of z. The constant a is called a *period* of the function. If there does *not* exist an integer n such that a/n is also a period, then a is said to be a *fundamental period*. If for each two distinct periods $a_1 \neq a_2$ it is true that $\text{Im}(a_1^* a_2) = 0$, then $f(z)$ is called *simply periodic*, since the functional values will be repeated in only *one direction* of the z plane. If, however, there exist two distinct periods $a_1 \neq a_2$ such that $\text{Im}(a_1^* a_2) \neq 0$, then $f(z)$ is called *doubly periodic*, since the functional values will be repeated in the *two directions* of a_1 and a_2. Since there are no more than two independent directions in the z plane, it is clear that there are *no* triply periodic analytic functions but only simply or double periodic ones. Examples of simply periodic functions are $\sin z$, $\tan z$, e^z and $\cosh z$. The doubly periodic analytic functions are called *elliptic functions*, and examples of these will be constructed below.

One notes that if $g(z)$ is any analytic function for which the series

$$\sum_{n=-\infty}^{\infty} g(z - na) = f(z) \tag{12.2}$$

converges uniformly, then $f(z)$ will be periodic of period a, since replacing z by $z + a$ does not change the series because all integer multiples of a

already occur as shifts in the variable. The same is true of

$$g(z) + \sum_{n \neq 0} [g(z - na) - g(na)] = F(z)$$

That is, $F(z + a) = F(z)$.

Also, if Ω is a sum of any integer multiples of a_1 and a_2 with $\mathrm{Im}(a_1^* a_2) \neq 0$, then

$$\sum_{\Omega} g(z - \Omega) = p(z)$$

and the summation over *all* possible Ω yields $p(z)$, a doubly periodic function under the same conditions. Also,

$$g(z) + \sum_{\Omega \neq 0} [g(z - \Omega) - g(\Omega)] = P(z) \tag{12.3}$$

is doubly periodic. For the special case that $g(z) = 1/z^2$ the *Weierstrass elliptic function* $\wp(z)$ is obtained:

$$\wp(z) = \frac{1}{z^2} + \sum_{\Omega \neq 0} \left\{ \frac{1}{(z - \Omega)^2} - \frac{1}{\Omega^2} \right\} \tag{12.4}$$

It should be understood that $\wp(z)$ is only a definite function *after* the fundamental periods a_1 and a_2 have been fixed. Any particular choice yields [with $\mathrm{Im}(a_1^* a_2) \neq 0$] a double family of parallel lines dividing the z plane into a doubly infinite set of congruent *fundamental parallelograms* with sides of magnitudes $|a_1|$ and $|a_2|$ and angle given by $\sin \theta = \mathrm{Im}(a_1^* a_2)/|a_1 a_2|$. To generally discuss the properties of $\wp(z)$ without specifying a_1 and a_2 means that the class of all such possible functions is being treated. Since this class is almost as easily handled as any special function in it, this procedure will be followed.

Before entering into a discussion of the detailed properties of $\wp(z)$ it is useful to deduce some general properties of elliptic (doubly periodic) functions.

A nontrivial elliptic function *must* possess singularities within its fundamental parallelogram, for if it were bounded there, it would (by the double periodicity) be bounded everywhere, and hence it would be a constant by Liouville's theorem. By convention the term elliptic function excludes doubly periodic functions with essential singularities in a fundamental parallelogram.

The number of poles of an elliptic function in any fundamental parallelogram (each being counted according to its multiplicity) is called the *order* of the function. Since

$$f'(z) = \lim_{\varepsilon \to 0} \left[\frac{f(z + \varepsilon) - f(z)}{\varepsilon} \right]$$

$f(z + \Omega) = f(z)$ implies $f'(z + \Omega) = f'(z)$, and the derivative of an elliptic function is also an elliptic function. If $F(z)$ is the integral of an elliptic function $f(z)$, it is readily seen that

$$F(z + \Omega) - F(z) = \int_{z}^{z+\Omega} f(z)\, dz$$

is not generally zero, so that the integral of an elliptic function is not necessarily an elliptic function. The sum of the residues of an elliptic function within a fundamental parallelogram must be zero, since

$$\oint f(z)\, dz = 0 \tag{12.5}$$

because of the double periodicity of $f(z)$ producing cancellation on opposite sides. An elliptic function must be of at least second order since it could not have a single simple pole without violating the residue theorem (12.5).

The number of zeros (counting multiplicity) is equal to the number of poles (counting multiplicity) in a fundamental parallelogram for any elliptic function. This follows because if $f(z)$ is elliptic, so is $f'(z)/f(z)$ with the same double periodicity. Hence,

$$\oint \frac{f'(z)}{f(z)}\, dz = \oint d[\ln f(z)] = 0$$

where the last integral gives $2\pi i$ times the difference in number of poles and number of zeros by the principle of argument.

With this general information the specific properties of $\wp(z)$ can now be derived. First it will be shown that $\wp(z)$ satisfies the differential equations

$$6\wp^2 - \ddot{\wp} = 30k_4 \tag{12.6}$$

and

$$4\wp^3 - \dot{\wp}^2 - 60k_4\wp = 140k_6 \tag{12.7}$$

where $\dot{\wp}$ and $\ddot{\wp}$ are the first and second derivatives of $\wp$ and

$$k_4 = \sum_{\Omega \neq 0} \frac{1}{\Omega^4} \qquad k_6 = \sum_{\Omega \neq 0} \frac{1}{\Omega^6} \tag{12.8}$$

with the summations over all vertices of the doubly infinite replications of a fundamental parallelogram excepting only that at the origin.

To establish the second-order equation note that by definition

$$\wp(z) = \frac{1}{z^2} + \sum_{\Omega \neq 0} \left\{ \frac{1}{(z - \Omega)^2} - \frac{1}{\Omega^2} \right\}$$

Hence,

$$-\dot{\wp}(z) = 2\sum_{\Omega}\frac{1}{(z-\Omega)^3} = 2\left\{\frac{1}{z^3} + \sum_{\Omega\neq 0}\frac{1}{(z-\Omega)^3}\right\} \tag{12.9}$$

$$\ddot{\wp}(z) = 6\sum_{\Omega}\frac{1}{(z-\Omega)^4} = 6\left\{\frac{1}{z^4} + \sum_{\Omega\neq 0}\frac{1}{(z-\Omega)^4}\right\} \tag{12.10}$$

where summations $\sum_\Omega$ include the origin ($\Omega = 0$).

One notes that $\wp$ has a double pole at the origin and $\ddot{\wp}$ has a fourth-order pole at the origin, and so the fourth-order pole of $\wp^2$ at the origin can be eliminated by subtracting $\ddot{\wp}/6$ from $\wp^2$. Thus, writing (12.4) as

$$\wp(z) = \frac{1}{z^2} + \sum\nolimits_2 \tag{12.11}$$

one has

$$\wp^2 = \frac{1}{z^4} + \frac{2}{z^2}\sum\nolimits_2 + \sum\nolimits_2^2 \tag{12.12}$$

and for small z

$$\frac{1}{(z-\Omega)^2} \cong \frac{1}{\Omega^2}\left\{1 + 2\left(\frac{z}{\Omega}\right) + 3\left(\frac{z}{\Omega}\right)^2\right\} \tag{12.13}$$

and

$$\sum\nolimits_2 = \sum_{\Omega\neq 0}\left[\frac{1}{(z-\Omega)^2} - \frac{1}{\Omega^2}\right] \cong 2z\sum_{\Omega\neq 0}\frac{1}{\Omega^3} + 3z^2\sum_{\Omega\neq 0}\frac{1}{\Omega^4}$$

and since

$$\sum_{\Omega\neq 0}\frac{1}{\Omega^m} = 0 \qquad \text{for} \quad m = \text{odd integer}$$

$$\lim_{z\to 0}\sum\nolimits_2 = 0 \tag{12.14}$$

$$\lim_{z\to 0}\frac{1}{z^2}\sum\nolimits_2 = 3k_4 \tag{12.15}$$

$$\sum\nolimits_2 \cong 3k_4 z^2 \tag{12.16}$$

Hence by (12.10) and (12.12)

$$6\wp^2 - \ddot{\wp} = \frac{12}{z^2}\sum\nolimits_2 + 6\sum\nolimits_2^2 - 6\sum\nolimits_4 \tag{12.17}$$

where

$$\sum\nolimits_4 = \sum_{\Omega\neq 0}\frac{1}{(z-\Omega)^4} \tag{12.18}$$

and since $6\wp^2 - \ddot{\wp}$ is an elliptic function (the only possible singularities of which lie at the lattice points $z = \Omega$), it will follow from its double periodicity that it is in fact a constant if only it can be shown that

$$\lim_{z\to 0}[6\wp^2 - \ddot{\wp}]$$

exists. By (12.14), (12.15), and (12.18) the right side of (12.17) becomes (as $z \to 0$)

$$6\wp_0^2 - \ddot{\wp}_0 = 36k_4 - 6k_4 = 30k_4$$

and since $z = 0$ was the only eligible location for a singularity of $6\wp^2 - \ddot{\wp}$, it follows that

$$6\wp^2 - \ddot{\wp} = 30k_4$$

is this constant value for all z. Thus (12.6) is established. To prove (12.7), (12.6) may be multiplied by $2\dot{\wp}$ and integrated to yield

$$4\wp^3 - \dot{\wp}^2 - 60k_4\wp = \text{constant} \tag{12.19}$$

The constant may be determined by evaluating the limit of the left side as $z \to 0$, since it is the same for all values of z.

Using the abbreviation $\sum_3$, (12.9) may be written

$$-\dot{\wp} = 2\left\{\frac{1}{z^3} + \sum\nolimits_3\right\}$$

whence

$$\dot{\wp}^2 = 4\left\{\frac{1}{z^6} + \frac{2}{z^3}\sum\nolimits_3 + \sum\nolimits_3^2\right\} \tag{12.20}$$

while (12.11) yields

$$4\wp^3 = 4\left\{\frac{1}{z^6} + \frac{3}{z^4}\sum\nolimits_2 + \frac{3}{z^2}\sum\nolimits_2^2 + \sum\nolimits_2^3\right\} \tag{12.21}$$

so that (12.19) becomes

$$\frac{12}{z^4}\sum\nolimits_2 - \frac{8}{z^3}\sum\nolimits_3 - \frac{60k_4}{z^2} + \frac{12}{z^2}\sum\nolimits_2^2 + 4\sum\nolimits_2^3 - 4\sum\nolimits_3^2 - 60k_4\sum\nolimits_2$$

The last four terms approach zero as $z \to 0$ and the first two terms may be handled by noting that for small z

$$\frac{1}{(z-\Omega)^2} \cong \frac{1}{\Omega^2}\left\{1 + 2\left(\frac{z}{\Omega}\right) + 3\left(\frac{z}{\Omega}\right)^2 + 4\left(\frac{z}{\Omega}\right)^3 + 5\left(\frac{z}{\Omega}\right)^4\right\}$$

so that

$$\sum\nolimits_2 = \sum_{\Omega\neq 0}\left[\frac{1}{(z-\Omega)^2} - \frac{1}{\Omega^2}\right] \cong 3k_4z^2 + 5k_6z^4 \tag{12.22}$$

since

$$\sum_{\Omega\neq 0}\frac{1}{\Omega^3}=\sum_{\Omega\neq 0}\frac{1}{\Omega^5}=0$$

and

$$\frac{1}{(z-\Omega)^3}\cong-\frac{1}{\Omega^3}\left\{1+3\left(\frac{z}{\Omega}\right)+6\left(\frac{z}{\Omega}\right)^2+10\left(\frac{x}{\Omega}\right)^3\right\}$$

so that

$$-\sum_3=-\sum_{\Omega\neq 0}\frac{1}{(z-\Omega)^3}\cong 3k_4z+10k_6z^3 \tag{12.23}$$

Thus the constant in (12.19) becomes

$$\lim_{z\to 0}\left\{\frac{36k_4}{z^2}+60k_6+\frac{24k_4}{z^2}+80k_6-\frac{60k_4}{z^2}\right\}=140k_6$$

so that (12.19) may be written as

$$4\wp^3-\dot{\wp}^2-60k_4\wp=140k_6$$

and (12.7) is proved.

The integration of this equation yields

$$\int_\infty^\wp\frac{d\wp}{\sqrt{4\wp^3-60k_4\wp-140k_6}}=z \tag{12.24}$$

The symbol $\wp$ has already been defined as a function of z, so that this equation gives z as a function of $\wp$, i.e., the *inverse function* to the elliptic functions $\wp(z)$. This inverse function is called an *elliptic integral*. If the value of $\wp$ corresponding to a value of z is denoted by w, one has

$$w=\wp(z)$$

defined by (12.4)

$$z=\wp^{-1}(w)=\int_\infty^w\frac{dw}{\sqrt{4w^3-60k_4w-140k_6}} \tag{12.25}$$

Such an integral cannot be expressed in closed form in terms of algebraic, trigonometric, inverse trigonometric, logarithmic, and exponential (elementary) functions unless the cubic has repeated roots. In the latter case the integral is elementary:

$$\int\frac{dw}{2(w-w_1)\sqrt{w-w_2}}=\frac{1}{\sqrt{w_2-w_1}}\left[\arctan\sqrt{\frac{w-w_2}{w_2-w_1}}-\frac{\pi}{2}\right]$$

In the former case [nonrepeated roots of $(4w^3-60k_4w-140k_6)$] the value of the integral may be expressed in terms of the inverse of the Weierstrass elliptic $\wp$ function.

Similarly, the evaluation of any integral of the form

$$\int \frac{dw}{\sqrt{w^3 + aw^2 + bw + c}}$$

may be reduced to the evaluation of the previous integral by the substitution $w = \alpha v + \beta$ with α,β so chosen that the quadratic term in v^2 has a zero coefficient.

The criterion for repeated roots of

$$Q(w) = 4w^3 - 60k_4 w - 140k_6 = 0 \tag{12.26}$$

is that $Q'(w)$ should vanish for the same value of w. Thus,

$$Q'(w) = 12w^2 - 60k_4 = 0 \tag{12.27}$$

so that $w^2 = 5k_4$, which yields upon substitution into (12.26)

$$-2k_4 w = 7k_6$$

so that

$$4k_4^2(5k_4) = 20k_4^3 = 49k_6^2 \tag{12.28}$$

remains as the criterion for

$$\int \frac{dw}{\sqrt{Q(w)}}$$

to be an elementary integral of the form (12.25) rather than an elliptic integral.

More generally, any integral of the form

$$\int F[w,p(w)]\, dw \tag{12.29}$$

is called an *elliptic integral* if $p = aw^4 + bw^3 + cw^2 + dw + e$ is a cubic or quartic polynomial with distinct roots and $F[w,p]$ is a rational function of w and p, but not a rational function of w and p^2. Historically the terminology arose from the integral

$$\int \sqrt{\frac{a^2 - e^2x^2}{a^2 - x^2}}\, dx = a \int \sqrt{1 - e^2 \cos^2\theta}\, d\theta \tag{12.30}$$

representing the length of arc of an ellipse of eccentricity e, semimajor axis a, and equation

$$\frac{x^2}{a^2} + \frac{y^2}{a^2(1 - e^2)} = 1$$

for

$$x = a\cos\theta \qquad y = a\sqrt{1 - e^2}\sin\theta$$

The reader may refer to E. T. Copson's *Theory of Functions of a Complex Variable*, Oxford: Oxford University Press, 1935, p. 373, for a demonstration that any integral of the form (12.29) may be evaluated in terms of $\wp(z)$ and

$$\zeta(z) = -\int \wp(z)\, dz \tag{12.31}$$

$$\log \sigma(z) = \int \zeta(z)\, dz \tag{12.32}$$

Actually the integration of elliptic integrals may often be expressed more conveniently in terms of the Jacobian elliptic functions which will be introduced below.

Since

$$\wp(z) = \frac{1}{z^2} + \sum_{\Omega \neq 0} \left\{ \frac{1}{(z - \Omega)^2} - \frac{1}{\Omega^2} \right\}$$

for the case of a rectangular lattice, the Weierstrass elliptic function can be interpreted as a complex flow function for a quadrupole in a rectangular box centered about the origin.

The only singularity of $\wp(z)$ is a double pole at each lattice point $z = \Omega$ and the residue is naturally zero. Since $\wp(z) = \wp(-z)$ is even, $\dot{\wp}(z) = -\dot{\wp}(-z)$ is odd, and $\dot{\wp}(z + \Omega) = \dot{\wp}(z)$ is doubly periodic so that

$$\dot{\wp}\left(\frac{a_1}{2}\right) = -\dot{\wp}\left(-\frac{a_1}{2}\right) = -\dot{\wp}\left(\frac{a_1}{2}\right) = 0$$

$$\dot{\wp}\left(\frac{a_2}{2}\right) = -\dot{\wp}\left(-\frac{a_2}{2}\right) = -\dot{\wp}\left(\frac{a_2}{2}\right) = 0$$

Also, since

$$\dot{\wp}^2 = 4\wp^3 - 60k_4\wp - 140k_6$$

it is clear that the sum of the zeros of $\dot{\wp}$ must be zero as is the coefficient of $\wp^2$. Hence there must also be a zero of $\dot{\wp}$ at $z = (a_1 + a_2)/2$, and consequently also one at $z = (a_1 + a_2)/2$ because of the double periodicity. Hence,

$$\dot{\wp}\left(\frac{a_1}{2}\right) = \dot{\wp}\left(\frac{a_2}{2}\right) = \dot{\wp}\left(\frac{a_1 + a_2}{2}\right) = 0 \tag{12.33}$$

as indicated in Figure 12.1. The values taken by $\wp$ at these points are often designated by e_1, e_2, e_3:

$$e_1 = \wp\left(\frac{a_1}{2}\right) \qquad e_3 = \wp\left(\frac{a_2}{2}\right) \qquad e_2 = \wp\left(\frac{a_1 + a_2}{2}\right) \tag{12.34}$$

of which no two are equal; otherwise the differential equation (12.7) would

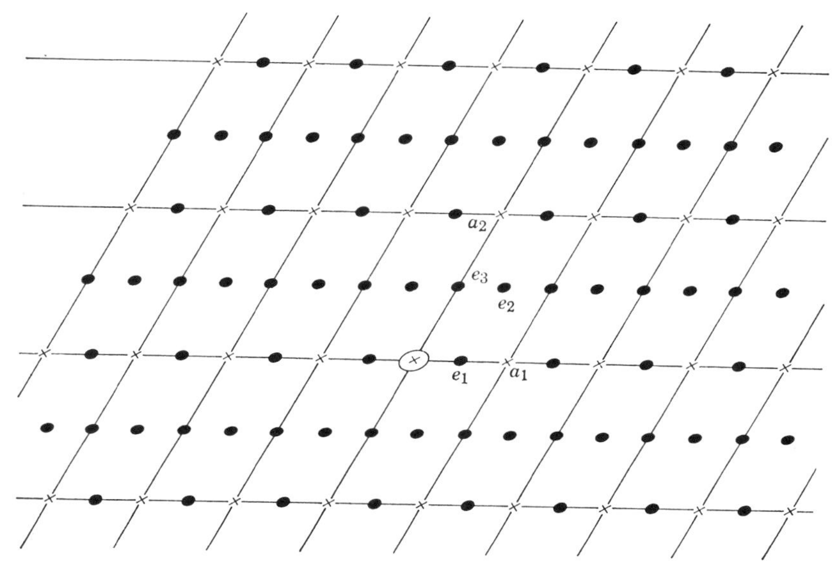

Figure 12.1. Zeros of $\dot{\wp}(z)$ are represented by ●; poles of $\wp(z)$ are represented by ×.

not integrate to $\wp(z)$ but to an elementary function. Also (12.7) can now be written

$$\dot{\wp}^2 = 4(\wp - e_1)(\wp - e_2)(\wp - e_3) \tag{12.35}$$

with

$$\begin{aligned} e_1 + e_2 + e_3 &= 0 \\ e_1e_2 + e_2e_3 + e_3e_1 &= -15k_4 \\ e_1e_2e_3 &= 35k_6 = e_G^3 \\ \frac{1}{e_1} + \frac{1}{e_2} + \frac{1}{e_3} &= -\frac{3k_4}{7k_6} = \frac{3}{e_H} \end{aligned} \tag{12.36}$$

where e_G and e_H are the (complex) geometric and harmonic means of e_1, e_2, e_3 and $e_H^3 + 4e_G^3 \neq 0$ in order that no double root occurs.

A conformal mapping associated with $\wp(z)$ can readily be obtained for the case where the lattice of poles of $\wp(z)$ is rectangular with $a_1 = a_1^*$ real and $a_2 = -a_2^*$ imaginary:

$$k_4 = \sum_{\Omega \neq 0} \frac{1}{\Omega^4} = \sum_{\Omega \neq 0} \frac{1}{(\Omega^*)^4} = k_4^*$$

$$k_6 = \sum_{\Omega \neq 0} \frac{1}{\Omega^6} = \sum_{\Omega \neq 0} \frac{1}{(\Omega^*)^6} = k_6^*$$

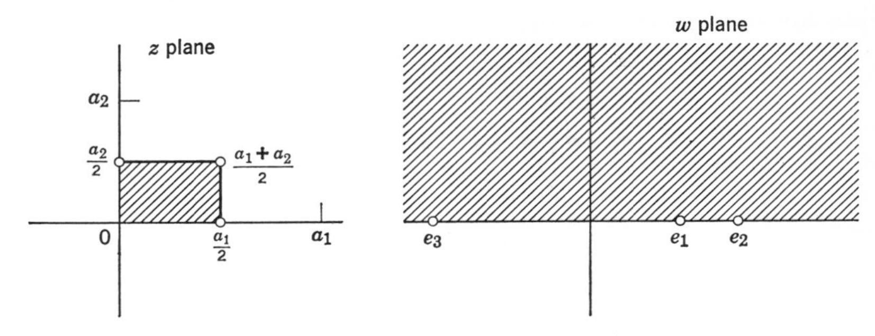

Figure 12.2. Mapping $w = \wp(z)$.

and if e_1, e_2, e_3 are *real* and distinct, $w = \wp(z)$ maps the rectangle

$$0 \leq 2\,\mathrm{Re}\, z < a_1 \qquad\qquad 0 \leq 2\,\mathrm{Im}\, z < -ia_2$$

onto the upper half-plane ($\mathrm{Im}\, w \geq 0$ is as shown in Fig. 12.2). This follows since then

$$\frac{dz}{dw} = 2[(w - e_1)(w - e_2)(w - e_3)]^{-1/2} \tag{12.37}$$

is a Schwarz–Christoffel mapping with interior angles $\pi/2$.

12.2 JACOBIAN ELLIPTIC FUNCTIONS sn z, cn z, dn z

As has been observed, $\wp(z)$ has as only singularities in the finite plane a double pole at each lattice point $z = \Omega$, and it takes the values e_1, e_2, e_3 at the zeros of $\dot{\wp}(z)$. The *Jacobian elliptic functions* expressed as sn z, cn z, and dn z in the notation of Gudermann may be defined as follows:

$$\begin{aligned} \mathrm{sn}\, z &= \left[\frac{e_1 - e_3}{\wp(z/\sqrt{e_1 - e_3}) - e_3}\right]^{1/2} \\ \mathrm{cn}\, z &= \left[\frac{\wp(z/\sqrt{e_1 - e_3}) - e_1}{\wp(z/\sqrt{e_1 - e_3}) - e_3}\right]^{1/2} \\ \mathrm{dn}\, z &= \left[\frac{\wp(z/\sqrt{e_1 - e_3}) - e_2}{\wp(z/\sqrt{e_1 - e_3}) - e_3}\right]^{1/2} \end{aligned} \tag{12.38}$$

That they are doubly periodic follows immediately from the same property of $\wp(z)$. It has been seen that the values of e_1, e_2, e_3 must be distinct to ensure the double periodicity of $\wp(z)$. Thus, the singularities of these three Jacobian elliptic functions must arise from the zeros of

$$\wp\left(\frac{z}{\sqrt{e_1 - e_3}}\right) - e_3$$

and these are double zeros, since according to (12.35) the derivative of the above function also vanishes at its zero. Thus, sn z, cn z, and dn z each have as only singularities in the finite plane *simple poles* at

$$\frac{z}{\sqrt{e_1 - e_3}} = \frac{a_2}{2} + ma_1 + na_2$$

according to (12.34).

Since the only poles of $\wp(z/\sqrt{e_1 - e_3})$ occur at

$$z = (ma_1 + na_2)\sqrt{e_1 - e_3}$$

these must also be the locations of the only *zeros* of sn z. Actually, it is customary to express these locations in terms of two parameters K and K' defined by

$$\begin{aligned} 2K &= a_1\sqrt{e_1 - e_3} \\ 2iK' &= a_2\sqrt{e_1 - e_3} \end{aligned} \tag{12.39}$$

In terms of these the zeros of sn z are located at $z = 2(mK + inK')$. Examination of (12.38) reveals that

$$\operatorname{sn} K = 1 \tag{12.40}$$

since $\wp(a_1/2) = e_1$. Also,

$$\operatorname{cn} K = 0 \tag{12.41}$$

$$\operatorname{dn} K = \left(\frac{e_1 - e_2}{e_1 - e_3}\right)^{1/2} = k' \tag{12.42}$$

which defines the *complementary modulus* k'. Also (12.38) implies that

$$\operatorname{sn}^2 z + \operatorname{cn}^2 z = 1 \tag{12.43}$$

$$1 - \operatorname{dn}^2 z = 1 - \frac{\wp - e_2}{\wp - e_3} = \left(\frac{e_2 - e_3}{e_1 - e_3}\right)\left(\frac{e_1 - e_3}{\wp - e_3}\right)$$

or

$$1 - \operatorname{dn}^2 z = k^2 \operatorname{sn}^2 z \tag{12.44}$$

where the *modulus* k is defined by

$$k = \left(\frac{e_2 - e_3}{e_1 - e_3}\right)^{1/2} \tag{12.45}$$

and comparison with (12.42) yields

$$k^2 + (k')^2 = 1 \tag{12.46}$$

Differentiation of (12.38) yields

$$d_z(\text{sn}\, z) = \frac{\dot{\wp}}{-2(\wp - e_3)^{3/2}}$$

which, with $\dot{\wp} = -2\sqrt{(\wp - e_1)(\wp - e_2)(\wp - e_3)}$, yields

$$d_z(\text{sn}\, z) = \frac{\sqrt{(\wp - e_1)(\wp - e_2)}}{(\wp - e_3)} = \text{cn}\, z\, \text{dn}\, z \tag{12.47}$$

Differentiation of (12.43) and substitution of (12.47) yields

$$d_z(\text{cn}\, z) = -\text{sn}\, z\, \text{dn}\, z \tag{12.48}$$

Similarly, differentiation of (12.44) leads to

$$d_z(\text{dn}\, z) = -k^2\, \text{sn}\, z\, \text{cn}\, z \tag{12.49}$$

12.3 ELLIPTIC INTEGRALS

Using (12.43) and (12.44) to substitute for cn z and dn z in (12.47), one obtains

$$d_z(\text{sn}\, z) = \sqrt{(1 - \text{sn}^2 z)(1 - k^2\, \text{sn}^2 z)} \tag{12.50}$$

Similarly,

$$d_z(\text{cn}\, z) = -\sqrt{(1 - \text{cn}^2 z)[(k')^2 + k^2\, \text{cn}^2 z]} \tag{12.51}$$

$$d_z(\text{dn}\, z) = -(1 - \text{dn}^2 z)[(k')^2 - \text{dn}^2 z] \tag{12.52}$$

With $w = \text{sn}\, z$, (12.50) can be written

$$z = \int_0^z dz = \int_0^w \frac{dw}{\sqrt{(1 - w^2)(1 - k^2 w^2)}} \tag{12.53}$$

which inverse function of sn z is called the *elliptic integral of the first kind.* The *elliptic integral of the second kind* is

$$\int_0^z \text{dn}^2 z\, dz = \int_0^w (1 - k^2\, \text{sn}^2 z)\left(\frac{dz}{dw}\right) dw$$

or

$$\int_0^z \text{dn}^2 z\, dz = \int_0^w \sqrt{\frac{1 - k^2 w^2}{1 - w^2}}\, dw \tag{12.54}$$

The *elliptic integral of the third kind* is

$$\int_0^z \frac{dz}{1 - \alpha^2 \operatorname{sn}^2 z} = \int_0^w \frac{1}{1 - \alpha^2 w^2}\left(\frac{dz}{dw}\right) dw$$

or

$$\int_0^z \frac{dz}{1 - \alpha^2 \operatorname{sn}^2 z} = \int_0^w \frac{dw}{(1 - \alpha^2 w^2)\sqrt{(1 - w^2)(1 - k^2 w^2)}} \tag{12.55}$$

Taking $w = \sin\theta$ in (12.53), one has*

$$z = \int_0^\phi \frac{d\theta}{\sqrt{1 - k^2 \sin^2\theta}} = F(\phi,k) \tag{12.56}$$

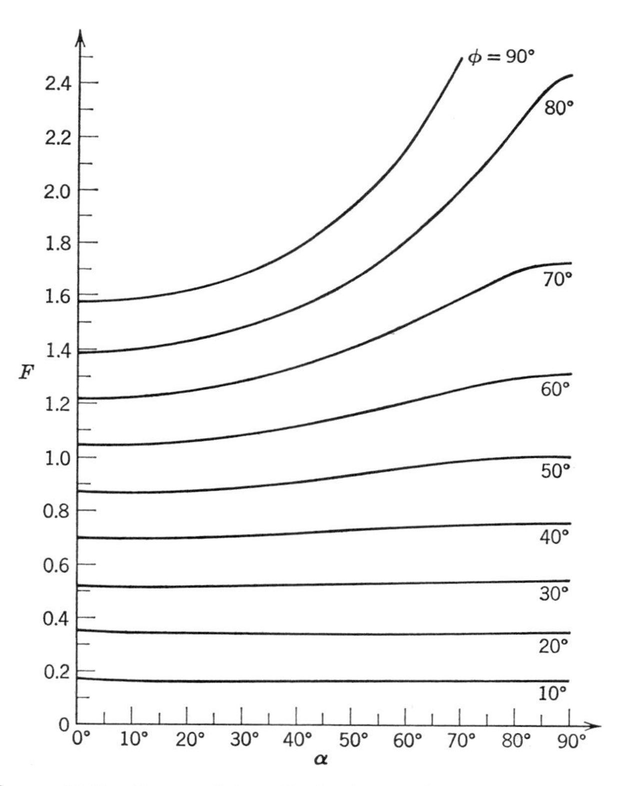

Figure 12.3. Incomplete elliptic integral of the first kind (12.56): $F(\phi,k)$, ϕ constant, $k = \sin\alpha$.

* Often the reverse order notation is used, e.g., $F(k,\phi)$ or $E(k,\phi)$.

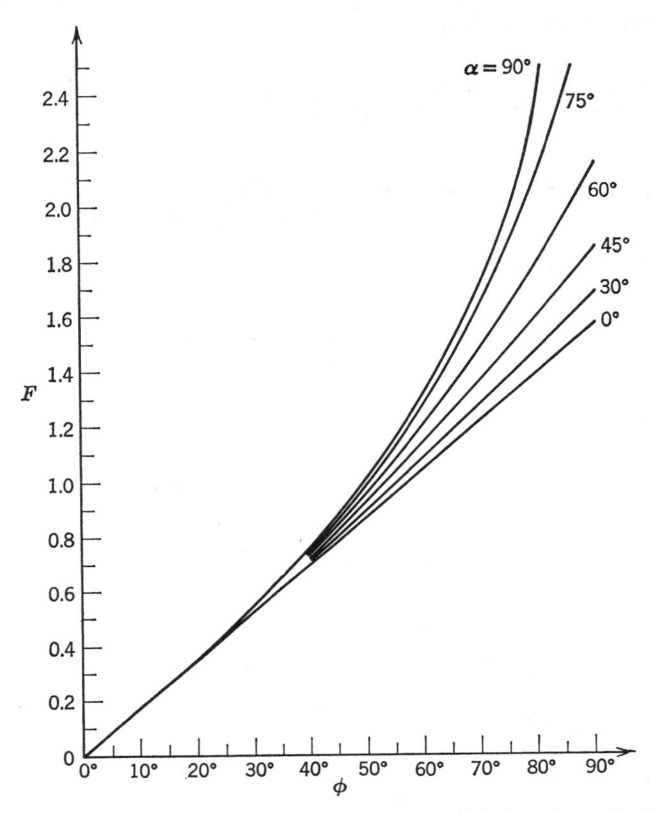

Figure 12.4. Incomplete elliptic integral of the first kind (12.56): $F(\phi,k)$, α constant, $k = \sin \alpha$.

(where $\sin \phi = w$), the Legendre notation for (12.53). Similarly, (12.54) becomes

$$\int_0^z \mathrm{dn}^2\, z\, dz = \int_0^\phi \sqrt{1 - k^2 \sin^2 \theta}\, d\theta = E(\phi,k) \qquad (12.57)$$

and (12.55) becomes

$$\int_0^z \frac{dz}{1 - \alpha^2 \,\mathrm{sn}^2\, z} = \int_0^\phi \frac{d\theta}{(1 - \alpha^2 \sin^2 \theta)\sqrt{1 - k^2 \sin^2 \theta}} = \Pi(\phi,\alpha^2,k) \qquad (12.58)$$

From (12.40) one concludes

$$\int_0^1 \frac{dw}{\sqrt{(1 - w^2)(1 - k^2 w^2)}} = F\left(\frac{\pi}{2},k\right) = K(k) \qquad (12.59)$$

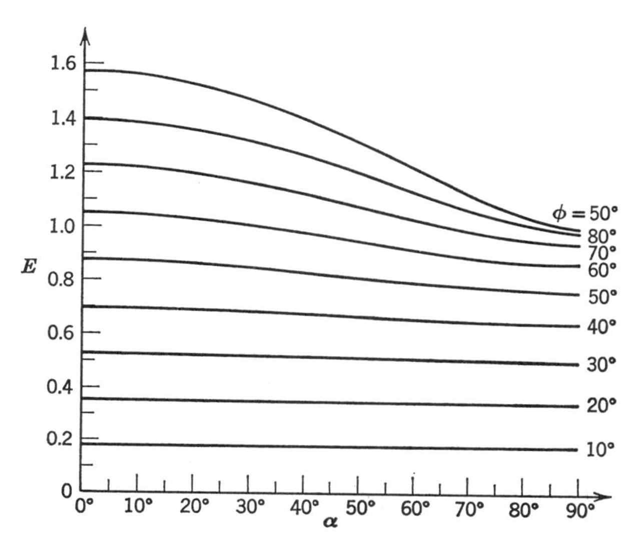

Figure 12.5. Incomplete elliptic integral of the second kind (12.57): $E(\phi,k)$, ϕ constant, $k = \sin\alpha$.

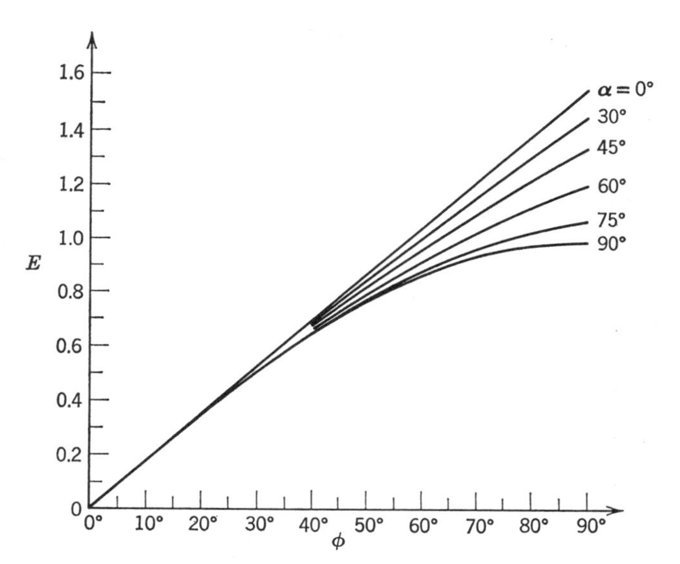

Figure 12.6. Incomplete elliptic integral of the second kind (12.57): $E(\phi,k)$, α constant, $k = \sin\alpha$.

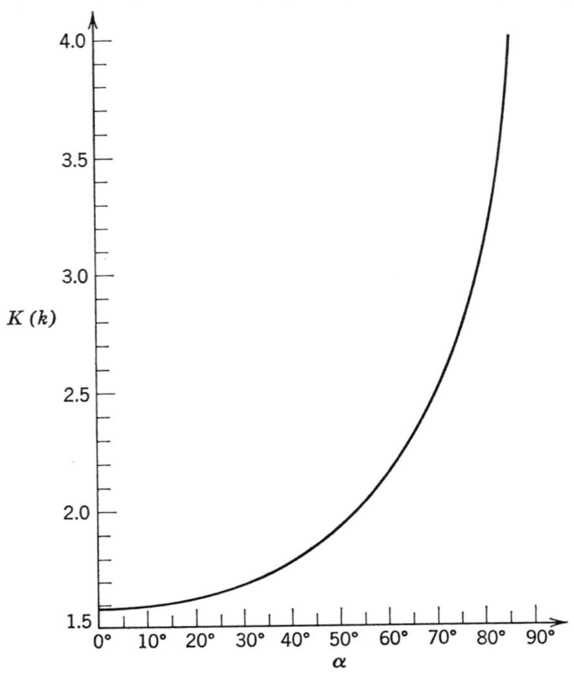

Figure 12.7. Complete elliptic integral of the first kind (12.59): $K(k)$, $k = \sin \alpha$.

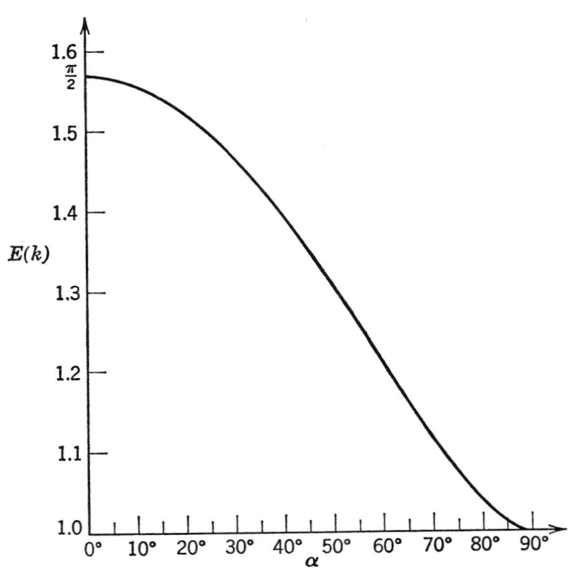

Figure 12.8. Complete elliptic integral of the second kind (12.60): $E(k)$, $k = \sin \alpha$.

while

$$\int_0^K \mathrm{dn}^2 z\, dz = \int_0^1 \sqrt{\frac{1 - k^2w^2}{1 - w^2}}\, dw = E\left(\frac{\pi}{2},k\right) = E(k) \qquad (12.60)$$

and

$$\int_0^K \frac{dz}{1 - \alpha^2 \mathrm{sn}^2 z} = \int_0^1 \frac{dw}{(1 - \alpha^2 w^2)\sqrt{(1 - w^2)(1 - k^2w^2)}} = \Pi\left(\frac{\pi}{2},\alpha^2,k\right) \qquad (12.61)$$

and (12.59)–(12.61) are called *complete elliptic integrals.*

12.4 PROPERTIES OF JACOBIAN ELLIPTIC FUNCTIONS

According to (12.59) K is a function of k alone or alternatively a function of $k' = \sqrt{1 - k^2}$ alone. From (12.34), (12.39), and (12.42) one sees that replacement of a_1 by $-a_2$ leads via

$$2K = a_1\sqrt{\wp\left(\frac{a_1}{2}\right) - \wp\left(\frac{a_2}{2}\right)}$$

to $K \to K'$, $e_3 \to e_1$, $e_1 \to e_3$, $e_2 \to e_2$ and hence $k' \to k$. Thus one has

$$K(k') = K'(k) \qquad (12.62)$$

It is clear from (12.53) that $\mathrm{sn}\, z = -\mathrm{sn}(-z)$ is an odd function of z since

$$\int_0^{-w} \frac{dw}{\sqrt{(1 - w^2)(1 - k^2w^2)}} = -\int_0^{w} \frac{dw}{\sqrt{(1 - w^2)(1 - k^2w^2)}}$$

This might also have been noticed from (12.38) if it is recognized that the sign of the square root $\sqrt{\wp - e_3}$ must be changed when z is replaced by $-z$. This is similar to recognizing $\sqrt{1 - \cos^2 z}$ as an odd function. Since $\mathrm{cn}\, z$ and $\mathrm{dn}\, z$ are ratios of such square roots, they are recognized as even functions of z.

The *addition theorem*

$$\mathrm{sn}(u + v) = \frac{\begin{vmatrix} \mathrm{sn}\, u & -\mathrm{sn}\, v \\ \mathrm{cn}\, u\, \mathrm{dn}\, u & \mathrm{cn}\, v\, \mathrm{dn}\, v \end{vmatrix}}{\begin{vmatrix} 1 & k\, \mathrm{sn}^2 v \\ k\, \mathrm{sn}^2 u & 1 \end{vmatrix}} \qquad (12.63)$$

may be established by calling the ratio of determinants $\phi(u,v)$ and the denominator determinant D and differentiating ϕ to obtain

$$D^2\partial_u\phi = \mathrm{cn}\, u\, \mathrm{dn}\, u\, \mathrm{cn}\, v\, \mathrm{dn}\, v(1 + k^2\, \mathrm{sn}^2 u\, \mathrm{sn}^2 v)$$
$$- \mathrm{sn}\, u\, \mathrm{sn}\, v(\mathrm{dn}^2 u + k^2\, \mathrm{cn}^2 u\, \mathrm{cn}^2 v)$$

This is symmetrical in u and v, and since ϕ is also symmetrical

$$\phi(u,v) = \phi(v,u)$$

$$\partial_u \phi = \partial_v \phi$$

which implies $\phi(u,v) = g(u + v)$ so that

$$\phi(u,0) = g(u) = \text{sn } u$$

and hence,

$$\phi(u,v) = g(u + v) = \text{sn}(u + v)$$

as was to be shown.

Also,

$$\text{cn}(u + v) = \frac{\begin{vmatrix} \text{cn } u & \text{sn } v \text{ dn } v \\ \text{sn } u \text{ dn } u & \text{cn } v \end{vmatrix}}{\begin{vmatrix} 1 & k \text{ sn}^2 v \\ k \text{ sn}^2 u & 1 \end{vmatrix}} \tag{12.64}$$

may be proved by writing

$$\text{cn}^2(u + v) = 1 - \text{sn}^2(u + v)$$

in terms of (12.63) and rearranging terms.

Similarly,

$$\text{dn}(u + v) = \frac{\begin{vmatrix} \text{dn } u & k \text{ sn } v \text{ cn } v \\ k \text{ sn } u \text{ cn } u & \text{dn } v \end{vmatrix}}{\begin{vmatrix} 1 & k \text{ sn}^2 v \\ k \text{ sn}^2 u & 1 \end{vmatrix}} \tag{12.65}$$

Recalling $\text{sn } K = 1$, $\text{cn } K = 0$, $\text{dn } K = k'$, one has from the addition formulae

$$\begin{aligned} \text{sn}(u + K) &= \frac{\text{cn } u}{\text{dn } u} \\ \text{cn}(u + K) &= -k' \frac{\text{sn } u}{\text{dn } u} \\ \text{dn}(u + K) &= \frac{k'}{\text{dn } u} \end{aligned} \tag{12.66}$$

For $u = K$ one has $\text{sn } 2K = 0$, $\text{cn } 2K = -1$, $\text{dn } 2K = 1$. Similarly, then

$$\begin{aligned} \text{sn}(u + 2K) &= -\text{sn } u \\ \text{cn}(u + 2K) &= -\text{cn } u \\ \text{dn}(u + 2K) &= \text{dn } u \end{aligned} \tag{12.67}$$

Hence, replacing u by $u + 2K$,

$$\begin{aligned} \operatorname{sn}(u + 4K) &= \operatorname{sn} u \\ \operatorname{cn}(u + 4K) &= \operatorname{cn} u \end{aligned} \tag{12.68}$$

so that sn z and cn z have period $4K$, and dn z has period $2K$.

It should be clearly understood that the functions designated by sn z, cn z, and dn z depend upon the particular value of k to which they correspond in addition to the particular value taken by z. To be more explicit about this dependence they are often written as $\operatorname{sn}(z,k)$, $\operatorname{cn}(z,k)$ and $\operatorname{dn}(z,k)$. In terms of this notation $\operatorname{sn}(iz,k)$, $\operatorname{cn}(iz,k)$, and $\operatorname{dn}(iz,k)$ can be expressed as follows:

$$\begin{aligned} \operatorname{sn}(iz,k) &= \frac{i \operatorname{sn}(z,k')}{\operatorname{cn}(z,k')} \\ \operatorname{cn}(iz,k) &= \frac{1}{\operatorname{cn}(z,k')} \\ \operatorname{dn}(iz,k) &= \frac{\operatorname{dn}(z,k')}{\operatorname{cn}(z,k')} \end{aligned} \tag{12.69}$$

which can be deduced from (12.38) by noting that interchange of e_3 and e_1 produces interchange of k and k'. Thus the addition theorems for functions of $(u + iv)$ become

$$\operatorname{sn}(u + iv,k) = \frac{\operatorname{sn}(u,k)\operatorname{dn}(v,k') + i \operatorname{cn}(u,k)\operatorname{dn}(u,k)\operatorname{sn}(v,k')\operatorname{cn}(v,k')}{1 - \operatorname{dn}^2(u,k)\operatorname{sn}^2(v,k')}$$

$$\operatorname{cn}(u + iv,k) = \frac{\operatorname{cn}(u,k)\operatorname{cn}(v,k') - i \operatorname{sn}(u,k)\operatorname{dn}(u,k)\operatorname{sn}(v,k')\operatorname{dn}(v,k')}{1 - \operatorname{dn}^2(u,k)\operatorname{sn}^2(v,k')}$$

$$\operatorname{dn}(u + iv,k) = \frac{\operatorname{dn}(u,k)\operatorname{cn}(v,k')\operatorname{dn}(v,k') - ik^2 \operatorname{sn}(u,k)\operatorname{cn}(u,k)\operatorname{sn}(v,k')}{1 - \operatorname{dn}^2(u,k)\operatorname{sn}^2(v,k')}$$

Substitution of $v = K'$ in the above yields

$$\operatorname{sn}(u + iK') = \frac{1}{k \operatorname{sn} u} \tag{12.70}$$

$$\operatorname{cn}(u + iK') = \frac{-i \operatorname{dn} u}{k \operatorname{sn} u} \tag{12.71}$$

$$\operatorname{dn}(u + iK') = \frac{-i \operatorname{cn} u}{\operatorname{sn} u} \tag{12.72}$$

Replacing u by $(u + iK')$ in (12.70) yields

$$\operatorname{sn}(u + 2iK') = \operatorname{sn} u \tag{12.73}$$

Replacing u by $(u + iK')$ in (12.71) yields

$$\text{cn}(u + 2iK') = \frac{-i\,\text{dn}(u + iK')}{k\,\text{sn}(u + iK')} = \frac{(-i)^2 k\,\text{sn}\,u\,\text{cn}\,u}{k\,\text{sn}\,u}$$

or

$$\text{cn}(u + 2iK') = -\text{cn}\,u$$

From (12.67) one then has

$$\text{cn}(u + 2K + 2iK') = \text{cn}\,u \tag{12.74}$$

Replacement of u by $(u + 3iK')$ in (12.72) results in

$$\text{dn}(u + 4iK') = -\text{dn}(u + 2iK') = \text{dn}\,u \tag{12.75}$$

Thus from (12.67), (12.68), (12.73), (12.74), and (12.75) one concludes that

$$\text{sn}(z + 4mK + 2inK') = \text{sn}\,z \tag{12.76}$$

$$\text{cn}(z + 4mK + 2nK + 2inK') = \text{cn}\,z \tag{12.77}$$

$$\text{dn}(z + 2mK + 4inK') = \text{dn}\,z \tag{12.78}$$

for all integer values of m and n.

One recalls that elliptic functions have the same number of poles and zeros in a fundamental parallelogram. Thus, $\text{sn}\,0 = 0$ by (12.38) while $\text{sn}\,2K = 0$ by (12.67). Since $\text{sn}\,K = 1$, $\text{cn}\,K = 0$ and $\text{cn}\,3K = 0$ by (12.67). Also by (12.72) $\text{dn}(K + iK') = 0$, since $\text{cn}\,K = 0$. Furthermore, by (12.72)

$$\text{dn}(K + 3iK') = -i\frac{\text{cn}(K + 2iK')}{\text{sn}(K + 2iK')} = \frac{i\,\text{cn}\,K}{\text{sn}(K + 2iK')} = 0$$

The location of a simple pole for each of $\text{sn}\,z$, $\text{cn}\,z$, $\text{dn}\,z$ may be deduced from (12.38) by noting that

$$z = \frac{a_2}{2}\sqrt{e_1 - e_3} = iK'$$

Table 12.1. Critical points and periods of Jacobian elliptic functions (m, n are integers)

	Zeros	Poles	Periods	
$\text{sn}\,z$	$2mK + 2inK'$	$2mK + (2n + 1)iK'$	$4K$	$2iK'$
$\text{cn}\,z$	$(2m + 1)K + 2inK'$	$2mK + (2n + 1)iK'$	$4K$	$2K + 2iK'$
$\text{dn}\,z$	$(2m + 1)K + (2n + 1)iK'$	$2mK + (2n + 1)iK'$	$2K$	$4iK'$

makes each denominator vanish. Since $\operatorname{sn} z$ is odd and $\operatorname{cn} z$ and $\operatorname{dn} z$ are even, they must also have simple poles at $z = -iK'$.

The critical lattice locations for the three functions may then be summarized in Table 12.1.

EXAMPLES 12

1. Show that the position of a particle $x(t)$ is a periodic function of time if

(1) $E = \dfrac{m\dot{x}^2}{2} + V$

(2) $\dot{x}(a) = \dot{x}(b) = 0$

(3) $\dfrac{\dot{x}^2}{(x-a)(x-b)}$ is continuous in $a \le x \le b$

(4) $\dfrac{\dot{x}^2}{(x-a)(b-x)} \ge \delta > 0 \quad \text{in} \quad a \le x \le b$

Ans. Writing the energy relation as

$$\dot{x}^2 = \frac{2(E-V)}{m} = f(x)$$

and substituting

$$x = \left(\frac{a+b}{2}\right) - \left(\frac{a-b}{2}\right)\sin\mu$$

where $-\pi \le \mu \le \pi$ corresponds to $a \le x \le b$ and

$$\dot{x}^2 = \left(\frac{a-b}{2}\right)^2 (\cos^2\mu)\dot{\mu}^2$$

so that if $G(\sin\mu)$ is defined by

$$f(x) = \left(\frac{a-b}{2}\right)^2 \cos^2\mu \; G(\sin\mu)$$

the differential equation becomes

$$\dot{\mu}^2 = G(\sin\mu) = \frac{f(x)}{(x-a)(b-x)} = \frac{\dot{x}^2}{(x-a)(b-x)}$$

or

$$\dot{\mu} = \sqrt{G(\sin\mu)}$$

with a *positive* square root, while the other equation with negative square root is satisfied by $(\pi - \mu)$ leading to exactly the same motion $x(t)$. Integration yields

$$\int_\tau^t dt = t - \tau = \int_0^\mu \frac{du}{\sqrt{G(\sin \mu)}} = H(\mu) = \int_{(a+b)/2}^x \frac{dx}{\sqrt{f(x)}}$$

Because of (3),

$$H'(\mu) = \frac{1}{\sqrt{G(\sin \mu)}} = \left[\frac{(x-a)(b-x)}{\dot{x}^2}\right]^{1/2} > 0$$

is never zero for $-\pi \le \mu \le \pi$ $(a \le x \le b)$, and because of (4), $H'(\mu)$ is never infinite in the same range. Thus a unique inverse function

$$\mu = h(t - \tau)$$

exists. Clearly, $H'(\mu) = H'(\mu + 2\pi)$ is periodic in μ, so

$$H(\mu + 2\pi) - H(\mu) = k = \text{constant}$$

and as μ increases by 2π the increase in t must be k. Hence,

$$x(t + k) - x(t) = \left[\frac{b-a}{2}\right][\sin(\mu + 2\pi) - \sin \mu] = 0$$

as was to be shown. The period k may be expressed as

$$k = 2\int_{-\pi/2}^{\pi/2} \frac{d\mu}{\sqrt{G(\sin \mu)}} = 2\int_a^b \frac{dx}{\sqrt{f(x)}}$$

since replacing μ by $(\pi - \mu)$ leads to a change in sign of $\dot{x}$ but no change in x, so that the motion from a to b is simply the reverse of that from b to a.

2. Verify that the Weierstrass zeta function

$$\zeta(z) = \frac{1}{z} + \sum_{\Omega \ne 0}\left(\frac{1}{z - \Omega} + \frac{1}{\Omega} + \frac{z}{\Omega^2}\right)$$

satisfies (12.31).

Ans. $$\wp(z) = -\zeta'(z) = \frac{1}{z^2} + \sum_{\Omega \ne 0}\left\{\frac{1}{(z-\Omega)^2} - \frac{1}{\Omega^2}\right\}$$

3. Show that the Weierstrass sigma function $\sigma(z)$ defined by

$$\sigma(z) = z \prod_{\Omega \ne 0}\left(1 - \frac{z}{\Omega}\right) e^{(z/\Omega) + (z^2/2\Omega^2)}$$

satisfies (12.32).

Ans. Logarithmic differentiation yields

$$\frac{\sigma'(z)}{\sigma(z)} = d_z \log \sigma(z) = \zeta(z)$$

4. Show that the constants η_1 and η_2 defined by

$$\zeta(z + a_1) - \zeta(z) = 2\eta_1$$

$$\zeta(z + a_2) - \zeta(z) = 2\eta_2$$

satisfy the (Legendre) relation

$$a_2\eta_1 - a_1\eta_2 = \pi i$$

Ans. Integrating $\zeta(z)$ over a fundamental parallelogram, one has

$$\oint \zeta(z)\, dz = \int_{a_1+a_2}^{a_2} 2\eta_2\, dz + \int_{a_1}^{a_1+a_2} 2\eta_1\, dz = 2\pi i$$

since $\zeta(z)$ has only a simple pole of residue one.

5. Show that the entire function $\sigma(z)$ is *quasi doubly periodic*, i.e., that

$$\sigma(z + a_1) = -e^{\eta_1(2z+a_1)}\sigma(z)$$

$$\sigma(z + a_2) = -e^{\eta_2(2z+a_2)}\sigma(z)$$

Ans. According to (12.32)

$$\zeta(z) = \frac{\sigma'(z)}{\sigma(z)} = d_z \log \sigma(z)$$

so substituting into the relations for $\zeta(z)$ in Example 4, one has after integration

$$\log \sigma(z + a_1) - \log \sigma(z) = 2\eta_1 z + k_1$$

$$\log \sigma(z + a_2) - \log \sigma(z) = 2\eta_2 z + k_2$$

or

$$\sigma(z + a_1) = \sigma(z)e^{2\eta_1 z}e^{k_1}$$

$$\sigma(z + a_2) = \sigma(z)e^{2\eta_2 z}e^{k_2}$$

and since $\sigma(z) = -\sigma(-z)$ is an odd function if $2z = -a_1$,

$$\sigma\left(\frac{a_1}{2}\right) = \sigma\left(-\frac{a_1}{2}\right)e^{-\eta_1 a_1}e^{k_1}$$

and

$$e^{k_1} = -e^{\eta_1 a_1}$$

and similarly for k_2.

6. Show that if the zeros and poles of an elliptic function $f(z)$ are located at $\{z_k\}$ and $\{p_k\}$ in a fundamental parallelogram with

$$\sum_k z_k = \sum_k p_k$$

then

$$f(z) = C \prod_k \frac{\sigma(z - z_k)}{\sigma(z - p_k)}$$

where C is constant.

Ans. Call the product $g(z)$. Then, according to the preceding example,

$$\sigma(z - z_k + a_1) = -\sigma(z - z_k)e^{2\eta_1(z-z_k)}e^{\eta_1 a_1}$$

$$\sigma(z - p_k + a_1) = -\sigma(z - p_k)e^{2\eta_1(z-p_k)}e^{\eta_1 a_1}$$

so that

$$g(z + a_1) = g(z)\exp\left[2\eta_1 \sum_k (p_k - z_k)\right] = g(z)$$

Similarly,

$$g(z + a_2) = g(z)$$

so $g(z)$ is an elliptic function. Then $f(z)/g(z)$ is an elliptic function without any singularities, and hence it must be constant.

7. Discuss the motion of a simple pendulum.

Ans. With θ the angle from the vertical (downward) position, l the length, m the mass, g the acceleration due to gravity, and ω_m the angular velocity at $\theta = 0$, the equation of motion is

$$\frac{ml^2}{2}\dot{\theta}^2 - mgl\cos\theta = \frac{ml^2\omega_m^2}{2} - mgl$$

or in terms of the small amplitude oscillation frequency ω_p for which $l\omega_p^2 = g$, one has

$$\dot{\theta}^2 = \omega_m^2 - 2\omega_p^2 + 2\omega_p^2\cos\theta = \omega_m^2 - 4\omega_p^2\sin^2(\theta/2)$$

so that, with $k = 2\omega_p/\omega_m$,

$$t = \int_0^\theta \frac{d\theta}{\sqrt{(\omega_m^2 - 2\omega_p^2) + 2\omega_p^2\cos\theta}} = \frac{2}{\omega_m}\int_0^{\theta/2} \frac{d(\theta/2)}{\sqrt{1 - k^2\sin^2(\theta/2)}}$$

there are then three possibilities. If $\omega_m/\omega_p > 2$, then $\dot{\theta}^2 > 0$ and

$$t = \frac{2}{\omega_m} F\left(\frac{2\omega_p}{\omega_m}, \frac{\theta}{2}\right)$$

with the pendulum performing complete revolutions in one direction. The period of a complete revolution is then $T = (4/\omega_m)K(2\omega_p/\omega_m)$. If $\omega_m/\omega_p < 2$, the former value for k becomes larger than unity, so the integral must be subjected to a change of variable. Let

$$\sin\phi = \frac{2\omega_p}{\omega_m}\sin\left(\frac{\theta}{2}\right)$$

$$\cos\phi\, d\phi = \frac{2\omega_p}{\omega_m}\cos\left(\frac{\theta}{2}\right)d\left(\frac{\theta}{2}\right)$$

so

$$d\left(\frac{\theta}{2}\right) = \frac{\cos\phi\, d\phi}{(2\omega_p/\omega_m)\sqrt{1-\sin^2(\theta/2)}}$$

$$d\left(\frac{\theta}{2}\right) = \frac{\cos\phi\, d\phi}{(2\omega_p/\omega_m)\sqrt{1-(\omega_m/2\omega_p)^2\sin^2\phi}}$$

while

$$\sqrt{1-(2\omega_p/\omega_m)^2\sin^2(\theta/2)} = \cos\phi$$

so

$$\omega_p t = \int_0^\phi \frac{d\phi}{\sqrt{1-(\omega_m/2\omega_p)^2\sin^2\phi}}$$

and

$$\omega_p t = F\left(\frac{\omega_m}{2\omega_p},\phi\right)$$

with a k value of $\omega_m/2\omega_p < 1$. The maximum angular amplitude θ_m of the pendulum occurs for $\dot{\theta} = 0$, so that

$$\sin\left(\frac{\theta_m}{2}\right) = \frac{\omega_m}{2\omega_p}$$

The corresponding value ϕ_m of ϕ is given by

$$\sin\phi_m = \frac{2\omega_p}{\omega_m}\sin\left(\frac{\theta_m}{2}\right) = 1$$

so $\phi_m = \pi/2$, and for $-\theta_m \le \theta \le \theta_m$ one has $-\pi/2 \le \phi \le \pi/2$, but for the negative range the sign must be reversed, since

$$-F(k,\phi) = F(k,-\phi)$$

The period of the motion will be

$$T = \frac{4}{\omega_p}F\left(\frac{\omega_m}{2\omega_p},\frac{\pi}{2}\right) = \frac{4}{\omega_p}K\left(\frac{\omega_m}{2\omega_p}\right)$$

If $\omega_m = 2\omega_p$, then $\dot{\theta} = 0$ for $\theta = \pi = \theta_m$ and the integral becomes elementary:

$$t = \frac{2}{\omega_m} F\left(1, \frac{\theta}{2}\right) = \frac{2}{\omega_m} \log\left[\tan\left(\frac{\theta}{2}\right) + \sec\left(\frac{\theta}{2}\right)\right]$$

But $t \to \infty$ as $\theta \to \pi$; hence it requires an infinite time for the pendulum to reach a vertical (upward) position. In all three cases the $(\omega_m/2\omega_p)^2$ is the ratio of maximum kinetic energy to potential energy at top.

8. Calculate the perimeter of the ellipse

$$\left(\frac{x}{a}\right)^2 + \left(\frac{y}{b}\right)^2 = 1 \qquad \text{for} \quad b > a$$

Ans. Equivalent parametric equations are

$$\begin{cases} x = a \cos \theta \\ y = b \sin \theta \end{cases} \qquad \begin{cases} dx = -a \sin \theta \, d\theta \\ dy = b \cos \theta \, d\theta \end{cases}$$

$$ds = \sqrt{a^2 \sin^2 \theta + b^2 \cos^2 \theta} \, d\theta$$

$$s = \int_0^\theta \sqrt{b^2 - (b^2 - a^2)\sin^2 \theta} \, d\theta$$

or

$$s = b \int_0^\theta \sqrt{1 - k^2 \sin^2 \theta} \, d\theta$$

with $k^2 = (b^2 - a^2)/b^2 = c^2/b^2 = e^2 < 1$. Hence $s(\theta) = bE(e,\theta)$ for the arc length corresponding to any θ. The perimeter is given by the complete integral

$$4s\left(\frac{\pi}{2}\right) = bE\left(e, \frac{\pi}{2}\right) = bE(e)$$

● PROBLEMS 12

1. What kind of function of time is $x(t)$ in Example 1 if $f(x)$ is of the form

   $f(x) = a_0x^4 + a_1x^3 + a_2x^2 + a_3x + a_4$ with $|a_0| + |a_1| > 0$?

2. Show that the Weierstrass zeta function $\zeta(z)$ defined by (12.31) is *not* an elliptic function.

3. Describe the singularities of the Weierstrass sigma function $\sigma(z)$.
4. Using the (canonical) product (in terms of the Weierstrass sigma functions) of Example 6 applied to $[\wp(z) - \wp(w)]$, where $w \neq \Omega$, derive the addition theorem for the Weierstrass $\zeta(z)$:

$$\zeta(z + w) = \zeta(z) + \zeta(w) + \frac{1}{2}\left[\frac{\wp'(z) - \wp'(w)}{\wp(z) - \wp(w)}\right]$$

5. Show that

$$\wp(z + w) = \wp(z) - \tfrac{1}{2}\partial_z\left[\frac{\wp'(z) - \wp'(w)}{\wp(z) - \wp(w)}\right]$$

is the addition theorem for $\wp(z)$.

6. Show that the theta functions defined by

$$\theta_1(z,q) = 2\sum_{n=0}^{\infty}(-1)^n q^{[n+(1/2)]^2}\sin[(2n + 1)z]$$

$$\theta_2(z,q) = 2\sum_{n=0}^{\infty} q^{[n+(1/2)]^2}\cos[(2n + 1)z]$$

$$\theta_3(z,q) = 1 + 2\sum_{n=1}^{\infty} q^{n^2}\cos(2nz)$$

$$\theta_4(z,q) = 1 + 2\sum_{n=1}^{\infty}(-1)^n q^{n^2}\cos(2nz)$$

where $q = e^{i\pi\tau}$ with Im $\tau > 0$ so $|q| < 1$, are entire functions and hence are *not* elliptic functions.

7. Show that for $k = 1, 2, 3, 4$, $z = x + iy$,

$$|\theta_k(z + \pi,q)| = |\theta_k(z,q)|$$

$$|q|\,|\theta_k(z + \pi\tau,q)| = e^{2y}\,|\theta_k(z,q)|$$

8. Show that for $k = 1, 2, 3, 4$,

$$D\partial_z^2\theta_k = \partial_\tau\theta_k$$

with $D = \pi/4i$.

9. Verify that the theta functions are solutions of the Schrödinger wave equation for a free particle

$$-\frac{\hbar^2}{2m}\partial_x^2\psi = -\frac{\hbar}{i}\partial_t\psi$$

moving in one space dimension.

10. How can the equations of a vibrating beam be related to the theta functions?

$$D^2\partial_x^4\phi + \partial_t^2\phi = 0$$

[Here $\phi(x,t)$ is transverse displacement, $D^2\rho = EI$ is flexural rigidity, and ρ is mass per unit length.]

11. Show that for $0 < k < 1$

$$|\phi| \leq |F(k,\phi)| \leq |\log(\tan\phi + \sec\phi)|$$

12. Show that for $0 < k < 1$

$$|\phi| \leq |E(k,\phi)| \leq |\sin\phi|$$

13. Find the perimeter of the ellipse $8x^2 + 9y^2 = 72$.
14. Find the midpoint of the arc of $x = 4\cos\theta$, $y = 2\sqrt{2}\sin\theta$ for $0 \leq \theta \leq \pi/2$.
15. Find the ratio of area to perimeter squared for $x = 2\cos\theta$, $y = \sin\theta$.
16. Show that

$$\int_{-1}^{0} \frac{dx}{\sqrt{x(x^2-1)}} = \sqrt{2}\, K\left(\frac{1}{\sqrt{2}}\right)$$

17. Show that if $a > b > c \geq m$,

$$\int_{-\infty}^{m} \frac{dt}{\sqrt{(a-t)(b-t)(c-t)}} = \frac{2}{\sqrt{a-c}} F(k,\phi)$$

where $k^2(a-c) = (a-b)$ and $\sin^2\phi = (a-c)/(a-m)$.

18. Show that if $m > 1$,

$$\int_{1}^{m} \frac{dt}{\sqrt{t^3-1}} = \frac{1}{3^{1/4}} F(k,\phi)$$

for $4k^2 = 2 - \sqrt{3}$ and $\cos\phi = (\sqrt{3}+1-m)/(\sqrt{3}-1+m)$.

19. Show that for $m < 1$,

$$\int_{m}^{1} \sqrt{1-t^3}\, dt = \frac{3^{3/4}}{5} F(k,\phi) - \frac{2}{5} m\sqrt{1-m^3}$$

for $4k^2 = 2 + \sqrt{3}$ and $\cos\phi = (\sqrt{3}-1+m)/(\sqrt{3}+1-m)$.

20. Show that for $b < 2a^2$, $0 \leq m < \infty$,

$$\int_{m}^{\infty} \frac{dt}{\sqrt{t^4+bt^2+a^4}} = \frac{1}{2a} F(k,\phi)$$

for $4a^2k^2 = 2a^2 - b$ and $\cos\phi = (m^2-a^2)/(m^2+a^2)$.

21. Show that

$$2k^3 \int \text{sn}^2 z \,\text{cn}\, z\, dz = \arcsin(k\,\text{sn}\, z) - k\,\text{sn}\, z\,\text{dn}\, z$$

22. Show that

$$\int \frac{dz}{1 + \operatorname{cn} z} = z - E(z) + \frac{\operatorname{sn} z \operatorname{dn} z}{1 + \operatorname{cn} z}$$

23. Show that the transformation $(K = K^*, K' = (K')^*)$

$$w = \frac{\operatorname{sn} z}{1 + \operatorname{cn} z}$$

maps the interior of the rectangle $|\operatorname{Re} z| < K$, $|\operatorname{Im} z| < K'$ onto the interior of the unit circle $|w| < 1$.

24. Show that $[K = K^*, K' = (K')^*]$

$$w = \operatorname{sn} z$$

maps $|\operatorname{Re} z| \leq K$, $0 \leq \operatorname{Im} z \leq K'$ onto $\operatorname{Im} w \geq 0$.

25. Show that the potential of a circular ring of uniformly distributed charge q and radius a at points in the plane of the ring inside it is given by

$$V(r) = \frac{2q}{\pi(a + r)} K\left(\frac{2\sqrt{ar}}{a + r}\right)$$

in terms of the complete elliptic integral of the first kind.

26. Show that $f(z) = \ln(\operatorname{sn} z)$ represents a flow inside a rectangle of corners $z = 0$, $z = iK'$, $z = 2K$, $z = 2K + iK'$ (for real K and K') from a source at $z = iK'$ to a sink at $z = 0$.

27. Prove the Legendre relation

$$EK' + E'K - K'K = \pi/2$$

between the complete elliptic integrals of the first and second kinds, by considering

$$\int_0^{\pi/2} \int_0^{\pi/2} \frac{[k^2 \cos^2 \theta + (k')^2 \cos^2 \phi] d\theta\, d\phi}{\sqrt{1 - k^2 \sin^2 \theta}\sqrt{1 - (k')^2 \sin^2 \phi}}$$

28. Show that the mapping

$$z = \int_0^w \frac{dt}{\sqrt{1 - t^4}}$$

takes $|w| \leq 1$ into the interior of the square with corners at $K/\sqrt{2}$, $iK/\sqrt{2}$, $-K/\sqrt{2}$, $-iK/\sqrt{2}$, the latter being the images of the fourth roots of unity in the w plane (with $k = 1/\sqrt{2}$).

29. Show that the inverse mapping to that of Problem 28 is

$$w\sqrt{2} = \frac{\operatorname{sn}(z\sqrt{2})}{\operatorname{dn}(z\sqrt{2})}$$

CHAPTER 13:

Differential Equations and Special Functions

13.1 SINGULARITIES OF SECOND-ORDER HOMOGENEOUS DIFFERENTIAL EQUATIONS

In this chapter attention will be restricted to ordinary linear homogeneous differential equations of second order. These are of the form

$$[d_z^2 + p(z)d_z + q(z)]w(z) = 0 \tag{13.1}$$

If $(z - a)p(z) = P(z)$ and $(z - a)^2q(z) = Q(z)$ are analytic (possessing Taylor expansions) at $z = a$, this point is said to be a *regular singularity* of (13.1) if

$$|P(a)| + |Q(a)| + |Q'(a)| > 0 \tag{13.2}$$

and $z = a$ is said to be an *ordinary point* of (13.1) if

$$|P(a)| + |Q(a)| + |Q'(a)| = 0 \tag{13.3}$$

(i.e., if all three terms vanish). If $P(z)$ or $Q(z)$ have a singularity at $z = a$, this point is said to be an *irregular singularity* of (13.1). Notice that these terms apply to the differential equation. This classification turns out to be useful when solving (13.1) by methods of series substitutions for $w(z)$. To simplify the notation it will often be supposed that $a = 0$ with the understanding that the same kind of argument may be applied to $z = a \neq 0$ by expanding the series about $z = a$.

In the case of a regular singularity at $z = 0$ (13.1) can be solved in the neighborhood of $z = 0$ by the substitution

$$w(z) = z^{\alpha} \sum_{k=0}^{\infty} w_k z^k \tag{13.4}$$

with

$$P(z) = zp(z) = \sum_{k=0}^{\infty} P_k z^k \tag{13.5}$$

$$Q(z) = z^2 q(z) = \sum_{k=0}^{\infty} Q_k z^k \tag{13.6}$$

Actually it is somewhat more convenient to multiply (13.1) by z^2 and substitute $\zeta = \ln z$ so that (13.1) becomes

$$[(zd_z)^2 - (zd_z) + P(z)(zd_z) + Q(z)]w(z) = 0 \tag{13.7}$$

since

$$z^2 d_z^2 = (zd_z)^2 - (zd_z)$$

and finally

$$[d_\zeta^2 - d_\zeta + P(e^\zeta)d_\zeta + Q(e^\zeta)]w(e^\zeta) = 0 \tag{13.8}$$

while (13.4)–(13.6) become

$$w(e^\zeta) = \sum_{k=0}^{\infty} w_k e^{(k+\alpha)\zeta} \tag{13.9}$$

$$P(e^\zeta) = \sum_{k=0}^{\infty} P_k e^{k\zeta} \tag{13.10}$$

$$Q(e^\zeta) = \sum_{k=0}^{\infty} Q_k e^{k\zeta} \tag{13.11}$$

and

$$d_\zeta w(e^\zeta) = \sum_{k=0}^{\infty} (k + \alpha) w_k e^{(k+\alpha)\zeta} \tag{13.12}$$

$$d_\zeta^2 w(e^\zeta) = \sum_{k=0}^{\infty} (k + \alpha)^2 w_k e^{(k+\alpha)\zeta} \tag{13.13}$$

Thus multiplying (Cauchy) (13.10) with (13.12) and (13.9) with (13.11), one has

$$\sum_{k=0}^{\infty} \left\{ (k + \alpha)^2 w_k - (k + \alpha) w_k + \sum_{m=0}^{k} [(m + \alpha) P_{k-m} w_m + Q_{k-m} w_m] \right\} e^{(k+\alpha)\zeta} = 0$$

so that for $k \geq 1$

$$[(k + \alpha)^2 + (P_0 - 1)(k + \alpha) + Q_0] w_k = -\sum_{m=0}^{k-1} [(m + \alpha) P_{k-m} + Q_{k-m}] w_m \tag{13.14}$$

while for $k = 0$

$$[\alpha^2 + (P_0 - 1)\alpha + Q_0]w_0 = 0$$

Since (13.1) is a second-order differential equation, there should be two *arbitrary* constants in the general solution. If w_0 is one of these, then, since it is arbitrary,

$$\alpha^2 + (P_0 - 1)\alpha + Q_0 = 0 \tag{13.15}$$

which is called the *indicial equation.* This determines two values of α which are called α_1, α_2, the *exponents* of the regular singularity. Representing the left side of (13.15) by $F(\alpha)$, (13.14) becomes

$$F(k + \alpha)w_k = -\sum_{m=0}^{k-1} [(m + \alpha)P_{k-m} + Q_{k-m}]w_m \tag{13.16}$$

so that any w_k can be determined in terms of $\{P_k\}$, $\{Q_k\}$ and the *preceding* w_m $(m < k)$ provided that

$$F(k + \alpha) \neq 0 \qquad \text{for} \quad k > 0 \tag{13.17}$$

If the indicial equation does *not* have a double root ($\alpha_1 \neq \alpha_2$) and (13.17) holds, (13.16) will yield a distinct solution for each root of (13.15).

If $w_0 = 0$ instead of being arbitrary, (13.14) yields

$$[(1 + \alpha)^2 + (P_0 - 1)(1 + \alpha) + Q_0]w_1 = 0$$

and for w_1 arbitrary the same indicial equation (13.15) is satisfied by $(1 + \alpha)$. Continuing in this way, it is seen that the indicial equation must hold if there are any arbitrary constants in the series solution (13.4). Under these conditions there will be one arbitrary constant in each distinct solution; hence there will be two arbitrary constants in the general solution of (13.1).

When $4Q_0 = (P_0 - 1)^2$ the roots of the indicial equation (13.15) become

$$\alpha = \frac{1 - P_0}{2} \tag{13.18}$$

and for this single value of α (13.16) can be used to successively determine the coefficients in (13.4) for one solution with one arbitrary constant.

Again if the roots of the indicial equation (13.15) differ by an integer, (13.16) can be used (with α equal to the root with the larger real part) to determine successively the coefficients in (13.4) for one solution.

In each of these cases (13.16) cannot be used to find a second solution of (13.1)—in the case of (13.18) because there is no second value of α, and in the case of integer difference of roots because the successive determinations will become impossible when $F(k + \alpha) = 0$. In both these cases a first solution can be found [call it $w_f(z)$] and then a second solution $w_s(z)$ can be found as follows: Take $w(z) = h(z)w_f(z)$ in (13.1), which is already satisfied by $w_f(z)$, the solution found from (13.16). Substitution yields

$$(d_z^2 + [2d_z \ln w_f(z) + p(z)]d_z)h = 0$$

so that

$$-\frac{d_z(d_z h)}{(d_z h)} = -d_z(\ln d_z h) = 2d_z[\ln w_f(z)] + p(z)$$

Hence,

$$-\ln(d_z h) = \ln w_f^2(z) + \int p(z)\, dz - \ln B$$

and

$$h(z) = A + B\int e^{-\int p\, dz} \frac{dz}{w_f^2(z)}$$

with A and B integration constants; $A = 1$, $B = 0$ yields the original solution, while $A = 0$, $B = 1$ yields a new (second) solution $w_s(z)$ with

$$w_s(z) = w_f(z)\int e^{-\int p\, dz} \frac{dz}{w_f^2(z)} \tag{13.19}$$

Since

$$p(z) - \frac{P_0}{z} = \frac{P(z) - P_0}{z}$$

is free of singularity at $z = 0$,

$$\int p\, dz - P_0 \ln z$$

is also analytic and free of singularity at $z = 0$. If in (13.4) $w_0 \neq 0$, then

$$w_s(z) = w_f(z)\int \frac{g(z)\, dz}{z^{P_0+2\alpha}} \tag{13.20}$$

where $g(z)$ is analytic and free of singularity at $z = 0$. Hence,

$$g(z) = \sum_{k=0}^{\infty} g_k z^k \tag{13.21}$$

If α and $\alpha - n$ (n = nonnegative integer) are the roots of the indicial equation

$$P_0 - 1 = -(2\alpha - n) = -2\alpha + n$$

so, $P_0 + 2\alpha = n + 1$ and (13.21) may be inserted into (13.20) yielding

$$\begin{aligned} w_s(z) &= w_f(z)\sum_{k=0}^{\infty} g_k \int z^{k-n-1}\, dz \\ w_s(z) &= w_f(z)\left\{\sum_{k\neq n} g_k \frac{z^{k-n}}{k-n} + g_n \log z\right\} \end{aligned} \tag{13.22}$$

For the case of double root α to the indicial equation this becomes ($n = 0$)

$$w_s(z) = w_f(z)\left\{\sum_{k=1}^{\infty} g_k \frac{z^k}{k} + g_0 \log z\right\} \tag{13.23}$$

The condition for absence of the logarithmic term in (13.22) is then $g_n = 0$. The function $g(z)$ in (13.21) is readily recognized to be

$$g(z) = \frac{z^{P_0+2\alpha} e^{-\int p(z)\,dz}}{w_f^2(z)} \tag{13.24}$$

If $z = 0$ is an ordinary point of (13.1), then $w(z)$ may be taken to be analytic and free of singularity there, since with $\alpha = 0 = P_0 = Q_0 = Q_1$ (13.15) is automatically satisfied and (13.14) becomes ($k \geq 1$)

$$k(k-1)w_k = -\sum_{m=0}^{k-1} (mP_{k-m} + Q_{k-m})w_m \tag{13.25}$$

Aside from the trivial case $k = 1$, one has for $k = 2$

$$-2w_2 = Q_2 w_0 + P_1 w_1$$

and for $k = 3$

$$-3 \cdot 2w_3 = Q_3 w_0 + (P_2 + Q_2)w_1 + 2P_1 w_2$$

for $k = 4$

$$-4 \cdot 3w_4 = Q_4 w_0 + (P_3 + Q_3)w_1 + (2P_2 + Q_2)w_2 + 3P_1 w_3$$

and so on. It is readily seen that by successive substitutions all the coefficients w_k can be expressed in terms of w_0 and w_1 which may then be interpreted as the two arbitrary constants in the general solution of (13.1).

To determine the nature of solutions of (13.1) for large values of $|z|$ it is convenient to use the inversion $z = 1/u$. Then (13.1) becomes

$$\left\{d_u^2 + \left[\frac{2}{u} - \frac{1}{u^2}\, p\left(\frac{1}{u}\right)\right] d_u + \frac{1}{u^4}\, q\left(\frac{1}{u}\right)\right\} w(u) = 0 \tag{13.26}$$

and *the point* ($z = \infty$) *at infinity* is called an *ordinary* point if the origin ($u = 0$) is an ordinary point of (13.26). Equivalently the point at infinity in the z plane is said to be an ordinary point if $2z - z^2 p(z)$ and $z^4 q(z)$ are analytic and free of singularities at $z = \infty$. Again this can be expressed as

$$\begin{aligned} p(z) &= \sum_{k=0}^{\infty} p_{-k} z^{-k} \\ q(z) &= \sum_{k=0}^{\infty} q_{-k} z^{-k} \end{aligned} \tag{13.27}$$

valid for $|z|$ sufficiently large.

The *point at infinity* ($z = \infty$) is called a *regular singular* point if the origin ($u = 0$) is a regular singularity of (13.26). Equivalently the point at infinity is said to be a regular singularity of (13.1) if $zp(z)$ and $z^2q(z)$ are analytic and free of singularities at $z = \infty$. Again this can be expressed by setting $p_0 = q_0 = q_{-1} = 0$ in (13.27). A solution of (13.1) valid for large $|z|$ may be found by substituting a solution of the form

$$w = z^{-\alpha}\sum_{k=0}^{\infty} w_{-k}z^{-k} \tag{13.28}$$

into (13.1) whence an indicial equation for α is found and a second solution is correspondingly found from the second root of this equation provided the roots do *not* differ by a nonnegative integer. Otherwise the second solution is found as before by (depression of order) using the knowledge of the first solution yielded from the indicial equation.

13.2 THE GAUSSIAN HYPERGEOMETRIC FUNCTION

Fortunately most of the second-order linear homogeneous differential equations occurring in practice possess no more than three regular singularities. The archetype of such equations is that of the *hypergeometric differential equation* which has regular singularities with exponents α_1 and α_2 of values $0, 1 - c$ at $z = 0$; a, b at $z = \infty$; $0, c - a - b$ at $z = 1$. This equation is

$$\{z(1 - z)d_z^2 + [c - (a + b + 1)z]d_z - ab\}w(z) = 0 \tag{13.29}$$

Here

$$P(z) = zp(z) = \frac{c - (a + b + 1)z}{1 - z} = \sum_{k=0}^{\infty} P_k z^k$$

with $P_0 = c$ and $P_k = c - (a + b + 1)$ for $k > 0$, while

$$Q(z) = z^2q(z) = \frac{-abz}{1 - z} = \sum_{k=0}^{\infty} Q_k z^k$$

with $Q_0 = 0$ and $Q_k = -ab$ for $k > 0$. Thus the indicial equation (13.15) becomes

$$F(\alpha) = \alpha^2 + (c - 1)\alpha = \alpha(\alpha + c - 1)$$

while (13.16) becomes

$$(k + \alpha)(k + \alpha + c - 1)w_k = -\sum_{m=0}^{k-1}\{(m + \alpha)[c - (a + b + 1)] - ab\}w_m$$

so that

$$(k + 1 + \alpha)(k + \alpha + c)w_{k+1} - (k + \alpha)(k + \alpha + c - 1)w_k = \{ab - (k + \alpha)[c - (a + b + 1)]\}w_k$$

whence

$$(k + 1 + \alpha)(k + c + a)w_{k+1} = (k + a + \alpha)(k + b + \alpha)w_k \tag{13.30}$$

and in particular

$$(\alpha + 1)(\alpha + c)w_1 = (\alpha + a)(\alpha + b)w_0$$
$$(\alpha + 2)(\alpha + c + 1)w_2 = (\alpha + a + 1)(\alpha + b + 1)w_1$$
$$(\alpha + 3)(\alpha + c + 2)w_3 = (\alpha + a + 2)(\alpha + b + 2)w_2$$

etc. so that by successive substitutions one has $(k \geq 1)$

$$w_k = \frac{(\alpha + a + k - 1)\cdots(\alpha + a)(\alpha + b + k - 1)\cdots(\alpha + b)}{(\alpha + c + k - 1)\cdots(\alpha + c)(\alpha + k)\cdots(\alpha + 1)} w_0 \tag{13.31}$$

or

$$w_k = \frac{g(\alpha + a + k)g(\alpha + b + k)}{g(\alpha + c + k)g(\alpha + k + 1)} w_0 \tag{13.32}$$

where $g(x + k) = x(x + 1)\cdots(x + k - 1)$. Furthermore, $g(x + k + 1) = (x + k)g(x + k)$, or with $x + k = y$, $g(y + 1) = yg(y)$ so that by (9.48) $g(y)$ is proportional to $\Gamma(y)$. Hence, (13.32) becomes

$$w_k = \frac{\Gamma(\alpha + a + k)\Gamma(\alpha + b + k)}{\Gamma(\alpha + c + k)\Gamma(\alpha + k + 1)} w_0$$

The two roots of the indicial equation are $\alpha = 0$ and $\alpha = 1 - c$. The solution of (13.29) corresponding to $\alpha = 0$ and $w_0 = \Gamma(c)/\Gamma(a)\Gamma(b)$ is called a (Gaussian) *hypergeometric function* and is denoted by $F(a,b,c,z)$ with

$$\frac{\Gamma(a)\Gamma(b)}{\Gamma(c)} F(a,b,c,z) = \sum_{k=0}^{\infty} \frac{\Gamma(a + k)\Gamma(b + k)}{\Gamma(c + k)k!} z^k \tag{13.33}$$

as its power series convergent within $|z| < 1$. If c is not an integer, the other solution (for $\alpha = 1 - c$) may be taken to be

$$z^{1-c}F(1 + a - c, 1 + b - c, 2 - c, z)$$

The general solution of (13.29) is then

$$AF(a,b,c,z) + Bz^{1-c}F(1 + a - c, 1 + b - c, 2 - c, z)$$

Since the hypergeometric function contains so many parameters in addition to the variable z, it might be supposed that many different types of functions could be obtained for special choices of these. Thus for $b = c$ in (13.33), one has

$$F(a,b,b,z) = \sum_{k=0}^{\infty} \frac{\Gamma(a + k)}{\Gamma(a)} \frac{z^k}{k!}$$

and by (9.48)

$$\frac{\Gamma(a + k)}{\Gamma(a)} = a(a + 1) \cdots (a + k - 1)$$

$$\frac{\Gamma(a + k)}{\Gamma(a)} = (-1)^k(-a)(-a - 1) \cdots (-a - k + 1)$$

Hence,

$$F(a,b,b,z) = \frac{1}{(1 - z)^a}$$

Again for $a = b = 1$, $c = 2$ one has

$$F(1,1,2,z) = \sum_{k=0}^{\infty} \frac{\Gamma(1 + k)\Gamma(1 + k)}{\Gamma(2 + k)k!} z^k$$

$$F(1,1,2,z) = \sum_{k=0}^{\infty} \frac{z^k}{k + 1} = -\frac{1}{z} \log(1 - z)$$

Also,

$$F\left(a,b,b,\frac{z}{a}\right) = \sum_{k=0}^{\infty} \frac{\Gamma(a + k)}{a^k\Gamma(a)} \frac{z^k}{k!}$$

and

$$\lim_{a \to \infty} \frac{\Gamma(a + k)}{a^k\Gamma(a)} = 1$$

so

$$\lim_{a \to \infty} F\left(a,b,b,\frac{z}{a}\right) = e^z$$

Most of the named functions of analysis can be derived from $F(a,b,c,z)$ in one way or another.

The derivative of $F(a,b,c,z)$ may be conveniently expressed in terms of the hypergeometric function as follows:

$$\frac{\Gamma(a)\Gamma(b)}{\Gamma(c)} d_z F(a,b,c,z) = \sum_{k=1}^{\infty} \frac{\Gamma(a + k)\Gamma(b + k)}{\Gamma(c + k)} \frac{z^{k-1}}{(k - 1)!}$$

$$\frac{\Gamma(a)\Gamma(b)}{\Gamma(c)} d_z F(a,b,c,z) = \sum_{m=0}^{\infty} \frac{\Gamma(a + 1 + m)\Gamma(b + 1 + m)}{\Gamma(c + 1 + m)} \frac{z^m}{m!}$$

Multiplying both sides by ab/c, one has

$$d_z F(a,b,c,z) = \frac{ab}{c} F(a + 1,b + 1,c + 1,z) \tag{13.34}$$

Clearly the hypergeometric function

$$F(a,b,c,z) = F(b,a,c,z) \tag{13.35}$$

is symmetric in the first two parameters. Also, if b (or a) is a negative integer $-m$, then

$$\frac{\Gamma(b+k)}{\Gamma(b)} = b(b+1)\cdots(b+k-1)$$

will be zero for $k > m$; otherwise it is evaluated from the residues of $\Gamma(z)$ so (13.33) becomes a polynomial

$$\frac{\Gamma(a)}{\Gamma(c)} F(a,-m,c,z) = \sum_{k=0}^{m} \frac{\Gamma(a+k)m!\,(-z)^k}{\Gamma(c+k)(m-k)!\,k!}$$

or with $a = m + q$

$$F(m+q,-m,c,z) = \frac{\Gamma(c)}{\Gamma(a)} \sum_{k=0}^{m} \binom{m}{k} \frac{\Gamma(a+k)}{\Gamma(c+k)} (-z)^k \tag{13.36}$$

is called $J_m(q,c,z)$ a *Jacobi polynomial.* It will usually be assumed that neither a nor b are negative integers.

It is readily verified that another [besides (13.33)] representation for $F(a,b,c,z)$ is given by the *Barnes integral*

$$\frac{\Gamma(a)\Gamma(b)}{\Gamma(c)} F(a,b,c,z) = \frac{1}{2\pi i} \int_{-i\infty}^{i\infty} \frac{\Gamma(a+s)\Gamma(b+s)\Gamma(-s)(-z)^s\,ds}{\Gamma(c+s)} \tag{13.37}$$

where the path of integration passes to the left of the poles of $\Gamma(-s)$ and to the right of the poles of $\Gamma(a+s)\Gamma(b+s)$ as shown in Figure 13.1. Such a

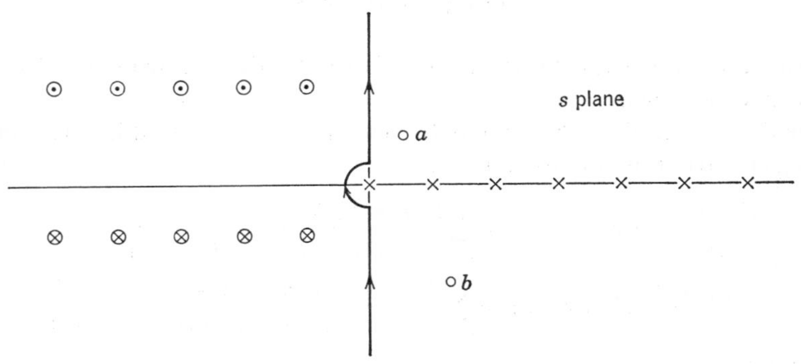

Figure 13.1. Path of integration for Barnes integral (13.37): $\times$ denotes poles of $\Gamma(-s)$; $\otimes$ denotes poles of $\Gamma(a+s)$; $\odot$ denotes poles of $\Gamma(b+s)$.

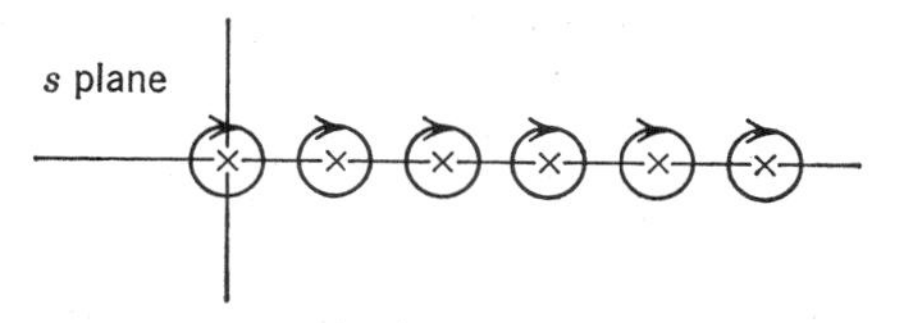

Figure 13.2. Equivalent contour for Barnes integral.

separation of poles is always possible (with excursion loops) if neither a nor b is a negative integer. Since the only singularities of the integrand of (13.37) are simple poles, the Barnes integral is equivalent to a sum of such integrals about small circular contours, each encircling a pole of $\Gamma(-s)$ as shown in Figure 13.2. The residue for $s = m$ is

$$\frac{\Gamma(a+m)\Gamma(b+m)}{\Gamma(c+m)}\left[\frac{-(-1)^m}{m!}\right]\frac{(-1)^m z^m}{1}$$

Taking account of the direction of integration, this yields (13.33).

The advantage of (13.37) over (13.33) is that the former is analytic and free of singularities in the z plane *cut* from $z = 0$ to $z = \infty$ along the nonnegative real axis, whereas the series (13.33) is valid only inside $|z| < 1$.*

Another integral for $F(a,b,c,z)$ is given by

$$\frac{\Gamma(b)\Gamma(c-b)}{\Gamma(c)} F(a,b,c,z) = \int_0^1 t^{b-1}(1-t)^{c-b-1}(1-zt)^{-a}\,dt \tag{13.38}$$

for Re $c >$ Re $b > 0$, which may be verified by noting that

$$\frac{1}{(1-zt)^a} = \sum_{k=0}^{\infty}\frac{\Gamma(a+k)}{\Gamma(a)}\frac{(zt)^k}{k!} \tag{13.39}$$

(obtained by differentiating $1/(1-zt)$ with respect to (zt) m times, replacing $m+1$ by a, and replacing the factorials by gamma functions) and then multiplying (13.39) by $t^{b-1}(1-t)^{c-b-1}$ and integrating. Thus the right-hand side of (13.38) becomes

$$\frac{1}{\Gamma(a)}\sum_{k=0}^{\infty}\frac{\Gamma(a+k)}{k!} z^k \int_0^1 t^{b-1+k}(1-t)^{c-b-1}\,dt$$

which by (9.59) is

$$\frac{1}{\Gamma(a)}\sum_{k=0}^{\infty}\frac{\Gamma(a+k)}{k!} z^k B(b+k,c-b)$$

* See E. T. Copson, *Theory of Functions of a Complex Variable*. Oxford: Oxford University Press, 1935, p. 253.

or, expressing the beta function in terms of gamma functions,

$$\frac{\Gamma(c-b)}{\Gamma(a)}\sum_{k=0}^{\infty}\frac{\Gamma(a+k)\Gamma(b+k)}{\Gamma(c+k)}\frac{z^k}{k!}$$

which is [by (13.33)] recognized as the left side of (13.38).

The behavior of the solution of (13.29) for large $|z|$ may be obtained by first observing that for large $|z|$ (13.29) becomes approximately

$$[z^2d_z^2 + (a+b+1)zd_z + ab]w(z) = 0 \tag{13.40}$$

which has solutions of the form z^α for

$$\alpha(\alpha-1) + (a+b+1)\alpha + ab = (\alpha+a)(\alpha+b) = 0$$

the indicial equation of (13.29) for large $|z|$. Substitution of

$$w(z) = z^{-a}\sum_{k=0}^{\infty} w_{-k}z^{-k}$$

into (13.29) yields for the coefficients

$$w_{-k-1} = \frac{(k+a)(k+a-c+1)}{(k+1)(k+a-b+1)}w_{-k}$$

or

$$z^{-a}F\left(a,a-c+1,a-b+1,\frac{1}{z}\right) \tag{13.41}$$

The corresponding substitution of

$$w(z) = z^{-b}\sum_{k=0}^{\infty} w_{-k}z^{-k}$$

would yield

$$z^{-b}F\left(b,b-c+1,b-a+1,\frac{1}{z}\right) \tag{13.42}$$

Hence, the general solution for large $|z|$ is

$$w = Az^{-a}F\left(a,a-c+1,a-b+1,\frac{1}{z}\right) + Bz^{-b}F\left(b,b-c+1,b-a+1,\frac{1}{z}\right) \tag{13.43}$$

Clearly the series converge for $|z| > 1$. For $|z| > 1$ in the cut z plane there is the relation [more explicit than (13.43)]

$$F(a,b,c,z) = \frac{\Gamma(b-a)\Gamma(c)}{\Gamma(b)\Gamma(c-a)}(-z)^{-a}F\left(a,1+a-c,1+a-b,\frac{1}{z}\right)$$
$$+ \frac{\Gamma(a-b)\Gamma(c)}{\Gamma(a)\Gamma(c-b)}(-z)^{-b}F\left(b,1+b-c,1+b-a,\frac{1}{z}\right)$$

proved by the calculus of residues applied to Barnes integral (13.37).*

13.3 THE CONFLUENT HYPERGEOMETRIC FUNCTION

The differential equation (13.29) has regular singularities at $z = 0$, $z = 1$, and $z = \infty$ as previously discussed. The corresponding equation with z replaced by z/b is

$$\{z(b-z)d_z^2 + [bc - (a+b+1)z]d_z - ab\}w(z) = 0$$

which has regular singularities at $z = 0$, $z = b$, and $z = \infty$. If $b \to \infty$, this becomes

$$\{zd_z^2 + (c-z)d_z - a\}w(z) = 0 \tag{13.44}$$

which is called the *confluent hypergeometric differential equation* and has a regular singularity at $z = 0$ with exponents (indicial roots) 0 and $(1-c)$, while at $z = \infty$ it has an irregular singularity formed by the "confluence" of the regular singularities at $z = b$ and $z = \infty$ of the differential equation preceding (13.44) when $b \to \infty$.

The solutions of (13.44) (for c noninteger) are denoted by $F(a,c,z)$ and $z^{1-c}F(a-c+1,2-c,z)$ with the series converging for all finite values of $|z|$ so that $F(a,c,z)$ is seen to be an entire function. An obvious result of this argument is

$$F(a,c,z) = \lim_{b\to\infty} F\left(a,b,c,\frac{z}{b}\right)$$

This may also be established by carrying out the limiting process on the series solution $F(a,b,c,z/b)$ of the differential equation preceding (13.44).

By using

$$\lim_{b\to\infty} \frac{\Gamma(b+s)}{b^s\Gamma(b)} = 1 \tag{13.45}$$

* See E. T. Copson, *Theory of Functions of a Complex Variable*. Oxford: Oxford University Press, 1935, p. 256.

in (13.37), a Barnes integral for the *confluent hypergeometric* function $F(a,c,z)$ is obtained:

$$\frac{\Gamma(a)}{\Gamma(c)} F(a,c,z) = \frac{1}{2\pi i} \int_{-i\infty}^{i\infty} \frac{\Gamma(a+s)\Gamma(-s)(-z)^s \, ds}{\Gamma(c+s)} \tag{13.46}$$

Since (13.38) [by virtue of the symmetry of $F(a,b,c,z)$ in a and b] can be written as

$$\frac{\Gamma(a)\Gamma(c-a)}{\Gamma(c)} F\left(a,b,c,\frac{z}{b}\right) = \int_0^1 t^{a-1}(1-t)^{c-a-1}\left(1-\frac{zt}{b}\right)^{-b} dt$$

where z has been replaced by z/b and since

$$\lim_{b\to\infty} \left(1-\frac{zt}{b}\right)^{-b} = e^{zt}$$

one has

$$\frac{\Gamma(a)\Gamma(c-a)}{\Gamma(c)} F(a,c,z) = \int_0^1 e^{zt}t^{a-1}(1-t)^{c-a-1}\, dt \tag{13.47}$$

From this it may be noted that

$$\frac{\Gamma(c-a)\Gamma(a)}{\Gamma(c)} F(c-a,c,-z) = \int_0^1 e^{-zt}t^{c-a-1}(1-t)^{a-1}\, dt$$

by replacing a by $(c-a)$ and z by $-z$ in (13.47). With $u = (1-t)$ the right side of this becomes

$$e^{-z}\int_0^1 e^{zu}u^{a-1}(1-u)^{c-a-1}\, du = e^{-z}\frac{\Gamma(c-a)\Gamma(a)}{\Gamma(c)} F(a,c,z)$$

Hence,

$$F(c-a,c,-z) = e^{-z}F(a,c,z) \tag{13.48}$$

Arguing heuristically that (13.47) can be regarded as a Laplace transform with $z = -s$ and

$$f(t) = \begin{cases} t^{a-1}(1-t)^{c-a-1} & \text{for } 0 \le t \le 1 \\ 0 & \text{for } 1 \le t \end{cases}$$

then for Re s large and positive, i.e., Re z large and negative, one may use the binomial theorem

$$t^{a-1}(1-t)^{c-a-1} = \sum_{k=0}^{\infty} \frac{\Gamma(c-a)(-1)^k t^{k+a-1}}{\Gamma(k+1)\Gamma(c-a-k)}$$

in (13.47), obtaining (by Watson's lemma) the asymptotic expansion

$$F(a,c,-s) \sim \frac{\Gamma(c)}{\Gamma(a)} \sum_{k=0}^{\infty} \frac{\Gamma(k+a)(-1)^k}{\Gamma(c-a-k)\Gamma(k+1)s^{a+k}}$$

or

$$F(a,c,z) \sim (-z)^{-a} \frac{\Gamma(c)}{\Gamma(a)} \sum_{k=0}^{\infty} \frac{\Gamma(k+a)}{\Gamma(c-a-k)\Gamma(k+1)} \frac{1}{z^k} \tag{13.49}$$

valid for Re z large and negative. The corresponding result for Re z large and positive can be obtained from (13.48):

$$F(a,c,z) \sim z^{a-c} e^z \frac{\Gamma(c)}{\Gamma(c-a)} \sum_{k=0}^{\infty} \frac{\Gamma(c-a+k)}{\Gamma(a-k)\Gamma(k+1)(-z)^k} \tag{13.50}$$

The leading terms of (13.49) and(13.50) are then

$$F(a,c,z) \sim \begin{cases} z^{a-c} e^z \Gamma(c)/\Gamma(a) & \text{for} \quad \text{Re}\, z > 0 \\ (-z)^{-a} \Gamma(c)/\Gamma(c-a) & \text{for} \quad \text{Re}\, z < 0 \end{cases}$$

Inspection of the Taylor series for $F(a,c,z)$ about $z = 0$,

$$\frac{\Gamma(a)}{\Gamma(c)} F(a,c,z) = \sum_{k=0}^{\infty} \frac{\Gamma(a+k)}{\Gamma(c+k)} \frac{z^k}{k!} \tag{13.51}$$

shows that if a is a negative integer, (13.51) will reduce to a polynomial.

13.4 RELATED SPECIAL FUNCTIONS

A commonly occurring relative of (13.44) is the Whittaker confluent hypergeometric differential equation,

$$[4z^2 d_z^2 + (4kz - z^2 + 1 - 4m^2)] M_{k,m}(z) = 0 \tag{13.52}$$

which for noninteger $2m$ has solutions

$$M_{k,m}(z) = z^{m+(1/2)} e^{-z/2} F(\tfrac{1}{2} - k + m, 1 + 2m, z)$$
$$M_{k,-m}(z) = z^{-m+(1/2)} e^{-z/2} F(\tfrac{1}{2} - k - m, 1 - 2m, z)$$

The *Whittaker functions* are then defined as the linear combination

$$W_{k,m} = W_{k,-m} = \frac{\Gamma(-2m)}{\Gamma(\frac{1}{2} - m - k)} M_{k,m}(z) + \frac{\Gamma(2m)}{\Gamma(\frac{1}{2} + m - k)} M_{k,-m}(z)$$

The related entire *parabolic cylinder functions* are defined by

$$D_\nu(z) = 2^{(1/4)+(\nu/2)} z^{-1/2} W_{(1/4)+(\nu/2),-1/4}\left(\frac{z^2}{2}\right)$$

and satisfy the equation

$$[4d_z^2 + (4\nu + 2 - z^2)] D_\nu(z) = 0 \tag{13.53}$$

Various other special functions of considerable importance may also be derived from the confluent hypergeometric function.

The *generalized Laguerre polynomials* $\mathcal{L}_n^{(\alpha)}(z)$ introduced by Sonine ($\alpha \neq 0$) and Laguerre ($\alpha = 0$) may be defined by ($n \geq 0$)

$$\mathcal{L}_n^{(\alpha)}(z) = \frac{\Gamma(\alpha + n + 1)}{n!\,\Gamma(\alpha + 1)} F(-n,\alpha + 1,z) \tag{13.54}$$

corresponding to the differential equation

$$[zd_z^2 + (1 + \alpha - z)d_z + n]\mathcal{L}_n^{(\alpha)}(z) = 0 \tag{13.55}$$

It is alternatively common to define the *associated Laguerre polynomials*

$$L_n^m(z) = (-1)^m n! \binom{n}{m} F(m - n,m + 1,z) \qquad n \geq m \tag{13.56}$$

corresponding to the differential equation

$$[zd_z^2 + (1 + m - z)d_z + (n - m)]L_n^m = 0 \tag{13.57}$$

Special examples are $L_1^1(z) = -1$, $L_2^1(z) = 2z - 4$, $L_2^2(z) = 2$, $L_3^1(z) = -3z^2 + 18z - 18$, $L_3^2(z) = 18 - 6z$, $L_3^3(z) = -6$, $L_4^1(z) = 4z^3 - 48z^2 + 144z - 96$, $L_4^2(z) = 12z^2 - 96z + 144$, $L_4^3(z) = 24z - 96$, $L_4^4(z) = 24$, etc.

The associated Laguerre functions $\pounds_n^m(z)$ are defined by

$$\pounds_n^m(z) = z^{-m/2}e^{-z/2}L_n^m(z) \tag{13.58}$$

and satisfy the differential equation

$$\{4z^2d_z^2 + 4zd_z + [(4n + 2 - 2m)z - z^2 - m^2]\}\pounds_n^m(z) = 0 \tag{13.59}$$

For $m = 0$ (13.56) and (13.58) are called ordinary (i.e., unassociated) Laguerre polynomials and functions, respectively.

The *Hermite polynomials* $\mathcal{H}_n(z)$ are defined by

$$\mathcal{H}_{2n}(z) = (-1)^n 2^{2n} n!\, \mathcal{L}_n^{(-1/2)}(z^2) \tag{13.60}$$

$$\mathcal{H}_{2n+1}(z) = (-1)^n 2^{2n+1} n!\, z\mathcal{L}_n^{(1/2)}(z^2) \tag{13.61}$$

for nonnegative integers n, and they satisfy the differential equation of Hermite,

$$[d_z^2 - 2zd_z + 2n]\mathcal{H}_n(z) = 0 \tag{13.62}$$

The *Hermite functions* $H_n(z)$ are defined by

$$H_n(z) = e^{-z^2/2}\mathcal{H}_n(z) \tag{13.63}$$

and they satisfy the differential equation*

$$[d_z^2 + (1 + 2n - z^2)]H_n(z) = 0 \tag{13.64}$$

* A special case is the *quantum harmonic oscillator* with $[-(\hbar^2/2m)\, d_x^2 + m\omega^2x^2/2]u_n = E_n u_n$ so that $E_n = [n + (\frac{1}{2})]\hbar\omega$ and $u_n \sim H_n(x\sqrt{m\omega/\hbar})$.

Since all the above differential equations are homogeneous, it should be understood that the normalizations do not stem from the equations themselves which are each satisfied by any constant multiple of the function given here. Because the constant factors adopted above are not universally used by all authors, it should be kept in mind that the terminologies for the functions used by different authors often differ by constant factors. When using formulae from different sources, it is wise to check the normalization conventions.

All the above definitions are given in terms of $F(a,c,z)$, and the differential equations can all be obtained by appropriately modifying (13.44). From (13.46) and (13.47) corresponding integral representations can be derived.

Another extremely useful function is the *error function* Erf z, defined as

$$\operatorname{Erf} z = zF(\tfrac{1}{2},\tfrac{3}{2},-z^2) \tag{13.65}$$

so that substituting $2a = 1$, $2c = 3$ into (13.47) and replacing z by $-z^2$, one obtains

$$2F(\tfrac{1}{2},\tfrac{3}{2},-z^2) = \int_0^1 e^{-z^2 t}\frac{dt}{\sqrt{t}} = \frac{2}{z}\int_0^z e^{-u^2}\,du$$

and

$$\operatorname{Erf} z = \int_0^z e^{-u^2}\,du \tag{13.66}$$

It is readily verified that (13.66) satisfies

$$(d_z^2 + 2zd_z)w(z) = 0 \tag{13.67}$$

The *incomplete gamma function* for Re $\nu > 0$ is defined by

$$\gamma(\nu,z) = z^\nu F(\nu,\nu+1,-z) \tag{13.68}$$

Substitution into (13.47) yields

$$\gamma(\nu,z) = \int_0^z e^{-t}t^{\nu-1}\,dt \tag{13.69}$$

Clearly, $\gamma(\nu,\infty) = \Gamma(\nu)$, the (complete) gamma function. From (13.69) one concludes that $\gamma(\nu,z)$ satisfies

$$[zd_z^2 + (z+1-\nu)d_z]w(z) = 0 \tag{13.70}$$

13.5 BESSEL FUNCTIONS

The *Bessel function* $J_\nu(z)$ *of the first kind* of order ν may be defined by the (Schläfli) integral

$$2^\nu z^{-\nu}J_\nu(z) = \frac{1}{2\pi i}\int_L e^{(\sigma - z^2/4\sigma)}\sigma^{-\nu-1}\,d\sigma \tag{13.71}$$

where L is the same contour used in (9.44) to define the gamma function (counterclockwise about the σ negative real axis). Compare this with the definition of $J_n(z)$ in Example 4 of Chapter 6 by means of a generating function. The right side of (13.71) is an entire function of z which therefore has a MacLaurin expansion convergent for all finite values of $|z|$. Using the series about $z = 0$ for the exponential,

$$e^{-z^2/4\sigma} = \sum_{k=0}^{\infty} \frac{(-1)^k z^{2k}}{k!\,(4\sigma)^k}$$

in (13.71) termwise integration results in

$$2^\nu z^{-\nu} J_\nu(z) = \frac{1}{2\pi i} \sum_{k=0}^{\infty} \frac{(-1)^k z^{2k}}{4^k k!} \int_L e^\sigma \sigma^{-(\nu+k+1)}\, d\sigma$$

which by (9.44) may be written

$$J_\nu(z) = \left(\frac{z}{2}\right)^\nu \sum_{k=0}^{\infty} \frac{(-1)^k z^{2k}}{4^k k!\,\Gamma(\nu + k + 1)} \tag{13.72}$$

By termwise differentiation one has

$$zJ_\nu' = \sum_{k=0}^{\infty} \frac{(-1)^k (2k + \nu) z^{2k+\nu}}{2^\nu 4^k k!\,\Gamma(\nu + k + 1)} = zd_z J_\nu$$

and

$$z(zJ_\nu')' = zd_z(zd_z J_\nu) = \sum_{k=0}^{\infty} \frac{(-1)^k (2k + \nu)^2 z^{2k+\nu}}{2^\nu 4^k k!\,\Gamma(\nu + k + 1)}$$

$$zd_z(zd_z J_\nu) - \nu^2 J_\nu = \sum_{k=1}^{\infty} \frac{(-1)^k (k^2 + k\nu) z^{2k+\nu}}{4^{k-1} k!\,\Gamma(\nu + k + 1) 2^\nu}$$

or, with $k = m + 1$,

$$zd_z(zd_z J_\nu) - \nu^2 J_\nu = -z^2 \sum_{m=0}^{\infty} \frac{(-1)^m (m + 1)(m + 1 + \nu) z^{2m+\nu}}{2^\nu 4^m (m + 1)!\,\Gamma(\nu + m + 2)}$$

Using $\Gamma(\nu + m + 2) = (\nu + m + 1)\Gamma(\nu + m + 1)$ and $(m + 1)! = (m + 1)m!$ the series on the right side is recognized as (13.72). Hence, $J_\nu(z)$ satisfies the *Bessel differential equation*

$$[zd_z(zd_z) + (z^2 - \nu^2)]w(z) = 0 \tag{13.73}$$

For $\nu = n$ an integer (positive, negative, or zero) the integrand of (13.71) is single-valued and possesses as only singularity in the finite σ plane an essential singularity at $\sigma = 0$. Hence, the contour L of (13.71) may be continuously deformed into the circle $2\,|\sigma| = |z| \neq 0$ without passing over any singularity. Thus (13.71) becomes, with $2\sigma = zu$,

$$J_n(z) = \frac{1}{2\pi i} \oint e^{(z/2)[u-(1/u)]} \frac{du}{u^{n+1}} \tag{13.74}$$

which is recognized as the Laurent coefficient formula (6.8) for the function

$$e^{(z/2)[u-(1/u)]} = \sum_{-\infty}^{\infty} J_n(z)u^n \tag{13.75}$$

which is called the (Schlömilch) *generating function for the Bessel functions* of the first kind. On the unit circle $u = e^{i\theta}$ in the u plane (13.74) becomes $([u - (1/u)] = 2i \sin \theta)$

$$J_n(z) = \frac{1}{2\pi} \int_{-\pi}^{\pi} e^{-in\theta} e^{iz \sin \theta} \, d\theta \tag{13.76}$$

or *Bessel's integral*

$$J_n(z) = \frac{1}{\pi} \int_0^{\pi} \cos(n\theta - z \sin \theta) \, d\theta \tag{13.77}$$

because $(n\theta - z \sin \theta)$ is odd in θ.

Since

$$e^{(z/2)[u-(1/u)]} e^{(w/2)[u-(1/u)]} = e^{[(z+w)/2][u-(1/u)]}$$

Cauchy multiplication of the corresponding Laurent series yields

$$J_n(z + w) = \sum_{-\infty}^{\infty} J_m(z) J_{n-m}(w) \tag{13.78}$$

a result called the (Neumann) *addition theorem* for the Bessel functions $J_n(z)$.

The transformation $u = e^w$ takes the unit circle in the u plane into the line segment $(-i\pi, i\pi)$, applying this to (13.74) yields

$$J_n(z) = \frac{1}{2\pi i} \int_{-\pi i}^{\pi i} e^{z \sinh w - nw} \, dw \tag{13.79}$$

Since the integrand has no finite singularity in w, the path of integration may be any one joining the end points.

Substitution of $u = e^{i\theta}$ into (13.75) leads to the Fourier series

$$e^{iz \sin \theta} = \sum_{-\infty}^{\infty} J_n(z) e^{in\theta} \tag{13.80}$$

whence (13.76) may be interpreted as a Fourier coefficient formula.

From (13.76) one has also

$$J_n(-z) = \frac{1}{2\pi} \int_{-\pi}^{\pi} e^{-in\theta} e^{-iz \sin \theta} \, d\theta = J_{-n}(z) \tag{13.81}$$

while replacement of θ by $(\theta + \pi)$ in (13.80) leads to the conclusion that

$$J_n(-z) = (-1)^n J_n(z) \tag{13.82}$$

so that even-order functions are even and odd-order functions are odd. For $\theta = 0$, then, (13.80) yields

$$1 = \sum_{-\infty}^{\infty} J_n(z) = J_0(z) + 2\sum_{1}^{\infty} J_{2n}(z)$$

The function $J_\nu(z)$ is not the only solution of (13.73). The indicial equation for (13.73) is $\alpha^2 - \nu^2 = 0$, so that $\alpha = \pm\nu$. For ν a noninteger, $J_\nu(z)$ and $J_{-\nu}(z)$ are linearly independent solutions of (13.73) which then has a general solution which may be written

$$w(z) = AJ_\nu(z) + BJ_{-\nu}(z) \tag{13.83}$$

For ν an integer (including zero) $J_{-\nu}(z)$ is not linearly independent of $J_\nu(z)$ but is in fact a multiple of it, as (13.81) and (13.82) show. In this case recourse must be had to (13.19) to find a second linearly independent solution of (13.73).

This can be done in several ways giving rise to several different versions of *Bessel functions of the second kind*. They generally have the property [unlike $J_\nu(z)$] of becoming infinite at $z = 0$.

The most widely adopted of these is the Weber–Schläfli Bessel function of the second kind, which may be defined by

$$Y_\nu(z) = \frac{\cos \pi\nu J_\nu(z) - J_{-\nu}(z)}{\sin \pi\nu} \tag{13.84}$$

or its limit as ν approaches an integer.*

The general solution of (13.73) can then be written

$$w(z) = AJ_\nu(z) + BY_\nu(z) \tag{13.85}$$

If (13.73) is written as

$$\left[d_z^2 + \frac{1}{z}d_z + 1 - \frac{\nu^2}{z^2}\right]w = 0 \tag{13.86}$$

and it is supposed that $|z|$ is so large that (13.86) could be approximated by

$$(d_z^2 + 1)w = 0 \tag{13.87}$$

the general solution of the latter could be written as

$$w = A\cos z + B\sin z \tag{13.88}$$

* For a full discussion of the functions of the second kind one may refer to G. N. Watson, *A Treatise on the Theory of Bessel Functions*, Cambridge: Cambridge University Press, 1945, p. 57.

In some respects $J_\nu(z)$ and $Y_\nu(z)$ can be thought of as analogous to cos and sin z. They are oscillatory though not of constant amplitude or frequency and $Y_\nu(z)$ becomes infinite for $z = 0$, as mentioned above.

An equation closely related to that of Bessel is obtained by replacing z by iz in (13.86), which then becomes

$$\left[d_z^2 + \frac{1}{z}d_z - \left(1 + \frac{\nu^2}{z^2}\right)\right]w = 0 \tag{13.89}$$

with general solution

$$w = AI_\nu(z) + BK_\nu(z) \tag{13.90}$$

where $i^\nu I_\nu(z) = J_\nu(iz)$ and

$$-K_\nu(z) = \frac{\pi[I_\nu(z) - I_{-\nu}(z)]}{2 \sin \pi\nu}$$

The function $I_\nu(z)$ is called the *modified Bessel function* of the *first kind* and $K_\nu(z)$ is the *modified Bessel function* of the *second kind.* If for large values of $|z|$ (13.89) is approximated by

$$(d_z^2 - 1)w = 0 \tag{13.91}$$

with general solution

$$w = Ae^z + Be^{-z} \tag{13.92}$$

it can be seen that $I_\nu(z)$ and $K_\nu(z)$ can be expected to behave somewhat like positive and negative exponentials for large $|z|$. Actually it is readily discerned that $z^{-\nu}I_\nu(z)$, like $z^{-\nu}J_\nu(z)$, is entire (free of singularities in the finite z plane), while $K_\nu(z)$ has a singularity only at $z = 0$.

Just as it is often the case (especially in wave propagation problems) that the general solution of (13.87) is more conveniently expressed [than by (13.88)] as

$$w = Ae^{iz} + Be^{-iz} \tag{13.93}$$

so is it also often more convenient to take the general solution of (13.73) in the form

$$w(z) = AH_\nu^{(1)}(z) + BH_\nu^{(2)}(z) \tag{13.94}$$

where the *Bessel functions of the third kind*, or *Hankel functions*, of order ν are defined by

$$\begin{aligned} H_\nu^{(1)}(z) &= J_\nu(z) + iY_\nu(z) \\ H_\nu^{(2)}(z) &= J_\nu(z) - iY_\nu(z) \end{aligned} \tag{13.95}$$

analogous to the (Euler) relations

$$\begin{aligned} e^{iz} &= \cos z + i \sin z \\ e^{-iz} &= \cos z - i \sin z \end{aligned}$$

The Hankel functions occur most commonly in cylindrical wave propagation and related quantum problems were they play a role analogous to that of complex exponentials in the solutions of the one-dimensional wave equation.

Another way of defining the Hankel functions is as follows: It is readily verified that the substitution $2\sigma = ze^u$ applied to (13.71) yields

$$J_\nu(z) = \int_{£} e^{z \sinh u - \nu u}\, du \tag{13.96}$$

where £ is the image in the u plane of L under the transformation $2\sigma = ze^u$ for any fixed z. This suggests that there may be other solutions [than $J_\nu(z)$] of (13.73) with integrands of the same form as in (13.96) but with *different* paths of integration. If

$$w = \int_a^b e^{z \sinh u - \nu u}\, du \tag{13.97}$$

is substituted into (13.73) or (13.86), one has

$$[(zd_z)^2 + (z^2 - \nu^2)]w = \int_a^b e^{z \sinh u - \nu u}[z^2 \cosh^2 u + z \sinh u - \nu^2]\, du$$

$$0 = e^{z \sinh u - \nu u}[z \cosh u + \nu]_a^b$$

In other words, a and b must be so chosen that the result is zero in order for (13.97) to be a solution of Bessel's equation (13.73). Since the integrated function is entire with an essential singularity at $u = \infty$, it might be expected that it could be made to approach zero as u approaches ∞ appropriately. Taking $a = -N + \alpha i$, one has [with $\nu = \lambda + i\mu$ and $|\arg z - \alpha| \leq (\pi/2) - \delta$]

$$\operatorname{Re}(z \sinh u) \leq \frac{|z|}{2}(e^{-N} - e^N \sin \delta)$$

Hence, for fixed z,

$$\lim_{n \to \infty} |e^{z \sinh a - \nu a}(z \cosh a + \nu)| = 0$$

because of the factor $e^{-(|z|/2)e^N \sin \delta}$. For $b = N + \beta i$, with $\beta = \pm\pi - \alpha$, there is a similar argument. Hence, $a = -\infty + i\alpha$ and $b = \infty + i\beta$ are integration limits for which (13.97) is a solution of (13.73). This is the method given by Copson* which leads to the alternative definition of the Hankel functions:

$$\begin{aligned} i\pi H_\nu^{(1)}(z) &= \int_{-\infty + \alpha i}^{\infty + (\pi - \alpha)i} e^{z \sinh u - \nu u}\, du \\ -i\pi H_\nu^{(2)}(z) &= \int_{-\infty + \alpha i}^{\infty - (\pi + \alpha)i} e^{z \sinh u - \nu u}\, du \end{aligned} \tag{13.98}$$

* E. T. Copson, *Theory of Functions of a Complex Variable*. Oxford: Oxford University Press, 1935, Section 12.3.

13.6 LEGENDRE AND ASSOCIATED LEGENDRE FUNCTIONS

The *associated Legendre function of the first kind* is defined by the (Schläfli) integral

$$P_n^m(z) = \frac{(n+m)!}{2^n n!} \frac{(z^2-1)^{m/2}}{2\pi i} \oint \frac{(\sigma^2-1)^n \, d\sigma}{(\sigma - z)^{n+m+1}} \tag{13.99}$$

where the (simple) closed contour of integration encircles the point $\sigma = z$ in the σ plane. By differentiation ($n + m$ times) of the Cauchy integral theorem applied to the function $(z^2 - 1)^n$, one readily concludes

$$2^n n! \, P_n^m(z) = (z^2-1)^{m/2} d_z^{n+m}[(z^2-1)^n] \tag{13.100}$$

the *generalized Rodrigues formula*, whence it is seen that if $m > n$, $P_n^m(z) = 0$, which could also have been concluded from the definition, since the residue would then be zero. Also, if $m < -n$, the integrand in the definition would not possess a singularity, so again $P_n^m(z) = 0$. Here it is assumed that n is a nonnegative integer and m is an integer. Thus it has been concluded that $|m| > n$ implies $P_n^m(z) = 0$.

For $m = 0$ the function becomes $P_n^0(z)$, which is usually denoted simply by $P_n(z)$ and called a *Legendre polynomial*. The corresponding Rodrigues formula becomes

$$2^n n! \, P_n(z) = d_z^n (z^2-1)^n \tag{13.101}$$

Comparison of (13.100) and (13.101) yields

$$P_n^m(z) = (z^2-1)^{m/2} d_z^m P_n(z) \tag{13.102}$$

while (13.99) becomes (for $m = 0$)

$$P_n(z) = \frac{1}{2^{n+1} \pi i} \oint \frac{(\sigma^2-1)^n \, d\sigma}{(\sigma - z)^{n+1}} \tag{13.103}$$

Therefore,

$$\begin{aligned} 2\pi i d_z[(z^2-1) d_z P_n] &= \frac{(n+1)}{2^n} \oint \frac{(\sigma^2-1)^n}{(\sigma-z)^{n+3}} \\ &\quad \times [2z(\sigma - z) + (z^2-1)(n+2)] \, d\sigma \\ &= \frac{(n+1)}{2^n} \oint \frac{(\sigma^2-1)^n}{(\sigma-z)^{n+3}} \\ &\quad \times [n(\sigma-z)^2 - n\sigma^2 + 2n\sigma z + 2z\sigma - n - 2] \, d\sigma \end{aligned} \tag{13.104}$$

and since

$$d_\sigma\left[\frac{(\sigma^2-1)^{n+1}}{(\sigma-z)^{n+2}}\right] = [n\sigma^2 - 2n\sigma z - 2z\sigma + n + 2]\,\frac{(\sigma^2-1)^n}{(\sigma-z)^{n+3}}$$

(13.104) becomes (integral of derivative vanishes)

$$2\pi i d_z[(z^2-1)d_z P_n] = \frac{n(n+1)}{2^n}\oint \frac{(\sigma^2-1)^n\, d\sigma}{(\sigma-z)^{n+1}} = 2\pi i n(n+1)P_n$$

by (13.103). Hence the Legendre polynomial $P_n(z)$ is seen to satisfy the *Legendre differential equation*

$$d_z[(z^2-1)d_z w] = n(n+1)w \tag{13.105}$$

with $w = P_n(z)$. By using the Leibniz rule,*

$$d_z^m(ab) = \sum_{k=0}^{m}\binom{m}{k}(d_z^{m-k}a)(d_z^k b)$$

(13.105) may be differentiated m times, yielding

$$[(z^2-1)d_z^2 + 2(m+1)zd_z + m(m+1)]d_z^m w = n(n+1)d_z^m w \tag{13.106}$$

whence substitution from (13.102)

$$d_z^m w = (z^2-1)^{-m/2}P_n^m(z) = (z^2-1)^{-m/2}w$$

leads to the conclusion that the $P_n^m(z)$ satisfies the *associated Legendre equation* with $w = P_n^m(z)$:

$$(z^2-1)d_z[(z^2-1)d_z w] = [n(n+1)(z^2-1) + m^2]w \tag{13.107}$$

If the differential operator of (13.106),

$$(z^2-1)d_z^2 + 2(m+1)zd_z + m(m+1) - n(n+1)$$

is transformed by $z = 1 - 2\zeta$, one has $dz = -2d\zeta$ and $2d_z = -d_\zeta$ and $z^2 - 1 = 4\zeta(\zeta - 1)$, so that the operator may be written as

$$-[\zeta(1-\zeta)d_\zeta^2 + (m+1)(1-2\zeta)d_\zeta - (m-n)(n+m+1)]$$

which is recognized as the operator of the hypergeometric equation for $a = m - n$, $b = m + n + 1$, $c = m + 1$. Since the solution of (13.106) is

$$d_z^m w = (z^2-1)^{-m/2}P_n^m(z)$$

* P. Franklin, *Treatise on Advanced Calculus*. New York: Wiley, 1940, p. 111.

one conclues that $P_n^m(z)$ is given by

$$P_n^m(z) = \frac{(n+m)!\,(z^2-1)^{m/2}}{2^m(n-m)!\,m!} F\left(m-n, m+n+1, m+1, \frac{1-z}{2}\right) \qquad (13.108)$$

where the constant factor is obtained by evaluating the integral of (13.99) for $z = 1$ and recalling that the hypergeometric function is unity at the origin. For $m = 0$ the Legendre polynomial $P_n(z)$ is expressed in terms of the hypergeometric function $F(-n,n+1,1,(1-z)/2)$.

If in (13.99) for fixed z the contour of integration is chosen to be the circle $|\sigma - z| = \sqrt{|z^2-1|}$ so that $\sigma = z + (z^2-1)^{1/2}e^{i\theta}$ (where the square root has its principal value and θ varies between $-\pi$ and π), then (13.99) becomes

$$P_n^m(z) = \frac{(n+m)!}{2\pi n!}\int_{-\pi}^{\pi}(z + \sqrt{z^2-1}\cos\theta)^n e^{im\theta}\,d\theta \qquad (13.109)$$

since

$$\sigma^2 - 1 = (z^2-1) + (z^2-1)e^{2i\theta} + 2ze^{i\theta}\sqrt{z^2-1}$$

may be written in the form

$$\sigma^2 - 1 = 2e^{i\theta}\sqrt{z^2-1}\,(z + \sqrt{z^2-1}\cos\theta)$$

Expressing $e^{im\theta} = \cos m\theta + i\sin m\theta$, (13.109) may be written as

$$P_n^m(z) = \frac{(n+m)!}{\pi n!}\int_0^{\pi}(z + \sqrt{z^2-1}\cos\theta)^n \cos m\theta\,d\theta \qquad (13.110)$$

which is called the Laplace first integral for $P_n^m(z)$.

For $m = 0$ (13.110) becomes

$$\pi P_n(z) = \int_0^{\pi}(z + \sqrt{z^2-1}\cos\theta)^n\,d\theta \qquad (13.111)$$

whence

$$\sum_{n=0}^{\infty} P_n(z)h^n = \frac{1}{\pi}\int_0^{\pi}\sum_{n=0}^{\infty}[h(z + \sqrt{z^2-1}\cos\theta)]^n\,d\theta$$

and

$$\sum_{n=0}^{\infty} P_n(z)h^n = \frac{1}{\pi}\int_0^{\pi}\frac{d\theta}{1 - h(z + \sqrt{z^2-1}\cos\theta)} \qquad (13.112)$$

which may be evaluated by substituting $w = e^{i\theta}$ and applying the residue theorem to yield

$$\frac{1}{\sqrt{1-2hz+h^2}} = \sum_{n=0}^{\infty} P_n(z)h^n \qquad (13.113)$$

the *generating function for the Legendre polynomials.* If this is differentiated m times with respect to z, one has

$$\frac{h^m(2m-1)!!}{(1-2hz+h^2)^{m+(1/2)}} = \sum_{n=0}^{\infty} d_z^m P_n(z) h^n \tag{13.114}$$

with

$$(2m-1)!! = \prod_{k=1}^{m}(2k-1)$$

Substituting from (13.102) into (13.114), one has

$$\frac{(2m-1)!!\,(z^2-1)^{m/2}h^m}{(1-2hx+h^2)^{m+(1/2)}} = \sum_{n=m}^{\infty} P_n^m(z)h^n \tag{13.115}$$

the *generating function for the associated Legendre functions of the first kind.* (For $n < m$ the terms are all zero.)

The recurrence formulae can then be deduced from the generating functions as previously done in examples on the Bessel functions. (See Chapter 6.)

Both (13.105) and (13.107) possess second solutions $Q_n(z)$ and $Q_n^m(z)$, respectively called *Legendre functions of the second kind* and *associated Legendre functions of the second kind.* These may be found explicitly by exploiting the $P_n(z)$ and $P_n^m(z)$ to reduce the equations to first order. These functions Q_n and Q_n^m also satisfy the Rodrigues relations. The reduction of order procedure leads to a second solution of (13.105):

$$w = P_n(z)\int \frac{dz}{(z^2-1)[P_n(z)]^2} \tag{13.116}$$

Since the zeros z_k of $P_n(z)$ are distinct from one another and from ± 1, the integrand may be expressed as

$$\frac{1}{(z^2-1)[P_n(z)]^2} = \frac{1}{2(z-1)} - \frac{1}{2(z+1)} + \sum_{k=1}^{n}\left\{\frac{A_k}{z-z_k} + \frac{B_k}{(z-z_k)^2}\right\}$$

Here the A_k are the residues at $z = a_k$, which can be verified to be zero by subtracting $B_m/(z-z_m)^2$ from each side, multiplying by $(z-z_m)$ and taking the limit as $z \to z_m$. It is readily seen that

$$B_k = \frac{1}{(z_k^2-1)[P_n(z_k)]^2} \tag{13.117}$$

Carrying out the integration, one then has

$$w = P_n(z)\left\{\frac{1}{2}\log\left(\frac{z-1}{z+1}\right) - \frac{B_k}{z-z_k}\right\}$$

The second solution $Q_n(z)$ is then chosen to be

$$Q_n(z) = \tfrac{1}{2}P_n(z)\log\left(\frac{z+1}{z+1}\right) + \frac{B_k P_n(z)}{z - z_k} \tag{13.118}$$

where the second term is a polynomial of degree $(n - 1)$, since z_k is a root of $P_n(z)$. This Legendre function of the second kind thus has logarithmic branch points at $z = \pm 1$ and is discontinuous on the cut between -1 and $+1$. By expanding $Q_n(z)$ in reciprocal powers of z for $|z| > 1$ one can establish*

$$Q_n(z) = \frac{n!\sqrt{\pi}\,F[(n+1)/2,(n/2)+1,n+(3/2),1/z^2]}{2^{n+1}\Gamma[n+(3/2)]z^{n+1}} \tag{13.119}$$

The first few Legendre functions are explicitly

$$P_0(z) = 1 \qquad Q_0(z) = \tfrac{1}{2}\log\left(\frac{z+1}{z-1}\right)$$

$$P_1(z) = z \qquad Q_1(z) = \frac{z}{2}\log\left(\frac{z+1}{z-1}\right) - 1$$

$$P_2(z) = \frac{3z^2 - 1}{2} \qquad Q_2(z) = \frac{P_2(z)}{2}\log\left(\frac{z+1}{z-1}\right) = \frac{3z}{2}$$

$$P_3(z) = \frac{5z^3 - 3z}{2} \qquad Q_3(z) = \frac{P_3(z)}{2}\log\left(\frac{z+1}{z-1}\right) - \frac{5z^2}{2} + \frac{2}{3}$$

By using the Legendre duplication formula and the integral for the beta function in the series for (13.119) it can be established that

$$-2^{n+1}Q_n(z) = \int_{-1}^{1}\frac{(\sigma^2 - 1)^n\,d\sigma}{(\sigma - z)^{n+1}} \tag{13.120}$$

in the z plane cut along the real axis from -1 to $+1$. By differentiating m times with respect to z and using the analog of (13.102) for $Q_n^m(z)$, one has

$$-2^{n+1}Q_n^m(z) = \frac{(n+m)!}{n!}(z^2 - 1)^{m/2}\int_{-1}^{1}\frac{(\sigma^2 - 1)^n\,d\sigma}{(\sigma - z)^{n+m+1}} \tag{13.121}$$

Since $Q_n(z)$ is free of singularities in the cut plane, Cauchy's integral theorem (for the exterior) yields $[Q_n(\infty) = 0]$

$$Q_n(z) = \frac{1}{2\pi i}\oint Q_n(\sigma)\frac{d\sigma}{z - \sigma} \tag{13.122}$$

* E. T. Copson, *Theory of Functions of a Complex Variable*. Oxford: Oxford University Press, 1935, p. 288.

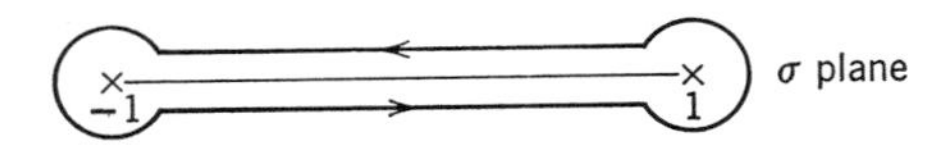

Figure 13.3. The σ plane contour of (13.122).

where the contour surrounds the cut $[-1,1]$. This is equivalent to a contour as shown in Figure 13.3.

Because of (13.118) the portions about the small circles go to zero as the radii do so. Thus,

$$Q_n(z) = \lim_{\varepsilon \to 0} \frac{1}{2\pi i} \int_{-1}^{1} [Q_n(x - i\varepsilon) - Q_n(x + i\varepsilon)] \frac{dx}{z - x}$$

and by (13.118) (change in $\arg[(z + 1)/(z - 1)]$)

$$\lim_{\varepsilon \to 0} [Q_n(x - i\varepsilon) - Q_n(x + i\varepsilon)] = \pi i P_n(x)$$

so that

$$2Q_n(z) = \int_{-1}^{1} \frac{P_n(x)\, dx}{z - x} \tag{13.123}$$

Neumann's integral for $Q_n(z)$.

13.7 STURM–LIOUVILLE EQUATION AND ORTHOGONAL SERIES

Most of the special functions considered in this chapter are solutions of the (self-adjoint) Sturm–Liouville differential equation

$$Lw = \{d_z[a(z)d_z] - [b(z) - \lambda\rho(z)]\}w(z) = 0 \tag{13.124}$$

Under appropriate boundary conditions there will exist solutions of this equation satisfying these boundary conditions only if λ takes certain discrete, characteristic, enumerable eigenvalues. If λ_n is such an eigenvalue, the corresponding eigenfunction would be denoted by $w_n(z)$. The operator which (acting on w) generates a scalar multiple λ of w is then

$$\frac{b}{\rho} - \frac{1}{\rho} d_z(ad_z) \tag{13.125}$$

or, in terms of $w\sqrt{\rho}$,

$$\left\{\frac{b}{\rho} - \frac{1}{\sqrt{\rho}} d_z\left[ad_z\left(\frac{1}{\sqrt{\rho}}\right)\right]\right\}(w\sqrt{\rho}) = \lambda(w\sqrt{\rho}) \tag{13.126}$$

If any two distinct eigenvalues $\lambda_m \neq \lambda_n$ correspond to two distinct eigenfunctions $w_m(z)$ and $w_n(z)$ of the Sturm–Liouville equation for the range $\alpha \leq z \leq \beta$ (with z real), one has

$$\int_\alpha^\beta [w_m L w_n - w_n L w_m] \, dz = 0$$

$$\int_\alpha^\beta \begin{vmatrix} w_m & d_z(ad_z w_m) \\ w_n & d_z(ad_z w_n) \end{vmatrix} dz + (\lambda_n - \lambda_m) \int_\alpha^\beta \rho w_m w_n \, dz = 0$$

or

$$\int_\alpha^\beta d\left[\begin{vmatrix} w_m & d_z w_m \\ w_n & d_z w_n \end{vmatrix} a\right] + (\lambda_n - \lambda_m) \int_\alpha^\beta \rho w_m w_n \, dz = 0 \quad (13.127)$$

so that if the boundary conditions are so chosen that the first integral vanishes (self-adjoint boundary conditions), then $\lambda_m \neq \lambda_n$ implies

$$\int_\alpha^\beta \rho w_m w_n \, dz = \int_\alpha^\beta (w_m \sqrt{\rho})(w_n \sqrt{\rho}) \, dz = 0 \quad (13.128)$$

which condition may be expressed by calling $w_n\sqrt{\rho}$ and $w_m\sqrt{\rho}$ orthogonal over the range (α,β) or by calling w_n and w_m *orthogonal with respect to the weighting factor* ρ over the range (α,β). Under these conditions the set of eigenfunctions will span a function space so that any function in this space can be represented as a linear combination of this set of mutually orthogonal (with respect to ρ) eigenfunctions. These (series) linear combinations of eigenfunctions then are analogous to Fourier series and indeed are often called *generalized Fourier series*. For any function $w(z)$ in the space spanned by $\{w_n(z)\}$ one has

$$w(z) = \sum_n c_n w_n(z)$$

so that

$$\int_\alpha^\beta w(z) w_m(z) \rho(z) \, dz = c_m \int_\alpha^\beta w_m^2(z) \rho(z) \, dz$$

by the orthogonality of w_m and w_n. For $w_m(z)$ real and $\rho(z) \geq 0$ for $\alpha < z < \beta$ and $\rho(z) > 0$ for some subinterval of (α,β) the existence of c_m may be concluded, and the formula for c_m is known as the *generalized Fourier coefficient formula*. The reciprocal formulae

$$w(z) = \sum_n c_n w_n(z) \quad (13.129)$$

$$c_n = \int_\alpha^\beta \rho(z) w(z) w_m(z) \, dz \Big/ \int_\alpha^\beta \rho(z) w_m^2(z) \, dz \quad (13.130)$$

Table 13.1

$$Lu_n = \lambda_n u_n$$

$$\langle u_k, \rho u_n \rangle = \int_\alpha^\beta u_k u_n \rho \, dz$$

Eigenfunction, u_n	Operator, L	Range, $\alpha \le z \le \beta$		Weight, ρ	Eigenvalue, λ_n	Normalization, $\langle u_k, \rho u_n \rangle$
$P_n(z)$	$d_z[(z^2-1)\, d_z]$	-1	1	1	$n(n+1)$	$2\delta_{kn}/(2n+1)$
$P_n^m(z)$	$d_z[(z^2-1)\, d_z] + \dfrac{m^2}{1-z^2}$	-1	1	1	$n(n+1)$	$2\delta_{kn}(n+m)!/(n-m)!\,(2n+1)$
$T_n(z)$	$-\sqrt{1-z^2}\, d_z[\sqrt{1-z^2}\, d_z]$	-1	1	$(1-z^2)^{-1/2}$	n^2	$\dfrac{\pi}{2}\delta_{kn}$ (for $n=0$, $\pi\delta_{k0}$)
$J_n(p,q,z)$	$-\dfrac{d_z[z^q(1-z)^{p-q+1}\, d_z]}{(1-z)^{p-q}z^{q-1}}$	-1	1	$(1-z)^{p-q}z^{q-1}$	$n(p+n)$	$\dfrac{n!\,\Gamma(n+q)\Gamma(n+p)\Gamma(n+p-q+1)}{(2n+p)\Gamma^2(2n+p)}\delta_{kn}$
$L_n(z)$	$-e^z\, d_z[ze^{-z}\, d_z]$	0	∞	e^{-z}	n	$(n!)^2\,\delta_{kn}$
$L_n^m(z)$	$-\dfrac{d^z[z^{m+1}e^{-z}\, d_z]}{z^m e^{-z}} + m$	0	∞	$z^m e^{-z}$	n	$(n!)^3\,\delta_{kn}/(n-m)!$
$z^{-m/2}e^{-z/2}L_n^m(z)$	$-d_z(z\, d_z) + \left(\dfrac{m^2}{4z} + \dfrac{z}{4} + \dfrac{m-1}{2}\right)$	0	∞	z^{2m+1}	n	$(n!)^3\,\delta_{kn}(2n-m+1)/(n-m)!$
$\mathscr{H}_n(z)$	$-e^{z^2}\, d_z(e^{-z^2}\, d_z)$	$-\infty$	∞	e^{-z^2}	$2n$	$2^n n!\,\delta_{kn}\sqrt{\pi}$
$H_n(z)$	$-d_z^2 + (z^2-1)$	$-\infty$	∞	1	$2n$	$2^n n!\,\delta_{kn}\sqrt{\pi}$

on the basis of a knowledge of $\{w_n(z)\}$, permit one to calculate $w(z)$ from a knowledge of $\{c_n\}$ or $\{c_n\}$ from a knowledge of $w(z)$. The information about the generalized Fourier coefficients is thus seen to be equivalent to the information about the function $w(z)$.

Because of the generality of this Fourier argument it helps to unify the properties and application of the special solutions of the Sturm–Liouville equation. Particular examples are shown in Table 13.1.

EXAMPLES 13

1. Prove that $\int_{-1}^{1} P_s(z)P_\sigma(z)\,dz = 0$ for $s \neq \sigma$.

Ans. $$I_{s\sigma} = \int_{-1}^{1} P_s P_\sigma\,dz = \frac{1}{s(s+1)}\int_{-1}^{1} P_\sigma d_z[(z^2-1)d_z P_s]\,dz$$

by the Legendre equation for P_s;

$$I_{s\sigma} = -\frac{1}{s(s+1)}\int_{-1}^{1}(z^2-1)d_z P_s d_z P_\sigma\,dz$$

by integration by parts;

$$I_{s\sigma} = \frac{1}{s(s+1)}\int_{-1}^{1} P_s d_z[(z^2-1)d_z P_\sigma]\,dz$$

by integration by parts. Invoking the Legendre equation for P_σ, one has

$$I_{s\sigma} = \frac{\sigma(\sigma+1)}{s(s+1)}\int_{-1}^{1} P_s P_\sigma\,dz = \frac{\sigma(\sigma+1)}{s(s+1)} I_{s\sigma}$$

For $s \neq \sigma$ this yields $I_{s\sigma} = 0$.

2. Prove that $\int_{-1}^{1} [P_s(z)]^2\,dz = 2/(2s+1)$.

Ans. $$I_s = \int_{-1}^{1} [P_s(z)]^2\,dz = \frac{1}{2^{2s}(s!)^2}\int_{-1}^{1} d_z^s(z^2-1)^s d_z^s(z^2-1)^s\,dz$$

by Rodrigues formula (13.92). Then s integrations by parts yields

$$I_s = \frac{(-1)^s}{2^{2s}(s!)^2}\int_{-1}^{1} d_z^{2s}[(z^2-1)^s](z^2-1)^s\,dz$$

but if $(z^2-1)^s$ is differentiated $2s$ times only, the leading term z^{2s} will lead to a nonvanishing result, and this will be $(2s)!$

Hence,

$$I_s = \frac{(-1)^s(2s)!}{2^{2s}(s!)^2}\int_{-1}^{1}(z^2 - 1)^s\,dz$$

The integral

$$\int_{-1}^{1}(z^2 - 1)^s\,dz = \int_{-1}^{1}(z - 1)^s(z + 1)^s\,dz$$

yields, after s integrations by parts,

$$-\frac{s}{s+1}\int_{-1}^{1}(z - 1)^{s-1}(z + 1)^{s+1}\,dz = \frac{(-1)^s[s!]^2}{(2s)!}\int_{-1}^{1}(z + 1)^{2s}\,dz$$

which is $(-1)^s[s!]^2 2^{2s+1}/(2s)!(2s + 1)$, so that $I_s = 2/(2s + 1)$.

3. Use the results,

$$\int_{-1}^{1}P_s(z)P_\sigma(z)\,dz = \frac{2\delta_{s\sigma}}{2s + 1}$$

of Examples 1 and 2 to determine f_k in the expansion of a given function $f(z)$ in terms of Legendre polynomials

$$f(z) = \sum_{k=0}^{\infty} f_k P_k(z)$$

Ans. Multiplying both sides by $P_m(z)$ and integrating, one has

$$\int_{-1}^{1}f(z)P_m(z)\,dz = \sum_{k=0}^{\infty} f_k\int_{-1}^{1}P_m(z)P_k(z)\,dz$$

which by the previous results is

$$\sum_k f_k\left(\frac{2\delta_{mk}}{2m + 1}\right) = \frac{2}{2m + 1}\sum_k f_k\delta_{mk} = \frac{2f_m}{2m + 1}$$

Hence,

$$f_k = \frac{2k + 1}{2}\int_{-1}^{1}f(z)P_k(z)\,dz$$

4. Establish the orthonormalization relations

$$I_{\sigma s}^{(m)} = \int_{-1}^{1}P_\sigma^m P_s^m\,dz = \frac{2}{2s + 1}\frac{(s + m)!}{(s - m)!}\delta_{s\sigma}$$

for the associated Legendre functions.

Ans. Since $P_\sigma^m = (z^2 - 1)^{m/2}d_z^m P_\sigma$,

$$I_{\sigma s}^{(m)} = \int_{-1}^{1}(z^2 - 1)^m d_z^m P_\sigma d_z^m P_s\,dz = -\int_{-1}^{1}d_z^{m-1}P_s d[(z^2 - 1)^m d_z^m P_\sigma]$$

Also $(m - 1)$-fold differentiation of the Legendre differential equation

$$d_z[(z^2 - 1)d_zP_\sigma] = \sigma(\sigma + 1)P_\sigma$$

using Leibniz rule yields

$$d_z[(z^2 - 1)^m d_z^m P_\sigma] = (\sigma + m)(\sigma - m + 1)(z^2 - 1)^{m-1} d_z^{m-1} P_\sigma$$

Hence,

$$I_{\sigma s}^{(m)} = (\sigma + m)(\sigma - m + 1) I_{\sigma s}^{(m-1)}$$

and, continuing to reduce m in this way,

$$I_{\sigma s}^{(m)} = \frac{(\sigma + m)!}{(\sigma - m)!} I_{\sigma s}^{(0)} = \frac{(\sigma + m)!}{(\sigma - m)!} \frac{2\delta_{\sigma s}}{(2\sigma + 1)}$$

from Examples 1 and 2.

5. Prove Lommel's identity:

$$(a^2 - b^2)\int_0^r rJ_\nu(ar)J_\nu(br)\,dr \equiv r\begin{vmatrix} J_\nu(ar) & J_\nu(br) \\ aJ'_\nu(ar) & bJ'_\nu(br) \end{vmatrix}$$

Ans. Differentiate, obtaining

$$(a^2 - b^2)rJ_\nu(ar)J_\nu(br)$$

$$\equiv \begin{vmatrix} J_\nu(ar) & J_\nu(br) \\ aJ'_\nu(ar) & bJ'_\nu(br) \end{vmatrix} + r\begin{vmatrix} J_\nu(ar) & J_\nu(br) \\ a^2J''_\nu(ar) & b^2J''_\nu(br) \end{vmatrix}$$

By the Bessel equation,

$$-a^2J''_\nu(ar) = \frac{a}{r} J'_\nu(ar) + \left(a^2 - \frac{\nu^2}{r^2}\right)J_\nu(ar)$$

so replacing the second derivatives, the determinants yield

$$(a^2 - b^2)rJ_\nu(ar)J_\nu(br) \equiv -rJ_\nu(ar)J_\nu(br)\begin{vmatrix} 1 & 1 \\ a^2 - \dfrac{\nu^2}{r^2} & b^2 - \dfrac{\nu^2}{r^2} \end{vmatrix}$$

which are identical, as was to be shown.

6. Prove Lommel's second identity

$$2a^2\int_0^r r[J_\nu(ar)]^2\,dr = (a^2r^2 - \nu^2)[J_\nu(ar)]^2 + [arJ'_\nu(ar)]^2$$

Ans. The first Lommel identity (Example 5) can be written

$$(a + b)\int_0^r rJ_\nu(ar)J_\nu(br)\,dr$$

$$= -rJ_\nu(ar)J_\nu(br)\left[\frac{bJ_\nu'(br)/J_\nu(br) - aJ_\nu'(ar)/J_\nu(ar)}{b - a}\right]$$

Taking the limit as $b \to a$, one has

$$2a\int_0^r r[J_\nu(ar)]^2\,dr = -r[J_\nu(ar)]^2\partial_a\left[\frac{aJ_\nu'(ar)}{J_\nu(ar)}\right]$$

The result then follows by carrying out the differentiation and substituting for the second derivative from the Bessel equation.

7. Show how to determine f_k in the Bessel series for $f(r)$

$$f(r) = \sum_k f_k J_\nu(a_k r)$$

if $J_\nu(a_k) = 0$.

Ans. For $a \neq b$ zeros of the νth-order Bessel function, the Lommel identities imply

$$\int_0^1 rJ_\nu(ar)J_\nu(br)\,dr = \frac{\delta_{ab}}{2}[J_{\nu+1}(a)]^2$$

Hence, multiplying both sides of the series by $J_\nu(a_m r)$ and integrating,

$$f_m = \frac{2}{[J_{\nu+1}(a_m)]^2}\int_0^1 rJ_\nu(a_m r)f(r)\,dr$$

8. Show that $\zeta(z)$, the Riemann zeta function defined by

$$\zeta(-z) = \frac{e^{i\pi z}\Gamma(1+z)}{2\pi i}\int_L \frac{d\sigma}{(e^\sigma - 1)\sigma^{z+1}}$$

[where L is the same contour as in (9.44) "encircling" the negative real σ axis], has a simple pole at $z = 1$ as only singularity in the finite z plane.

Ans. Since the integral is an entire function of z, the only eligible singularities are those of $\Gamma(1 + z)$, which has simple poles at $1 + z = -n$ $(n = 0,1,2,\ldots)$, but for $1 + z = -n$

$$\int_L \frac{d\sigma}{(e^\sigma - 1)\sigma^{z+1}} = \oint \frac{\sigma^n\,d\sigma}{e^\sigma - 1} = \begin{cases} 2\pi i & n = 0 \\ 0 & n > 0 \end{cases}$$

Thus the integral has a zero at each pole of $\Gamma(1 + z)$ except for $z = -1$. Hence, $\zeta(-z)$ has a simple pole at $z = -1$ with residue 1. Then clearly the only singularity of $\zeta(z)$ in the finite z plane is a simple pole at $z = 1$ with residue 1.

9. Show that

$$\zeta(z) = \frac{1}{\Gamma(z)} \int_0^\infty \frac{t^{z-1}\, dt}{e^t - 1}$$

for $\operatorname{Re} z > 1$.

Ans. From Example 8

$$\zeta(z) = \frac{e^{-i\pi z}\Gamma(1-z)}{2\pi i} \int_{£} \frac{d\sigma}{(e^\sigma - 1)\sigma^{1-z}}$$

where £ "encircles" the positive real σ axis, and

$$\zeta(z) = \frac{e^{-i\pi z}\Gamma(1-z)}{2\pi i}\left[\int_\infty^0 \frac{x^{z-1}\, dx}{e^x - 1} + \int_0^\infty \frac{x^{z-1}e^{2\pi i z}\, dx}{e^x - 1}\right]$$

$$\zeta(z) = \frac{\Gamma(1-z)}{2\pi i}(e^{\pi i z} - e^{-\pi i z}) \int_0^\infty \frac{x^{z-1}\, dx}{e^x - 1}$$

Using (9.65),

$$\zeta(z) = \frac{1}{\Gamma(z)} \int_0^\infty \frac{x^{z-1}\, dx}{e^x - 1}$$

as was to be shown.

10. Establish the *Dirichlet series* for the zeta function

$$\zeta(z) = \sum_{n=1}^\infty \frac{1}{n^z} \qquad \text{for} \quad \operatorname{Re} z > 0$$

Ans. In Example 9,

$$\frac{1}{e^x - 1} = \frac{e^{-x}}{1 - e^{-x}} = \sum_{m=0}^\infty e^{-(m+1)x}$$

so

$$\zeta(z) = \frac{1}{\Gamma(z)} \sum_{m=0}^\infty \int_0^\infty e^{-(m+1)x} x^{z-1}\, dx = \sum_{n=1}^\infty \frac{1}{n^z}$$

with $n = m + 1$, provided $\operatorname{Re} z > 0$.

11. Show that (the product over the prime numbers p)

$$\zeta(z) = \frac{1}{\prod_p [1 - (1/p^z)]} \qquad \text{for} \quad \operatorname{Re} z \geq 1 + \delta > 1$$

Ans. From the series $\zeta(z) = \sum_{n=1}^\infty 1/n^z$ the series for $[1 - (1/2^z)]\, \zeta(z)$ has lost all multiples of 2, and $[1 - (1/2^z)][1 - (1/3^z)]\zeta(z)$ has lost all multiples of 2 and 3 in the series. Continuing in this way, only the first term will remain for

$$\prod_p \left[1 - \frac{1}{p^z}\right]\zeta(z) = 1$$

which was to be shown.

● PROBLEMS 13

Establish the recurrence relations for the following Legendre polynomials:

1. $nP_n - (2n-1)zP_{n-1} + (n-1)P_{n-2} = 0$
2. $zd_zP_n - d_zP_{n-1} = nP_n$
3. $d_zP_n - zd_zP_{n-1} = nP_{n-1}$
4. $d_zP_{n+1} - d_zP_{n-1} = (2n+1)P_n$
5. $(z^2-1)d_zP_n = nzP_n - nP_{n-1}$
6. Show that $P_n(1) = 1$, $P_n(-1) = (-1)^n$.
7. Show that $P_{2n}(0) = (-1)^n(2n!/2^{2n}(n!)^2$, $P_{2n+1}(0) = 0$.
8. Assuming that it is possible to expand

$$\frac{1}{w-z} = \sum_{n=0}^{\infty} c_n P_n(z)$$

show that $c_n = (2n+1)Q_n(z)$.

9. Show formally by the Cauchy integral formula that the coefficients in the Legendre series for an analytic function $f(z)$ are given by

$$f_n = \frac{2n+1}{2\pi i}\oint f(z)Q_n(z)\,dz$$

where the contour encircles the two branch points of $Q_n(z)$.

10. Show that

$$\int_{-1}^{1} \frac{P_n(x)}{\sqrt{1-x^2}}\,dx = \pi[P_n(0)]^2$$

11. Prove that

$$\csc\theta = \frac{\pi}{2}\sum_{n=0}^{\infty}(4n+1)\left[\frac{(2n-1)!!}{(2n)!!}\right]^2 P_{2n}(\cos\theta)$$

Establish the recurrence relations for the following Bessel functions:

12. $z(J_{\nu-1} + J_{\nu+1}) = 2\nu J_\nu$
13. $2d_zJ_\nu = J_{\nu-1} - J_{\nu+1}$
14. $\nu J_\nu = z(J_{\nu-1} - d_zJ_\nu)$
15. $\nu J_\nu = z(J_{\nu+1} + d_zJ_\nu)$

Find the Bessel series coefficients a_n in $\sum_{n=0}^{\infty} a_n J_n(z)$ for the following:

16. z 17. z^3

18. Show that $\sqrt{2}\, e^{iz} = \sqrt{\pi z}[J_{-1/2}(z) + iJ_{1/2}(z)]$.
19. Show that $[J_\nu J_{1-\nu} + J_{-\nu}J_{\nu-1}]z\pi = 2 \sin \nu\pi$.
20. Show that

$$\int_{-1}^{1} e^{i\omega t} P_n(t)\, dt = \sqrt{\frac{2\pi}{\omega}}\, i^n J_{n+(1/2)}(\omega)$$

21. Show that

$$e^{i\omega t} = \sqrt{\frac{\pi}{2\omega}} \sum_{n=0}^{\infty} (2n + 1) i^n J_{n+(1/2)}(\omega) P_n(t)$$

22. Prove that

$$\int_{-\infty}^{\infty} J_{m+(1/2)}(x) J_{n+(1/2)}(x) \frac{dx}{x} = \frac{2\delta_{mn}}{2n + 1}$$

23. Deduce the *Hankel transform* formulae

$$F(k) = \int_0^\infty f(r) J_\nu(kr) r\, dr$$

$$f(r) = \int_0^\infty F(k) J_\nu(kr) k\, dk$$

24. The halves of a conducting spherical shell of radius R are insulated from each other along a great circle and maintained at potentials V_1 and V_2. Find the potential inside and outside.

Find the Legendre series for the following:

25. x^2 26. x^5 27. x^7

28. By using the Rodrigues formula and Rolle's theorem show that the roots of $P_n(x)$ are real, distinct, and all in $-1 \le x \le 1$.
29. Solve the harmonic Dirichlet problem in a cylindrical shell of radii a and $c > a$ and height b with zero values on all surfaces except the inner curved one where the potential is V. Show that the potential inside the shell is

$$\phi(r,z) = \frac{2V}{\pi} \sum_{n=1}^{\infty} \left[\frac{1 - (-1)^n}{n}\right] \frac{\begin{vmatrix} I_0\left(\frac{n\pi r}{b}\right) & I_0\left(\frac{n\pi c}{b}\right) \\ K_0\left(\frac{n\pi r}{b}\right) & K_0\left(\frac{n\pi c}{b}\right) \end{vmatrix}}{\begin{vmatrix} I_0\left(\frac{n\pi a}{b}\right) & I_0\left(\frac{n\pi c}{b}\right) \\ K_0\left(\frac{n\pi a}{b}\right) & K_0\left(\frac{n\pi c}{b}\right) \end{vmatrix}} \sin\left(\frac{n\pi z}{b}\right)$$

30. Show that the harmonic Dirichlet problem for the cylinder of Problem 29 with zero values on all surfaces except the top one where $\phi(r,b) = f(r)$ is solved by

$$\phi(r,z) = \sum_k \frac{\sinh(a_k z)}{\sinh(a_k b)} A_k \frac{\begin{vmatrix} J_0(a_k r) & Y_0(a_k r) \\ J_0(a_k c) & Y_0(a_k c) \end{vmatrix}}{Y_0(a_k c)}$$

or, defining Z_0 by,

$$\phi(r,z) = \sum_k \frac{\sinh(a_k z)}{\sinh(a_k b)} A_k Z_0(a_k r) \qquad Z_0(a_k a) = 0$$

with

$$A_k = \frac{2\int_a^c r f(r) Z_0(a_k r)\, dr}{c^2[Z_0'(a_k c)]^2 - a^2[Z_0'(a_k a)]^2}$$

31. Show that the harmonic Dirichlet problem is solved for a cylindrical surface of radius a and height b with top at potential V and all other surfaces grounded by

$$\phi(r,z) = 2V \sum_{k=1}^{\infty} \frac{1}{a_k a}\left[\frac{\sinh(a_k z)}{\sinh(a_k b)}\right]\frac{J_0(a_k r)}{J_1(a_k a)}$$

32. Find the potential of a grounded metal sphere placed in a uniform electric field.
33. Solve Problem 32 for a dielectric sphere showing the potential to be

$$\phi(r,\theta) = E_0 r \cos\theta \left[1 - \left(\frac{a}{r}\right)^3 \left(\frac{\kappa - 1}{\kappa + 2}\right)\right] \qquad \text{outside}$$

and

$$\phi(r,\theta) = \frac{3E_0 r \cos\theta}{\kappa + 2} \qquad \text{inside}$$

with κ the relative permittivity (ratio of dielectric constant outside to inside). For $\kappa \to \infty$ one recovers Problem 32.
34. Show that the velocity potential for irrotational flow past a sphere of radius a is

$$\phi = V_\infty r \cos\theta \left[1 + \frac{1}{2}\left(\frac{a}{r}\right)^3\right]$$

35. Show that the scattering amplitude

$$f(E,\cos\theta) = \sum_{l=0}^{\infty} (2l + 1) a(l,E) P_l(\cos\theta)$$

[with $a(l,E) = (1/\sqrt{E})e^{i\delta(l,E)} \sin \delta(l,E)$ for particles of energy E incident on a scattering center] can be written as

$$f(E,\cos\theta) = \frac{-i}{2}\int_C \frac{(2l+1)a(l,E)}{\sin \pi l} P_l(-\cos\theta)\, dl$$

where C "encircles" the real positive l axis and P_l is defined for complex l by (13.108). Hence show that if $a(l,E)$ has simple (Regge) poles at l_n with Re $l_n > -\frac{1}{2}[\gamma_n(E) =$ residue of $a(l,E)$ at $l_n]$,

$$f(E,\cos\theta) = \frac{i}{2}\int_{-(1/2)-i\infty}^{(1/2)+i\infty} \frac{(2l+1)a(l,E)}{\sin \pi l} P_l(-\cos\theta)\, dl + \frac{1}{2\pi i}\sum_{n=1}^{N} \frac{(2l_n+1)\gamma_n(E)}{\sin \pi l_n} P_{l_n}(-\cos\theta)$$

36. Show that the differential equation

$$(d_z^2 - z d_z + n)w = 0$$

has two independent series solutions representing entire functions given by

$$w = 1 - \frac{nz^2}{2!} + \frac{n(n-2)}{4!}z^4 - \frac{n(n-2)(n-4)}{6!}z^6 + \cdots$$

$$w = z - \frac{(n-1)}{3!}z^3 + \frac{(n-1)(n-3)}{5!}z^5 - \frac{(n-1)(n-3)(n-5)}{7!}z^7 - \cdots$$

Show that the coefficients satisfy the recurrence relation

$$(k+1)(k+2)a_{k+2} = (k-n)a_k$$

37. Show that

$$(4z^2 d_z^2 + 1 - z^2)w = 0$$

(which has a regular singularity at $z = 0$ and an indicial equation with equal roots) possesses two independent series solutions w_1 and w_2 of the forms

$$w_1 = \sqrt{z}\left(1 + \frac{z^2}{4^2} + \frac{z^4}{4^2 \cdot 8^2} + \cdots\right)$$

$$w_2 = w_1 \log z - \sqrt{z}\left(\frac{z^2}{16} + \cdots\right)$$

38. Find two independent power series solutions (about $z = 0$) of $(d_z^2 + z)w = 0$.

39. Solve by series: $(4z^2 d_z^2 + 4z d_z + 4z^2 - 1)w = 0$ (about $z = 0$).

TABLE I: Laplace Transforms

$f(t) = \dfrac{1}{2\pi i}\displaystyle\int_{\sigma-i\infty}^{\sigma+i\infty} e^{st}F(s)\,ds$	$F(s) = \displaystyle\int_0^\infty e^{-st}f(t)\,dt$
(1) $\delta(t-q)$	e^{-sq}
(2) t^α	$\dfrac{\Gamma(\alpha+1)}{s^{\alpha+1}}$
(3) $\dfrac{(c-b)e^{at}+(a-c)e^{bt}+(b-a)e^{ct}}{(a-b)(a-c)(c-b)}$	$\dfrac{1}{(s-a)(s-b)(s-c)}$
(4) $e^{-kt}\sin\omega t$	$\dfrac{\omega}{s^2+2ks+(k^2+\omega^2)}$
(5) $\cos^2\omega t$	$\dfrac{s^2+2\omega^2}{s(s^2+4\omega^2)}$
(6) $\cosh at - \cosh bt$	$\dfrac{(a^2-b^2)s}{(s^2-a^2)(s^2-b^2)}$
(7) $t\sin\omega t$	$\dfrac{2\omega s}{(s^2+\omega^2)^2}$
(8) $L_n(t)$ (re: 13.56, $m=0$)	$\dfrac{n!\,(s-1)^n}{s^{n+1}}$
(9) $e^{-t/2}L_n(t)$	$\dfrac{n!\,[s-(1/2)]^n}{[s+(1/2)]^{n+1}}$

(continued)

Table I *(continued)*

	$f(t) = \dfrac{1}{2\pi i}\displaystyle\int_{\sigma-i\infty}^{\sigma+i\infty} e^{st}F(s)\,ds$	$F(s) = \displaystyle\int_0^\infty e^{-st}f(t)\,dt$
(10)	$\displaystyle\sum_k \frac{e^{q_k t}}{Q'(q_k)}$	$\dfrac{1}{Q(s)}$ with $Q(s) = \prod_k (s - q_k)$, $Q'(q_k) \neq 0$
(11)	$J_\nu(t) \quad (\mathrm{Re}\,\nu > -1)$	$\dfrac{(\sqrt{s^2+1} - s)^\nu}{\sqrt{s^2+1}}$
(12)	$t^\nu J_\nu(qt) \quad (\mathrm{Re}\,\nu > -\frac{1}{2})$	$\dfrac{(2q)^\nu \Gamma(\nu + \frac{1}{2})}{\sqrt{\pi}\,(s^2 + q^2)^{\nu+(1/2)}}$
(13)	$\dfrac{\sin qt}{\sqrt{t}}$	$\sqrt{\dfrac{\pi}{2}\,\dfrac{\sqrt{s^2+q^2} - s}{(s^2+q^2)}}$
(14)	$\log t$	$-\left(\dfrac{\gamma}{s} + \dfrac{\log s}{s}\right) \quad \gamma = 0.5772157\ldots$
(15)	$\dfrac{\log t}{\sqrt{t}}$	$-\sqrt{\dfrac{\pi}{s}}\,(\log 4s + \gamma)$
(16)	$\displaystyle\int_0^\infty \frac{t^{v-1}\,dv}{\Gamma(v)}$	$\dfrac{1}{\log s}$
(17)	$\dfrac{e^{qt} - e^{bt}}{t}$	$\log\left(\dfrac{s-b}{s-q}\right)$
(18)	$t^{\nu/2} J_\nu(q\sqrt{t})$	$\dfrac{e^{-q^2/4s}}{s}\,\dfrac{q^\nu}{2^\nu s^{\nu+1}} \quad (\mathrm{Re}\,\nu > -1)$
(19)	$\dfrac{e^{-q^2/4t}}{\sqrt{\pi t}}$	$\dfrac{e^{-q\sqrt{s}}}{\sqrt{s}}$
(20)	$\dfrac{\cos(q\sqrt{t})}{\pi\sqrt{t}}$	$\dfrac{e^{-q^2/4s}}{\sqrt{\pi s}}$
(21)	$\dfrac{\sin(q\sqrt{t})}{\pi}$	$\dfrac{q}{2\sqrt{\pi}\,s^{3/2}}\,e^{-q^2/4s}$
(22)	$\dfrac{\sin qt \cos bt}{t}$	$\frac{1}{2}\arctan\left(\dfrac{2qs}{s^2 - q^2 + b^2}\right)$

Table I (*continued*)

$f(t) = \dfrac{1}{2\pi i}\int_{\sigma-i\infty}^{\sigma+i\infty} e^{st}F(s)\,ds$	$F(s) = \int_0^\infty e^{-st}f(t)\,dt$
(23) e^{-t^2}	$e^{s^2/2}\int_{s/2}^{\infty} e^{-v^2}\,dv$
(24) $u(t-q)$	$\dfrac{e^{-qs}}{s}$
(25) $u(t-q)u(b-t)$	$\dfrac{e^{-qs}-e^{-bs}}{s}$
(26) $u(t-2q)u(q+b-t) - u(t-q-b)u(2b-t)$	$\dfrac{(e^{-qs}-e^{-bs})^2}{s}$
(27) $(t-q)u(t-q)u(b-t) + (b-q)u(t-b)$	$\dfrac{e^{-qs}-e^{-bs}}{s^2}$
(28) $\left(1-\dfrac{t}{q}\right)u(t)u(q-t)$	$\dfrac{e^{-qs}+qs-1}{qs^2}$
(29) $\lvert\sin(qt)\rvert$	$\dfrac{q}{s^2+q^2}\left[\dfrac{1+e^{-(\pi/q)s}}{1-e^{-(\pi/q)s}}\right]$
(30) $(1+[t]-t)u(t)$ ($[t]$ = integer part of t)	$\dfrac{e^{-s}+s-1}{s^2(1-e^{-s})}$
(31) $u(t)\sum_{k=0}^{\infty} u(t-2kq)u[(2k+1)q-t]$	$\dfrac{1}{s(1+e^{-qs})}$
(32) $u(t)\sum_{k=0}^{\infty} u(t-kq)u[(k+\varepsilon)q-t]$	$\dfrac{1-e^{-\varepsilon qs}}{s(1-e^{-qs})}$
(33) $\left[\dfrac{t}{q}\right]$	$\dfrac{1}{s(e^{qs}-1)}$
(34) $q^{[t]}$	$\dfrac{1}{s}\left(\dfrac{e^s-1}{e^s-q}\right)$
(35) $J_0(mt)J_0(nt)$	$\dfrac{Q_{-1/2}}{\pi\sqrt{mn}}\left(\dfrac{s^2+m^2+n^2}{2mn}\right)$
(36) $J_1(mt)J_1(nt)$	$\dfrac{Q_{1/2}}{\pi\sqrt{mn}}\left(\dfrac{s^2+m^2+n^2}{2mn}\right)$

(*continued*)

Table I *(continued)*

	$f(t) = \frac{1}{2\pi i}\int_{\sigma-i\infty}^{\sigma+i\infty} e^{st}F(s)\,ds$	$F(s) = \int_0^\infty e^{-st}f(t)\,dt$
(37)	$\lvert J_0(mt)\rvert^2$	$\frac{2}{\pi\sqrt{s^2+4m^2}}K\left(\frac{2m}{\sqrt{s^2+4m^2}}\right)$
(38)	$\frac{I_0(t)}{\sqrt{t}}$	$\frac{2}{\sqrt{\pi(s+1)}}K\left(\frac{\sqrt{2}}{\sqrt{s+1}}\right)$
(39)	$\lvert I_0(mt)\rvert^2$	$\frac{2}{\pi s}E\left(\frac{2m}{s}\right)$
(40)	$t^{\mu-1}J_\nu(mt)$	$\frac{m^\nu\Gamma(\mu+\nu)}{2^\nu s^{\mu+\nu}\Gamma(\nu+1)} \times F\left(\frac{\mu+\nu}{2},\frac{\mu+\nu+1}{2},\nu+1,-\frac{m^2}{s^2}\right)$ for $\mathrm{Re}(s+im)>0<\mathrm{Re}(s-im)$
(41)	$K_\nu(t)$	$\left(\frac{\pi}{\sin\nu\pi}\right)\frac{\sinh(\nu\,\mathrm{arcosh}\,s)}{\sinh(\mathrm{arcosh}\,s)}$

TABLE II:

Fourier Transforms

$f(t) = \frac{1}{2\pi}\int_{-\infty}^{\infty} e^{-i\omega t}F(\omega)\,d\omega$	$F(\omega) = \int_{-\infty}^{\infty} e^{i\omega t}f(t)\,dt$
(1) $t^{\nu}u(t)u(a-t)$ $(\mathrm{Re}\,\nu > -1)$	$\frac{i^{\nu+1}}{\omega^{\nu+1}}\gamma(\nu+1,-ia\omega)$ (re: 13.69)
(2) $t^{\nu}u(t-a)$ $(\mathrm{Re}\,\nu < 0)$	$-\frac{i^{\nu+1}}{\omega^{\nu+1}}[\Gamma(\nu+1)-\gamma(\nu+1,-ia\omega)]$
(3) $(a+it)^{-\nu}(b+it)^{-1}$ $(\mathrm{Re}\,\nu > -1)$	$\frac{2\pi u(\omega)\gamma(\nu,a\omega-b\omega)}{\Gamma(\nu)(a-b)^{\nu}e^{b\omega}}$
(4) $[a^2+(t+b)^2]^{-\nu}$ $(\mathrm{Re}\,\nu > 0)$	$2e^{-ib\omega}\frac{\sqrt{\pi}}{\Gamma(\nu)}\left(\frac{\lvert\omega\rvert}{2a}\right)^{\nu-(1/2)}K_{\nu-(1/2)}(a\,\lvert\omega)$
(5) $P_n(t)u(1-\lvert t\rvert)$	$(-1)^n\sqrt{\frac{2\pi}{\omega}}\,J_{n+(1/2)}(\omega)$
(6) $\frac{u(1-\lvert t\rvert)\cos(n\arccos t)}{\sqrt{1-t^2}}$	$(-1)^n\pi J_n(\omega)$
(7) $e^{i\alpha t^2}$	$\sqrt{\frac{\pi}{2\alpha}}(1+i)e^{-i\omega^2/4\alpha}$
(8) $\frac{\sin\alpha t}{t}$	$\pi u(\alpha-\lvert\omega\rvert)$ $(\alpha > 0)$
(9) $e^{-\alpha t^2}$ $(\mathrm{Re}\,\alpha > 0)$	$\sqrt{\frac{\pi}{\alpha}}\,e^{-\omega^2/4\alpha}$

(*continued*)

Table II *(continued)*

	$f(t) = \dfrac{1}{2\pi}\displaystyle\int_{-\infty}^{\infty} e^{-i\omega t}F(\omega)\,d\omega$	$F(\omega) = \displaystyle\int_{-\infty}^{\infty} e^{i\omega t}f(t)\,dt$
(10)	$e^{-\alpha\|t\|}$	$\dfrac{2\alpha}{\alpha^2 + \omega^2}$
(11)	$\dfrac{\cosh(\alpha t)}{\cosh(\pi t)} \quad (\|\alpha\| < \pi)$	$2\left(\dfrac{\cos(\alpha/2)\cosh(\omega/2)}{\cosh\omega + \cos\alpha}\right)$
(12)	$\dfrac{\sinh(\alpha t)}{\sinh(\pi t)} \quad (\|\alpha\| < \pi)$	$\dfrac{\sin\alpha}{\cosh\omega + \cos\alpha}$
(13)	$\dfrac{u(\alpha - \|t\|)}{\sqrt{\alpha^2 - t^2}}$	$\pi J_0(\alpha\omega)$
(14)	$\dfrac{\cosh(\beta\sqrt{\alpha^2 - t^2})}{\sqrt{\alpha^2 - t^2}}\, u(\alpha - \|t\|)$	$\pi J_0(\alpha\sqrt{\omega^2 - \beta^2})$
(15)	$\dfrac{1}{\sqrt{\alpha^2 - t^2}}$	$\dfrac{\pi}{\alpha}\sin(\alpha\omega)$
(16)	$\dfrac{1}{\sqrt{\alpha^2 + t^2}}$	$2_{K_0}(\alpha\omega)$

TABLE III:

Conformal Mapping

Table III

z plane	$w = f(z)$	w plane
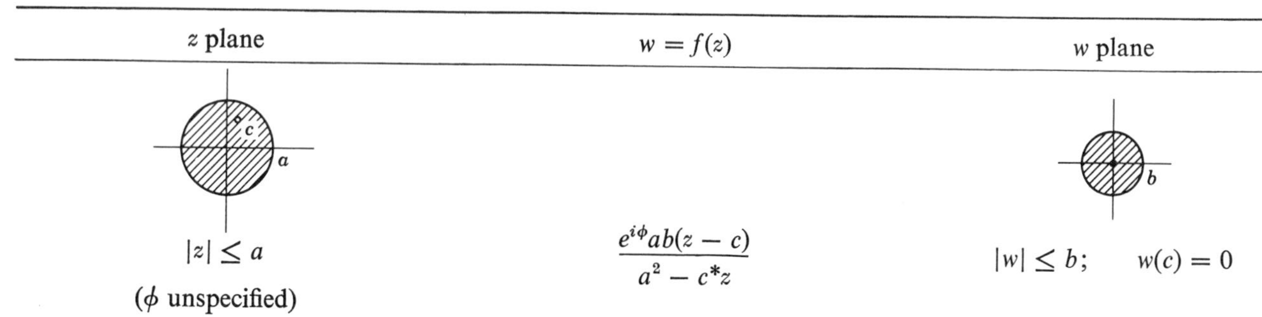 $\lvert z\rvert \leq a$ (ϕ unspecified)	$\dfrac{e^{i\phi}ab(z-c)}{a^2 - c^*z}$	$\lvert w\rvert \leq b; \quad w(c) = 0$
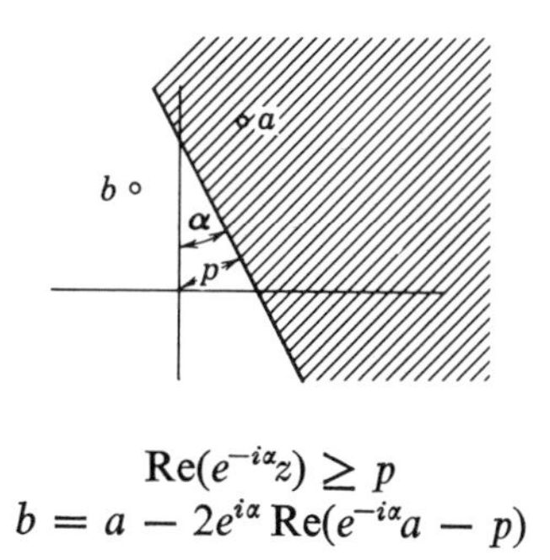 $\mathrm{Re}(e^{-i\alpha}z) \geq p$ $b = a - 2e^{i\alpha}\,\mathrm{Re}(e^{-i\alpha}a - p)$	$c + s\left(\dfrac{z-a}{z-b}\right)$	$\lvert w - c\rvert \leq s; \quad w(a) = c$

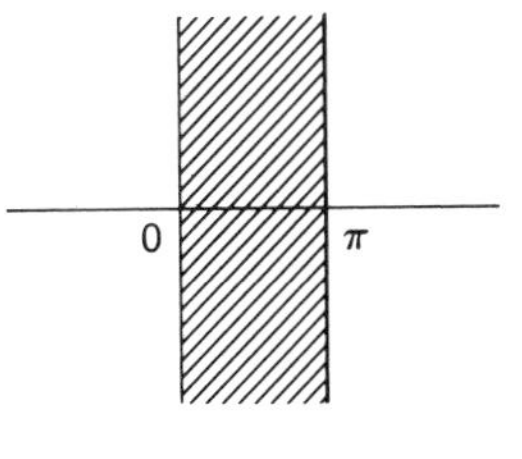

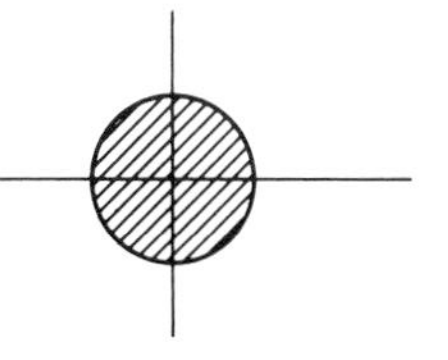

$\left\lvert \mathrm{Re}\left(z - \frac{\pi}{2}\right) \right\rvert \le \frac{\pi}{2}$	$\left(\frac{1 + ie^{iz}}{1 - ie^{iz}}\right)$	$\lvert w \rvert \le 1$

(*continued*)

Table III (*continued*)

z plane	$w = f(z)$	w plane
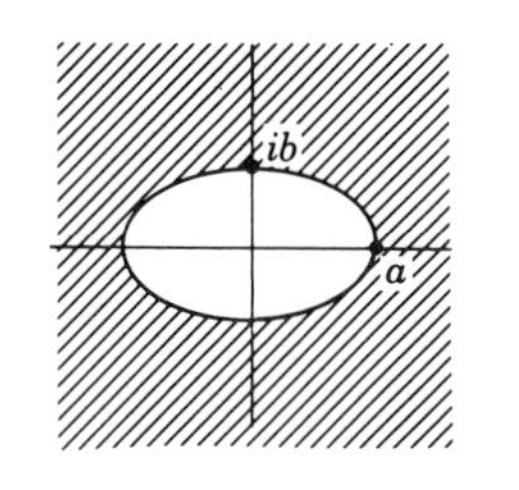 $1 \leq \mathrm{Re}\left[\left(\frac{1}{a^2} - \frac{1}{b^2}\right)z^2 + \left(\frac{1}{a^2} + \frac{1}{b^2}\right)z^*z\right]$ Ellipse	$\dfrac{z - \sqrt{z^2 + b^2 - a^2}}{a - b}$	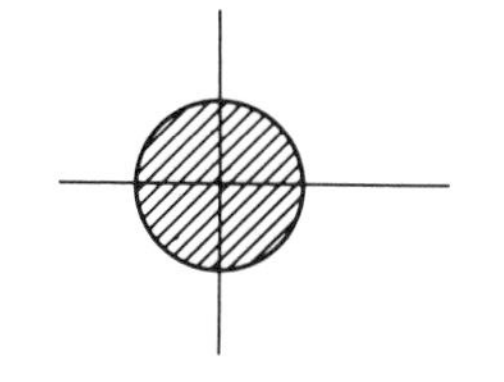 $\lvert w\rvert \leq 1; \quad w(\infty) = 0$

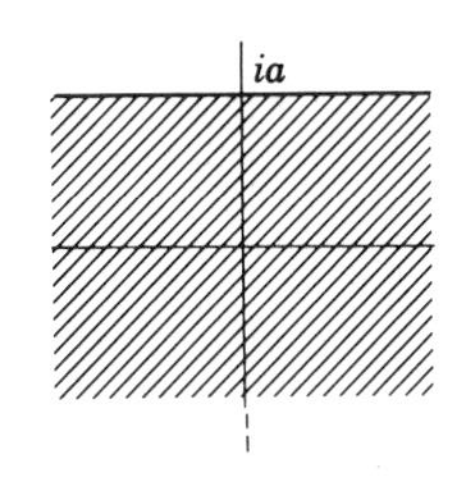		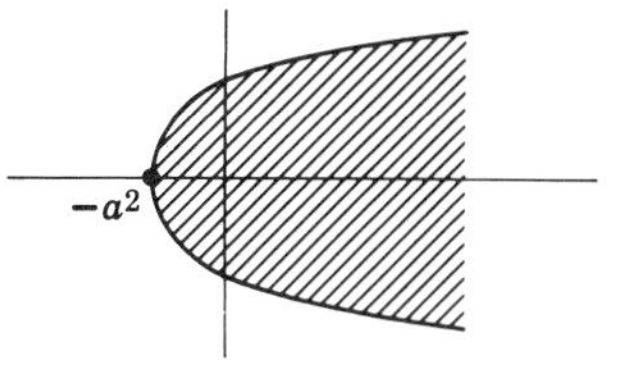
$\operatorname{Im} z \leq a$	z^2	$\operatorname{Re}(w^2 - w^*w + 8a^2w + 8a^4) \geq 0$ Parabola
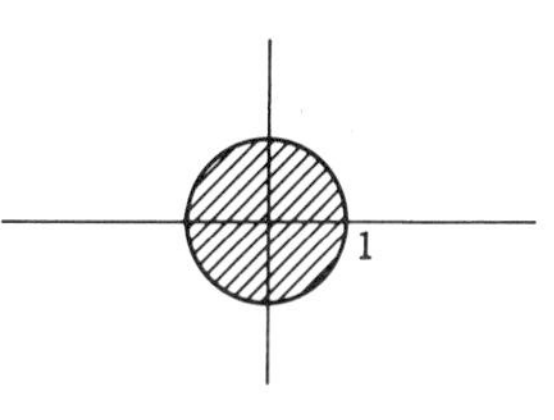	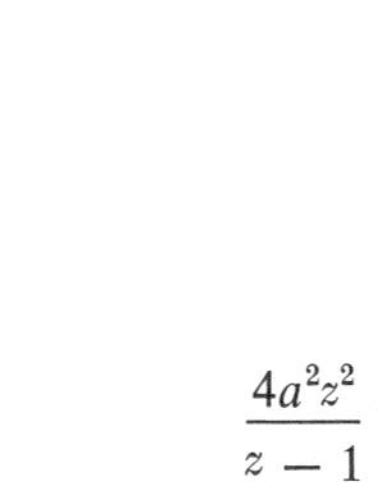	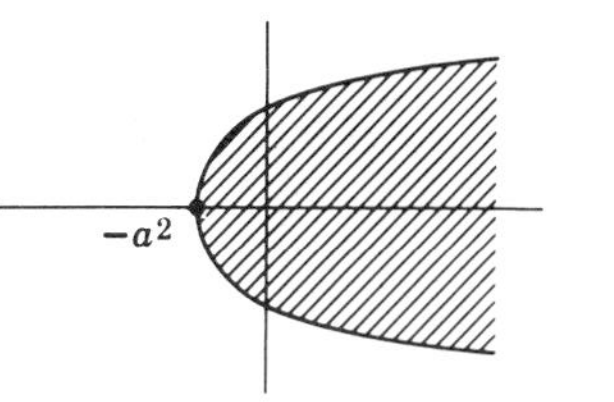
$\|z\| \leq 1$	$\dfrac{4a^2z^2}{z-1}$	$\operatorname{Re}(w^2 - w^*w + 8a^2w + 8a^4) \geq 0$ Parabola

(continued)

Table III *(continued)*

z plane	$w = f(z)$	w plane
$\lvert z - 1 \rvert \leq 1$	$\pm\sqrt{z}$	$\mathrm{Re}[2w^2 - (w^*w)^2] \geq 0$ Lemniscate
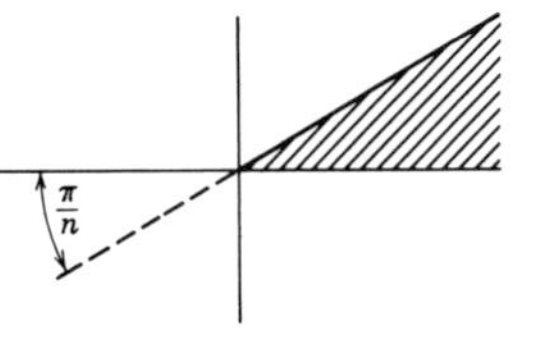 $0 \leq \arg z \leq \dfrac{\pi}{n}$	z^n	$\mathrm{Im}\, w \geq 0$

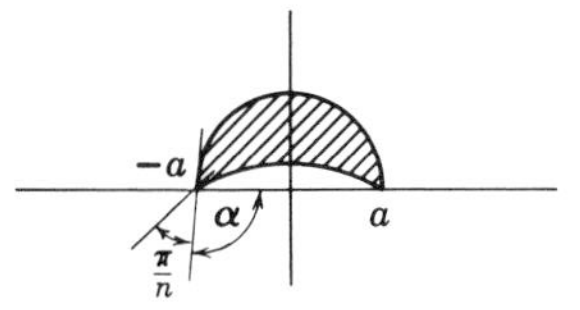 		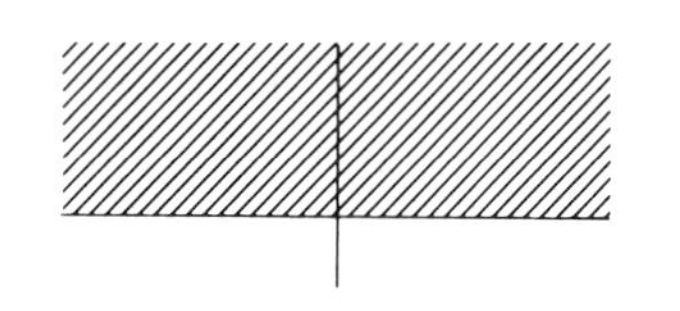
$\alpha \le \arg\left(\dfrac{z-a}{z+a}\right) \le \alpha + \dfrac{\pi}{n}$	$\left[\left(\dfrac{z-a}{z+a}\right)e^{-i\alpha}\right]^n$	$\text{Im}\, w \ge 0$
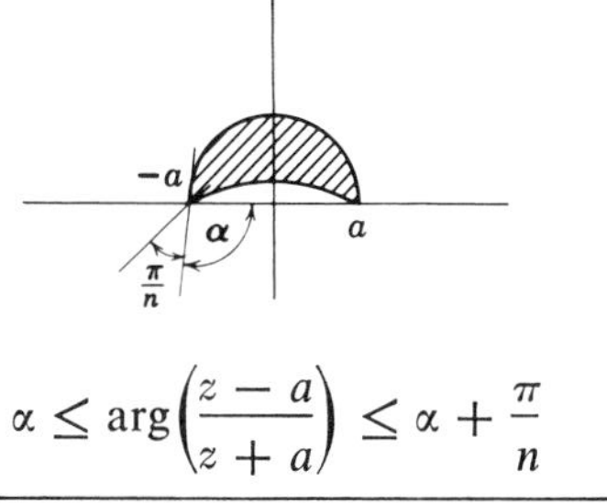 		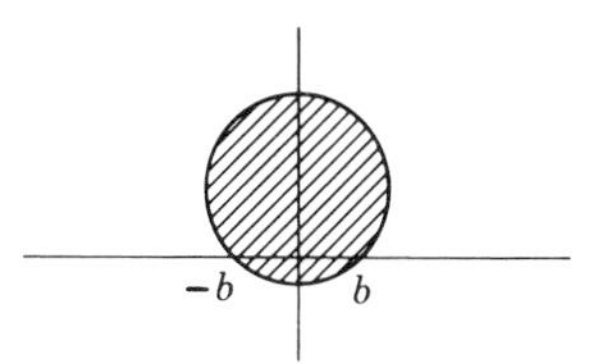
$\alpha \le \arg\left(\dfrac{z-a}{z+a}\right) \le \alpha + \dfrac{\pi}{n}$	$b\left[\dfrac{(z+a)^n + (z-a)^n}{(z+a)^n - (z-a)^n}\right]$	$-n\alpha \le \arg\left(\dfrac{w-b}{w+b}\right) \le n\alpha + \pi$

(continued)

Table III *(continued)*

z plane	$w = f(z)$	w plane
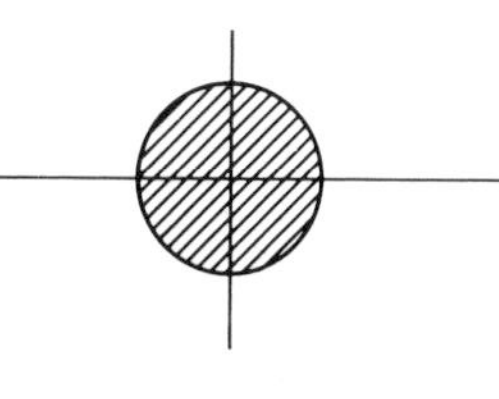		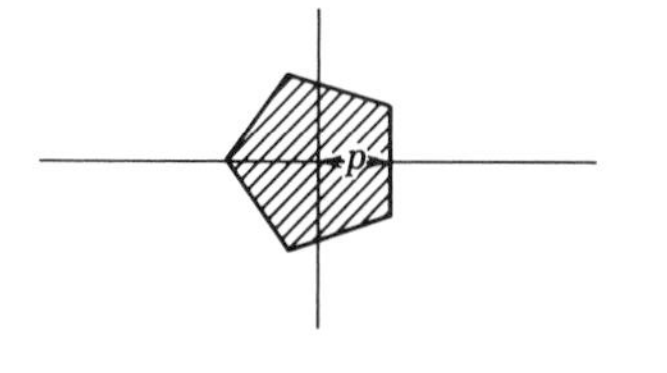
$\lvert z\rvert \leq 1$	$\int_0^z \frac{dt}{(1 - t^5)^{2/5}}$	$\mathrm{Re}(e^{-2\pi ik/5}z) \leq p$ for $k = 0, 1, 2, 3, 4$ Regular pentagon
$\lvert z\rvert \leq 1$	$\int_0^z \frac{dt}{(1 - t^n)^{2/n}}$	$\mathrm{Re}(e^{-2\pi ik/n}z) \leq p$ for $k = 0, 1, 2, \ldots, n - 1$ Regular n-gon

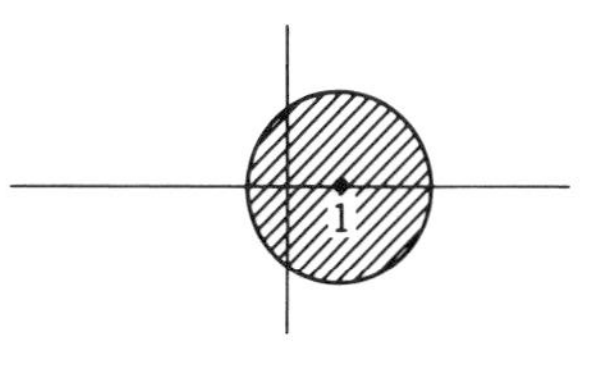		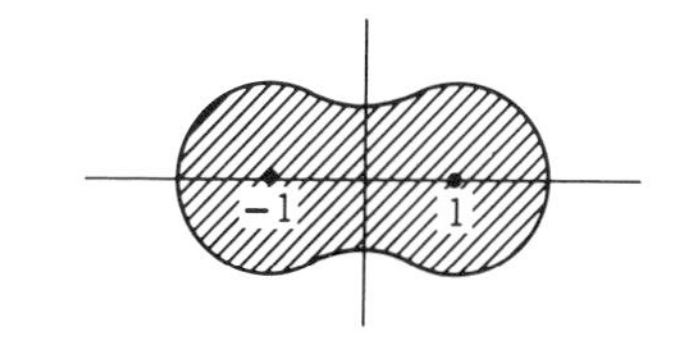
$\lvert z-1\rvert \leq R > 1$	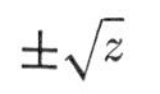 $\pm\sqrt{z}$	$\mathrm{Re}[(w^*w)^2 - 2w^2] \leq R^2 - 1$ Oval of Cassini: boundary $\lvert w-1\rvert\,\lvert w+1\rvert = R$

(*continued*)

Table III (*continued*)

z plane	$w = f(z)$	w plane
$\lvert z - 1\rvert \leq 1$	z^2	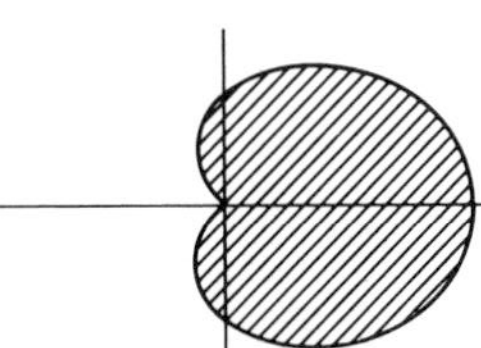 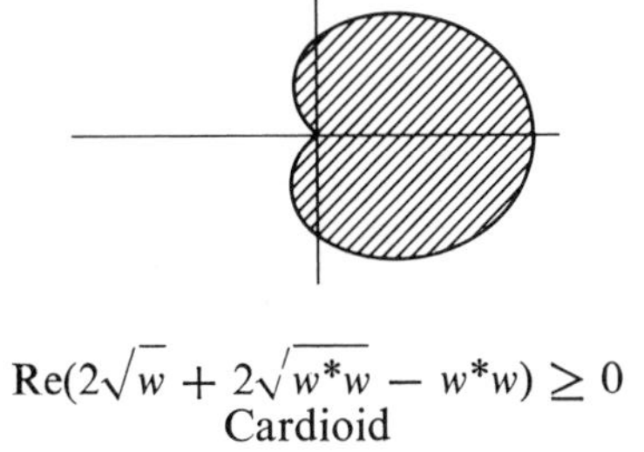 $\text{Re}(2\sqrt{w} + 2\sqrt{w^*w} - w^*w) \geq 0$ Cardioid
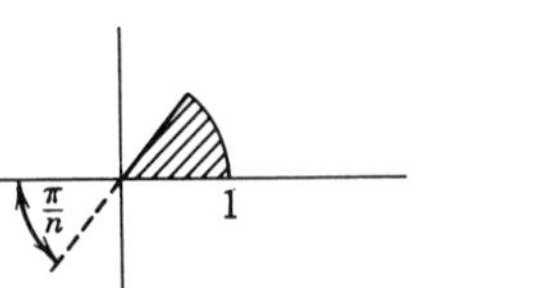 $\lvert z\rvert \leq 1 \geq \frac{n}{\pi} \arg z \geq 0$	$\left(\frac{z^n + 1}{z^n - 1}\right)^2$	$\text{Im}\, w \geq 0$

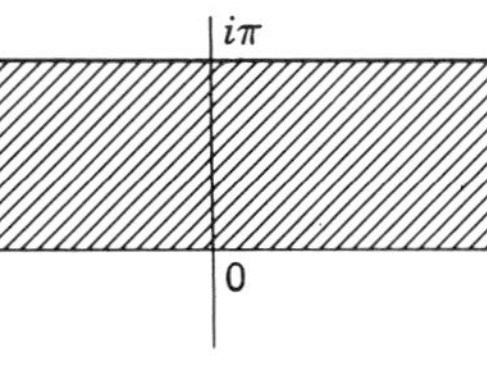

$0 \leq \operatorname{Im} z \leq \pi$

e^z

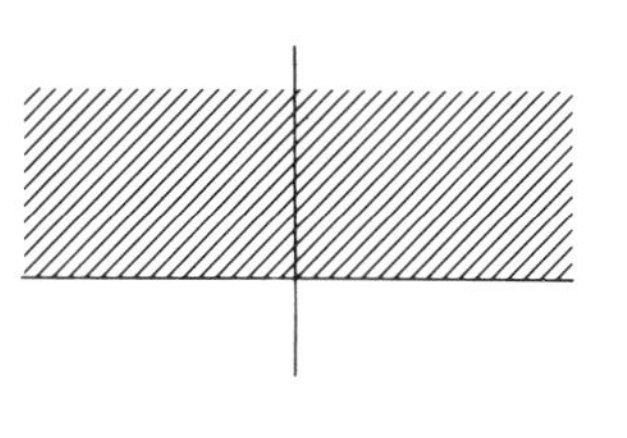

$\operatorname{Im} w \geq 0$

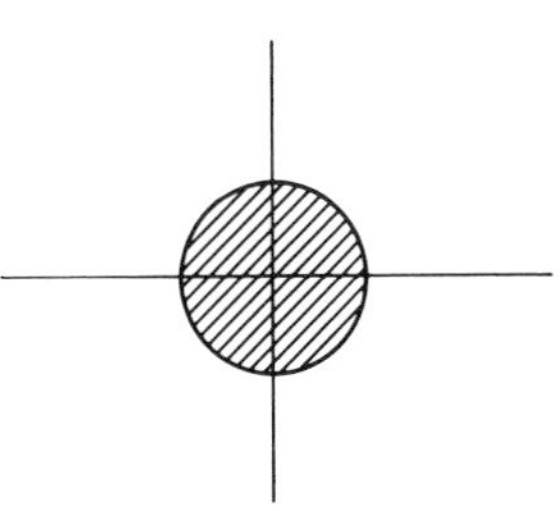

$|z| \leq 1$

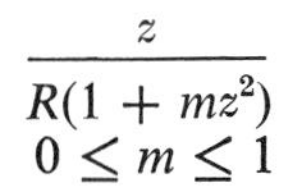

$\dfrac{z}{R(1 + mz^2)}$
$0 \leq m \leq 1$

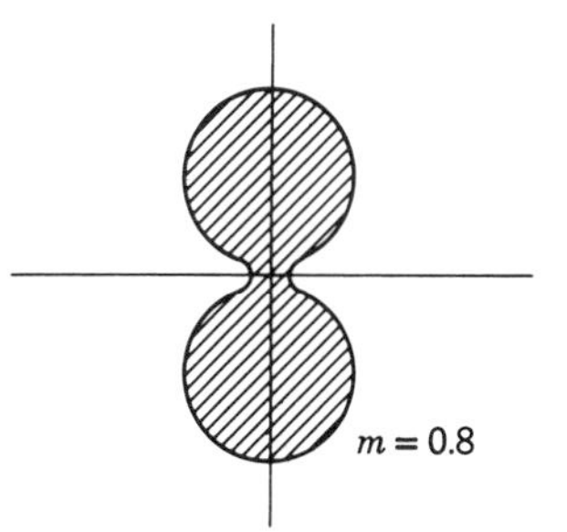

$|1 - \sqrt{1 - 4R^2w^2}| \leq 2mR\,|w|$

Lemniscate of Booth

(continued)

Table III *(continued)*

z plane	$w = f(z)$	w plane
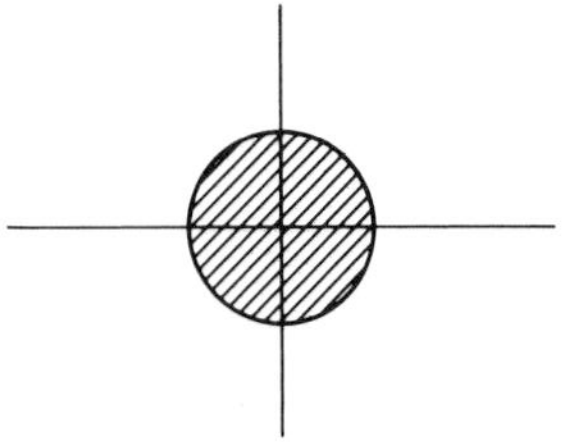 $\lvert z\rvert \le 1$	$R\left(\frac{1}{z} + mz^n\right)$ $0 \le m \le \frac{1}{n}$	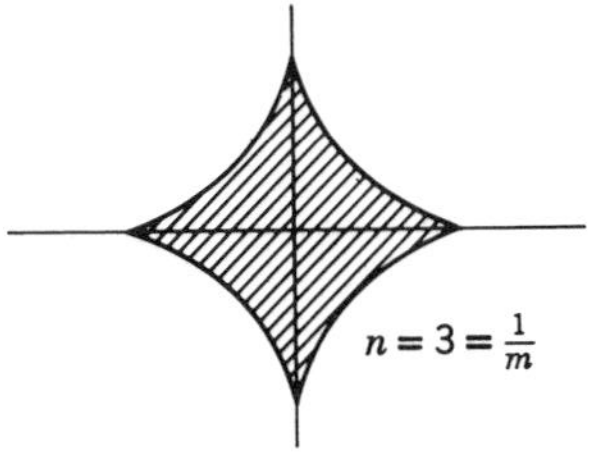 $n = 3 = \frac{1}{m}$ Hypocycloid

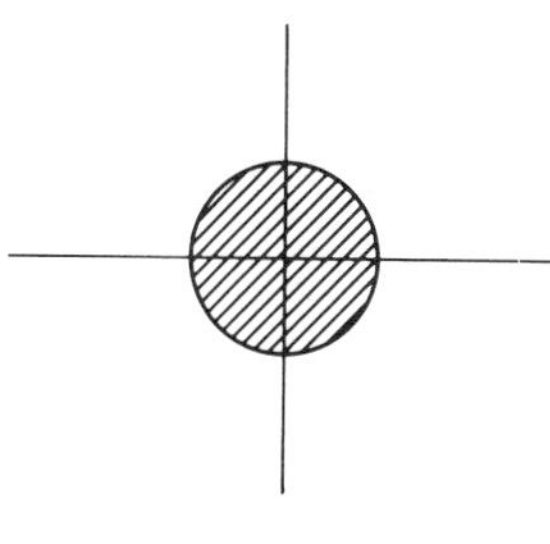	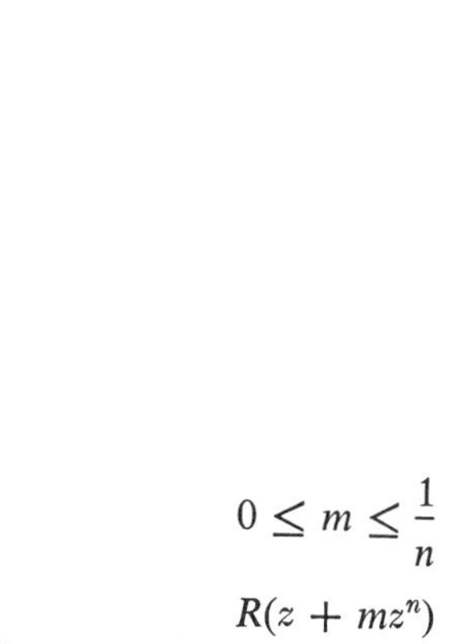	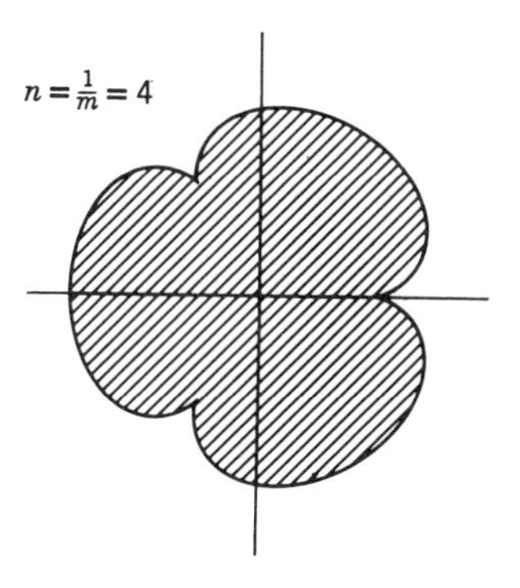
$\lvert z \rvert \leq 1$	$0 \leq m \leq \frac{1}{n}$ $R(z + mz^n)$	Epicycloid

(continued)

Table III *(continued)*

z plane	$w = f(z)$	w plane

$1 \geq \mathrm{Re}\left[\left(\frac{1}{a^2} - \frac{1}{b^2}\right)z^2 + \left(\frac{1}{a^2} + \frac{1}{b^2}\right)z^*z\right]$	$w = \sqrt{k}\,\mathrm{sn}\left[\frac{2K}{\pi}\arcsin\left(\frac{z}{\sqrt{a^2 - b^2}}\right)\right]$ $k = \left[\frac{\vartheta_2(\tau)}{\vartheta_3(\tau)}\right]^2$ $\tau = \frac{2i}{\pi}\log\left(\frac{a+b}{a-b}\right)$	$\lvert w\rvert \leq 1$

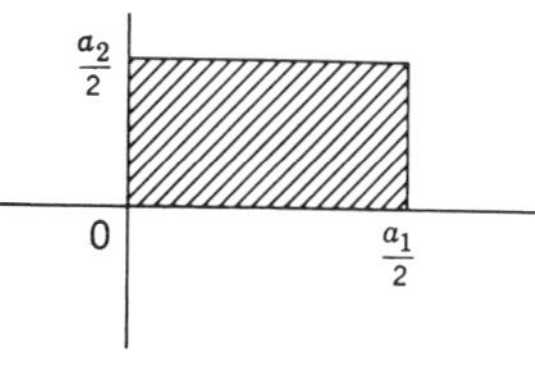

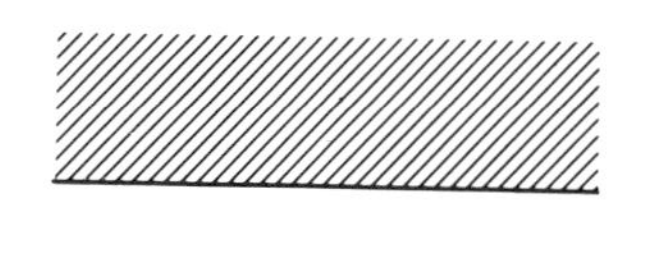

$$a_1 = a_1^* > 0; \qquad a_2 = -a_2^*$$

$$\left| \mathrm{Re}\left(z - \frac{a_1}{4}\right) \right| \leq \frac{a_1}{4}$$

$$\left| \mathrm{Im}\left(z - \frac{a_2}{4}\right) \right| \leq -\frac{ia_2}{4}$$

$$-ia_2 > 0$$

$\wp(z)$

$\mathrm{Im}\, w \geq 0$

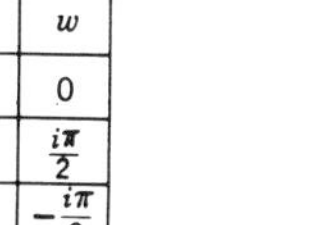

z	w
0	0
1	$\frac{i\pi}{2}$
−1	$-\frac{i\pi}{2}$

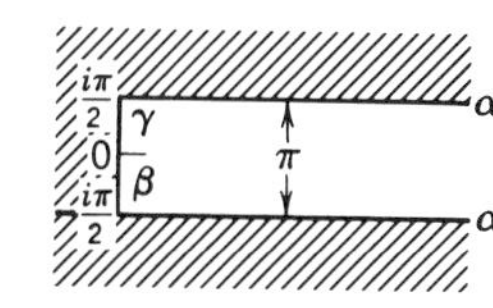

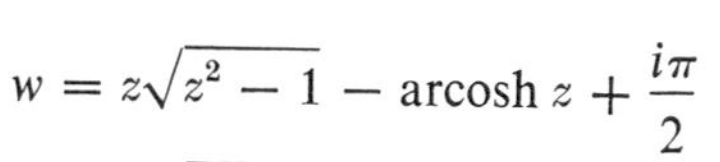

$$w = z\sqrt{z^2 - 1} - \mathrm{arcosh}\, z + \frac{i\pi}{2}$$

$$d_z w = 2\sqrt{z^2 - 1}$$

(continued)

Table III *(continued)*

z plane	$w = f(z)$	w plane

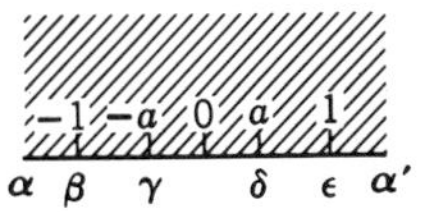

z	w
0	0
a	ih
$-a$	$-ih$

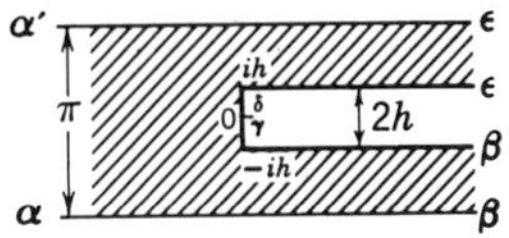

$$w = \frac{i\pi}{2} - \operatorname{arcosh}\left(\frac{z}{a}\right) + \frac{\sqrt{1-a^2}}{2}\operatorname{arcosh}\left[\frac{z^2(2-a^2)-a^2}{a^2(z^2-1)}\right]$$

$$d_z w = \frac{\sqrt{a^2-z^2}}{1-z^2} \qquad (0 < a < 1) \qquad a = \frac{2}{\pi}\sqrt{\pi h - h^2} \qquad 0 < 2h = \pi(1-\sqrt{1-a^2}) < \pi$$

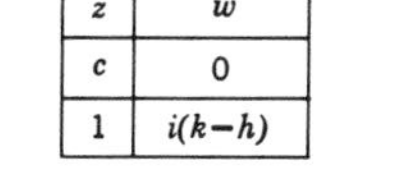

z	w
c	0
1	$i(k-h)$

$$w = \frac{k}{\pi}\operatorname{arcosh}\left(\frac{2z-c-1}{c-1}\right) - \frac{k}{\pi\sqrt{c}}\operatorname{arcosh}\left[\frac{(c+1)z-2c}{(c-1)z}\right]$$

$$d_z w = \frac{k\sqrt{z-1}}{\pi z\sqrt{z-c}} \qquad (c > 1;\ k > 0) \qquad k = h\sqrt{c}$$

z	w
0	0
1	∞
a	$\pi + i\pi\sqrt{a-1}$

$$w = \arccos\left(\frac{a-2z}{a}\right) + \sqrt{a-1}\,\operatorname{arcosh}\left[\frac{az+a-2z}{a(1-z)}\right]$$

$$d_z w = \frac{\sqrt{a-z}}{(1-z)\sqrt{z}} \qquad (a > 1)$$

(continued)

Table III *(continued)*

z plane	$w = f(z)$	w plane

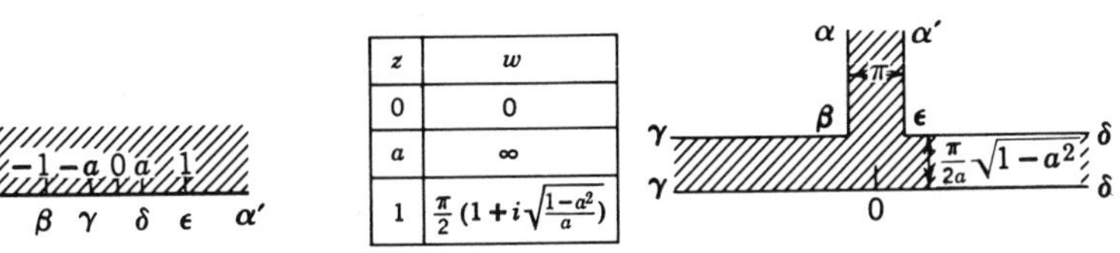

z	w
0	0
a	∞
1	$\frac{\pi}{2}\left(1 + i\sqrt{\frac{1-a^2}{a}}\right)$

$$w = \arcsin z + \frac{\sqrt{1 - a^2}}{2a} \operatorname{arcosh}\left(\frac{z^2 - 2a^2 z^2 + a^2}{a^2 - z^2}\right)$$

$$d_z w = \frac{\sqrt{1 - z^2}}{a^2 - z^2} \qquad (0 < a < 1)$$

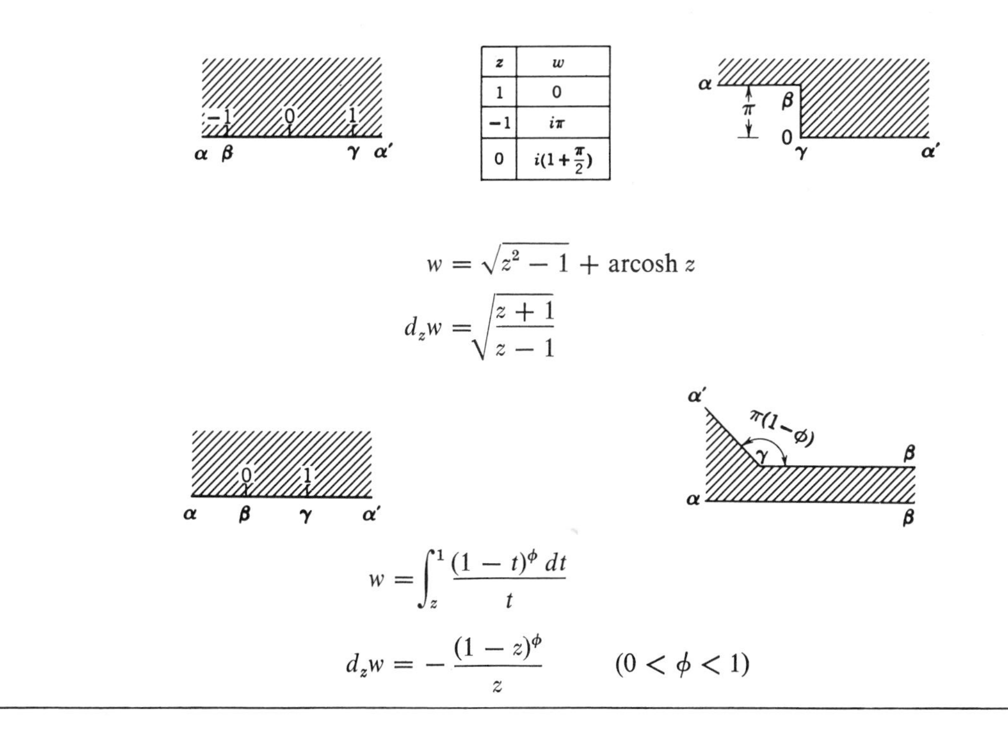

z	w
1	0
−1	$i\pi$
0	$i(1+\frac{\pi}{2})$

$$w = \sqrt{z^2 - 1} + \operatorname{arcosh} z$$

$$d_z w = \sqrt{\frac{z+1}{z-1}}$$

$$w = \int_z^1 \frac{(1-t)^\phi \, dt}{t}$$

$$d_z w = -\frac{(1-z)^\phi}{z} \qquad (0 < \phi < 1)$$

APPENDIX* A:

Riemann Mapping Theorem

Suppose that each function $f_n(z)$ of an infinite sequence is free of singularities for $|z| \leq R$ and that the image of $|z| \leq R$ under $w = f_n(z)$ is entirely contained within $|w| \leq M$ with M independent of n (and z). Also assume, at each point of an infinite set S of points in $|z| \leq R$, that $\lim_{n\to\infty} f_n(z)$ exists and that the set S has $z = 0$ as limit point (a point within an *arbitrarily small* distance of which there exists at least one point of S). Then $f_n(z)$ converges uniformly (independently of z) to an *analytic function* $f_\infty(z)$ for all $|z| \leq R - \varepsilon$ $(\varepsilon > 0)$. This is *Vitali's convergence theorem* for $|z| \leq R$. It is sometimes called Vitali's double series theorem. It may be proved as follows:

$$f_n(z) = \sum_{k=0}^{\infty} {}_nC_k z^k \qquad |z| \leq R$$

Also $|f_n(z) - f_n(0)| \leq 2M$ for $|z| \leq R$ and $|f_n(z) - f_n(0)| = 0$ for $z = 0$. Thus, by Schwarz's lemma,

$$|f_n(z) - f_n(0)| = |f_n(z) - {}_nC_0| \leq \frac{2M\,|z|}{R}$$

and by assumption M is independent of n. Let $\zeta \neq 0$ be in S. Then $f_\infty(\zeta)$ exists [i.e., $\lim f_n(\zeta)$ exists as $n \to \infty$] and

$$|{}_{n+m}C_0 - {}_nC_0| \leq |{}_{n+m}C_0 - f_{n+m}(\zeta)| + |f_{n+m}(\zeta) - f_n(\zeta)| + |f_n(\zeta) - {}_nC_0|$$

* The content of this appendix is based substantially on the presentation in E. C. Titchmarsh, *Theory of Functions*, 2nd ed., Oxford: Oxford University Press, 1939. The material is presented by courtesy of The Clarendon Press, Oxford.

so, applying Schwarz's lemma to the first and last terms on the right, one has

$$|_{n+m}C_0 - {}_nC_0| \le \frac{4M\,|\zeta|}{R} + |f_{n+m}(\zeta) - f_n(\zeta)|$$

Now since the origin is a limit point of the set S, the point ζ can be chosen so that $|\zeta|$ is arbitrarily small, and since $f_n(\zeta)$ approaches the limit $f_\infty(\zeta)$ as $n \to \infty$, the second term on the right can also be made arbitrarily small by choosing n sufficiently large. Hence, by the Cauchy criterion, the limit $f_\infty(0) = {}_\infty C_0$ of $f_n(0)$ as $n \to \infty$ exists.

Similarly,

$$g_n(z) = \frac{f_n(z) - {}_nC_0}{z} = \sum_{k=1}^{\infty} {}_nC_k z^{k-1} \qquad |z| \le R$$

is bounded with

$$|g_n(z)| \le \frac{2M}{R} \qquad \text{for} \quad |z| \le R$$

so that if $\zeta \ne 0$ is in S, one has

$$|_{n+m}C_1 - {}_nC_1| \le |_{n+m}C_1 - g_{n+m}(\zeta)| + |g_{n+m}(\zeta) - g_n(\zeta)| + |g_n(\zeta) - {}_nC_1|$$

and by Schwarz's lemma

$$|_{n+m}C_1 - {}_nC_1| \le \frac{4M\,|\zeta|}{R^2} + |g_{n+m}(\zeta) - g_n(\zeta)|$$

so that again the terms on the right can be made arbitrarily small by choosing $|\zeta|$ small and n large. Hence, again by the Cauchy criterion, the limit ${}_\infty C_1$ of ${}_nC_1$ as $n \to \infty$ exists.

Similarly,

$$\left|\frac{g_n(z) - {}_nC_1}{z}\right| \le \frac{3M}{R^2} \qquad \text{for} \quad |z| \le R$$

and ${}_\infty C_2$ can be shown to exist. Continuing in this way, ${}_\infty C_k$ can be shown to exist so that

$$f_\infty(z) = \sum_{k=0}^{\infty} {}_\infty C_k z^k$$

exists for all $|z| < R$ not just for z in S. Also for all $|z| \le R - \varepsilon$ (for ε a fixed small positive number) the series for $f_n(z)$ converges (uniformly) independently of z and n, since by the Weierstrass M test and the Cauchy inequality for ${}_nC_k$

$$\left|\sum_{k=0}^{\infty} {}_nC_k z^k\right| \le \sum_{k=0}^{\infty} \frac{M}{R^k}(R - \varepsilon)^k = \frac{MR}{\varepsilon}$$

with the bounds independent of z and n. By virtue of the uniform convergence the function $f_\infty(z)$ not only exists but is analytic and free of singularities in $|z| \leq R - \varepsilon$ however small ε may be chosen. Thus, Vitali's theorem is proved. The restriction to $|z| \leq R$ is not essential, since any simply connected closed set could be covered by a set of circular regions of convergence playing the role of $|z| \leq R$. Thus, the general statement of Vitali's convergence theorem would become: If each function $f_n(z)$ of an infinite sequence of functions is analytic and free of singularities in a region D and therein bounded independently of n and Z and if for a subset S of D possessing a limit point inside D, ζ in S implies the existence of $\lim_{n\to\infty} f_n(\zeta)$, then $f_n(z) \to f_\infty(z)$ uniformly for all z inside any contour interior to D (i.e., at minimal distance ε inside the boundary of D) and $f_\infty(z)$ is analytic and free of singularities in the latter region.

A corollary is that from any sequence of uniformly bounded functions $f_n(z)$ free of singularities in D a subsequence uniformly convergent to $f_\infty(z)$ inside any contour interior to D may be selected. To see this consider Σ to be an enumerable set with limit point inside D and members z_k $(1 \leq k < \infty)$. Then the image points $w_n = f_n(z_1)$ of the first point z_1 all lie within the circle $|w| \leq M$ [the uniform bound of all $f_n(z)$] and since there are an infinite number of such image points they must have a limit point. Hence, there exists a sequence of values $n_1, n_2, n_3 \to \infty$ such that the sequence of functions

$$f_{n_1}(z), f_{n_2}(z), f_{n_3}(z) \to f_\infty(z)$$

for $z = z_1$. Similarly, an infinite subsequence of functions for which

$$f_{m_1}(z), f_{m_2}(z), f_{m_3}(z) \to f_\infty(z)$$

at $z = z_2$ can be selected from the first sequence. Then from this subsequence an infinite subsequence of functions for which

$$f_{s_1}(z), f_{s_2}(z), f_{s_3}(z) \to f_\infty(z)$$

at $z = z_3$ can be selected, and so on. Finally, the subsequence

$$f_{n_1}(z), f_{m_2}(z), f_{s_3}(z) \to f_\infty(z)$$

in which successive terms are selected from successive subsequences converges to $f_\infty(z)$ at each of the points z_k of the set Σ, and Vitali's convergence theorem then implies uniform convergence to the analytic function $f_\infty(z)$ inside any contour interior to D.

As a final preparation for the fundamental theorem of conformal mapping the notion of a *schlicht* (simple) function will be introduced. This is a single-valued analytic function which does not take the same value twice in a region. It is the complex analog of a strictly monotonic function. For a function *schlicht* in a region its derivative cannot vanish there, otherwise an infinite number of points would be mapped into a single point via $dw = f'(z)\,dz$. Naturally, if $w = f(z)$ is *schlicht* in the region D of the z plane, the inverse function $z = g(w)$ will be *schlicht* in the corresponding (image) region of the w plane. If a uniformly convergent sequence of *schlicht* functions $f_n(z)$ approach a limiting function $f(z)$, the latter is either a constant or is itself *schlicht*. This may be seen by assuming $f'(z) \neq 0$. Then if $f(z)$ is not *schlicht*, there exist two distinct points $z_1 \neq z_2$ where $f(z_1) = f(z_2)$. Construct two nonoverlapping circles with z_1 and z_2 as centers for which $f(z) - f(z_1)$ is not zero on either circumference. Then if m is the least value of $|f(z) - f(z_1)|$ on *both* circumferences, n can be chosen so large that

$$|f(z) - f_n(z)| < m$$

Then the function

$$f_n(z) - f(z_1) = [f(z) - f(z_1)] + [f_n(z) - f(z)]$$

according to Rouché's theorem has the same number of zeros as $f(z) - f(z_1)$ inside the circles. This number is two; hence a contradiction to the *schlicht*-ness of $f_n(z)$. Thus it is concluded that $f(z)$ is *schlicht*, or constant.

A proof of the *Riemann mapping theorem* can now be given. The theorem states: Every simply connected closed domain D of the complex z plane with at least two boundary points can be mapped conformally onto the region $|w| \leq 1$ of the complex w plane by a *schlicht* function $w = f(z)$.

Let the two boundary points be $z = a$ and $z = b$ and consider the mapping

$$\xi = \sqrt{\frac{z - a}{z - b}}$$

where a definite branch B of the square root has been chosen. Each branch of the square root is a *schlicht* function, and the values taken by ξ for the branch chosen cover only a portion of the ξ plane (as *both* branches cover the whole ξ plane). Let ξ_0 be a point from the *other* branch $\tilde{B}$ at a distance at least $\delta > 0$ from any point of the chosen branch B in the ξ plane (see Fig. A.1). Then for any choice of k and d

$$w(\xi) = \frac{k}{\xi - \xi_0} + d$$

is *schlicht* and bounded for z in D. In particular, for any given interior point $z = c$ of D, k and d may be so chosen that

$$w[\xi(c)] = \frac{k}{\xi(c) - \xi_0} + d = 0$$

and

$$d_c w[\xi(c)] = -\frac{k}{[\xi(c) - \xi_0]^2}\, d_c \xi(c) = 1$$

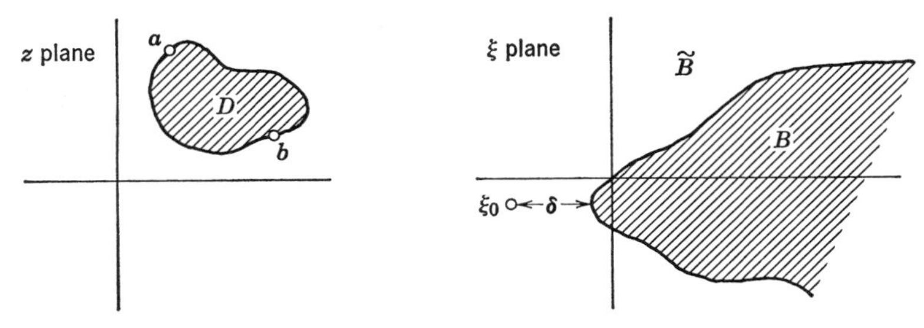

Figure A.1. Initial mapping diagram for Riemann mapping theorem.

Consider all functions $w[\xi(z)]$ which are *schlicht* and bounded in D and which have simple zeros of $w[\xi(z)]$ and $1 - d_z w[\xi(z)]$ at $z = c$ interior to D. Call any such function $g(z) = w[\xi(z)]$ and let $M(g)$ be the maximum modulus of $g(z)$ in D. Then a sequence of such functions may be found for which

$$\lim_{n \to \infty} M[g_n(z)] = \rho$$

where ρ is the greatest lower bound of the maximum moduli of the functions $g_n(z)$.

By Vitali's convergence theorem a subsequence $\{g_m(z)\}$ may be selected from $\{g_n(z)\}$ so that $g_m(z) \to f(z)$ uniformly. Then $f(z)$ is *schlicht* and bounded and $f(c) = 0 = 1 - f'(c)$, and so it is in the set of functions considered. Thus,

$$M[f(z)] \geq \rho$$

by the definition of ρ. Also,

$$M[g_m(z)] < \rho + \varepsilon \quad \text{for} \quad m > m_0; \; m \text{ sufficently large}$$

Hence also,

$$|g_m(z)| < \rho + \varepsilon \qquad \text{for} \quad m > m_0$$

and as $m \to \infty$

$$|f(z)| \leq \rho$$

so that $M[f(z)] \leq \rho$ and hence

$$Mf[(z)] = \rho$$

Thus, there exists a member $f(z)$ of the original set for which the limit is attained. Since $f'(c) = 1 \neq 0, f(z)$ is *not* constant and so $\rho > 0$.

So far it has only been shown that there exists a bounded *schlicht* function $w = f(z)$ with $f(c) = 0$ and $f'(c) = 1$ which maps D onto a region *contained in* a circle of radius ρ in the w plane. At this stage it is no loss of generality to assume $\rho = 1$ because whatever other value the maximum modulus of $f(z)$ might have it can be divided into $f(z)$ whence the new mapping function would have $\rho = 1$.

Let D' be the region of the w plane on which $w = f(z)$ maps D. Then D' is included in $|w| \leq 1$ and reaches $|w| = 1$ at least at one point. Suppose D' has a boundary point at $w = w_1$ with $|w_1| < 1$. By three successive mappings a function can be constructed which will contradict the condition $\rho = 1$.

Let a definite branch of

$$\zeta = \sqrt{\frac{w_1 - w}{1 - w_1^* w}}$$

be chosen. Then since ζ^2 maps the interior of the unit circle in the w plane into the interior of a unit circle in the ζ^2 plane, sending $w = w_1$ into the origin, one has

$$|\zeta| \leq 1 \qquad \text{for} \quad |w| \leq 1$$

while $\zeta = \sqrt{w_1}$ when $w = 0$. Now let

$$\eta = \frac{\zeta - \sqrt{w_1}}{1 - \sqrt{w_1^*}\zeta}$$

Since this maps $|\zeta| \leq 1$ into $|\eta| \leq 1$, it follows that

$$|\eta| \leq 1 \qquad \text{for} \quad |w| \leq 1$$

Also $\eta = 0$ when $w = 0$ and

$$d_w\eta = d_\zeta\eta \, d_w\zeta = \frac{d_\zeta\eta \, d_w\zeta^2}{2\zeta}$$

and at $w = 0$ ($\zeta = \sqrt{w_1}$)

$$d_w\zeta = \frac{d_w\zeta^2}{2\zeta} = \frac{\begin{vmatrix} 1 & -w_1 \\ w_1^* & -1 \end{vmatrix}}{2\zeta} = -\left(\frac{1 - |w_1|^2}{2\sqrt{w_1}}\right)$$

$$d_\zeta\eta = \frac{\begin{vmatrix} 1 & -\sqrt{w_1} \\ -\sqrt{w_1^*} & 1 \end{vmatrix}}{(1 - \sqrt{w_1^*}\zeta)^2} = \frac{1 - |w_1|}{(1 - |w_1|)^2}$$

so at $w = 0$

$$|d_w\eta| = \frac{1 + |w_1|}{2\sqrt{|w_1|}} > 1$$

Finally, let

$$-\phi = \frac{2\sqrt{w_1}}{1 + |w_1|}\eta$$

so that $\phi = 0$ for $w = 0$ $(z = c)$ and

$$- d_z\phi = \frac{2\sqrt{w_1}}{1 + |w_1|} d_w\eta \, d_z w$$

which for $w = 0$ $(z = c)$ yields

$$d_z\phi = 1$$

so that $\phi(z)$ is a function of the set considered. But then

$$M(\phi) = \frac{2\sqrt{|w_1|}}{1 + |w_1|} < 1$$

contrary to the assumption that $\rho = 1$. Hence, it must be that $|w_1| = 1$ and the map of D by $w = f(z)$ reaches $|w| = 1$ at *every* boundary point. The uniqueness of the mapping may be seen by assuming there are *two* such mappings $w = f(z)$ and $w = g(z)$. If the inverse mapping of the first is $z = F(w)$, then $w = g[F(w)]$ implies $g(z) = f(z)$.

This completes the proof of the Riemann mapping theorem. It is physically equivalent to the existence of a potential function for the field of an infinite straight charged wire inside a grounded conducting cylindrical container (with generators parallel to the wire) of arbitrary closed connected (non-self-intersecting) cross section. For those who hold that the existence of such a function is evident on physical grounds the Riemann mapping theorem may seem superfluous. For those not of this opinion it must remain a masterpiece of (nonconstructive) mathematical reasoning.

APPENDIX B:

Green's Theorem in the Plane

If $\partial_x u$ and $\partial_y v$ are single-valued and continuous in a simply connected (plane) region S^* and if S is a simply connected subregion of S^* bounded by a simple closed curve Γ, then

$$\iint_S (\partial_x u - \partial_y v)\, dx\, dy = \oint_\Gamma v\, dx + u\, dy$$

The proof follows: It suffices to consider S to be convex, since it can be decomposed into a set of convex regions (Fig. B.1). There exist at least four points A, B, C, D on Γ such that there is no point in S

(1) with a smaller ordinate than A;
(2) with a larger ordinate than C;
(3) with a smaller abscissa than D;
(4) with a larger abscissa than B.

The integration of $\partial_x u$ may be carried out on x first so that

$$\iint_S \partial_x u\, dx\, dy = \int \left[\int \partial_x u\, dx \right] dy = \int (u^+ - u^-)\, dy$$

where u^+ and u^- are values of u corresponding to the same ordinate but to points on ABC and CDA, respectively. Thus,

$$\int (u^+ - u^-)\, dy = \oint_{ABC} u\, dy - \oint_{ADC} u\, dy = \oint_{ABC} u\, dy + \oint_{CDA} u\, dy$$

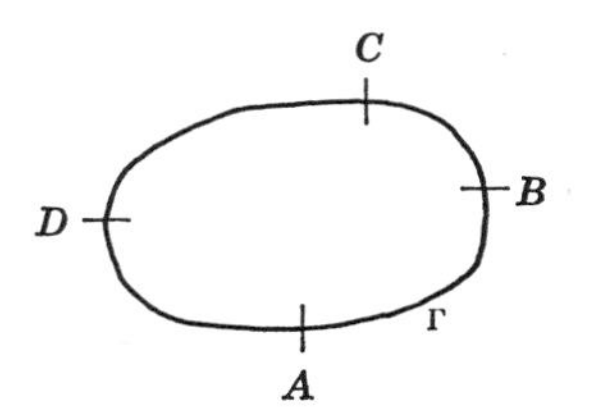

Figure B.1. Contour for Green's theorem in the plane.

As y increases, the portion ADC is traversed in a clockwise direction while the portion ABC is traversed in a counterclockwise direction. Finally, traversing Γ counterclockwise,

$$\iint_S \partial_x u \, dx \, dy = \oint_\Gamma u \, dy$$

Similarly,

$$\iint_S \partial_y v \, dx \, dy = \int \left[\int \partial_y v \, dy \right] dx = \int (v^+ - v^-) \, dx$$

where v^+ and v^- are values of v corresponding to the same abscissa but to points on DCB and DAB, respectively. Thus,

$$\int (v^+ - v^-) \, dx = \int_{DCB} v \, dx - \int_{DAB} v \, dx = \int_{DCB} v \, dx + \int_{BAD} v \, dx$$

As x increases, the portion DAB is traversed in a counterclockwise direction while the portion DCB is traversed in a clockwise direction. Finally, traversing Γ counterclockwise,

$$-\iint_S \partial_y v \, dx \, dy = \oint_\Gamma v \, dx$$

so that by addition the theorem follows.

Important corollaries are obtained by taking

$$u = \phi \partial_x \psi \qquad v = -\phi \partial_y \psi$$

whence

$$\iint_S [\phi \, \Delta \psi + (\partial_x \phi \partial_x \psi + \partial_y \phi \partial_y \psi)] \, dS = \oint \phi \partial_n \psi \, ds$$

with

$$dS = dx \, dy$$
$$\Delta = \partial_x^2 + \partial_y^2$$
$$\partial_n = \frac{dy}{ds} \partial_x - \frac{dx}{ds} \partial_y = d_s y \partial_x - d_s x \partial_y$$

the derivative in the (outward) direction normal to Γ. Interchanging ϕ and ψ and subtracting, one then has

$$\iint_S (\phi\,\Delta\psi - \psi\,\Delta\phi)\,dS = \oint [\phi\partial_n\psi - \psi\partial_n\phi]\,ds$$

For $\phi = 1$ this yields

$$\iint_S \Delta\psi\,dS = \oint \partial_n\psi\,ds$$

so that if ψ is harmonic ($\Delta\psi = 0$),

$$\oint \partial_n\psi\,ds = 0$$

which shows that the average value of the normal derivative of a harmonic function on a closed curve is zero.

APPENDIX* C:

Phragmén-Lindelöf Theorems

A sometimes useful extension of the maximum modulus theorem is the *Phragmén–Lindelöf* theorem which concludes that $|f(z)| \leq M$ at all points *inside* a simple closed contour C if

(1) $f(z)$ is analytic and free of singularities inside and on C *except at one point* P on C;
(2) $|f(z)| \leq M$ *on* C *except at* P;
(3) there exists a function $\omega(z)$ analytic and free of singularities and zeros inside C with $|\omega(z)| \leq 1$ inside C and a curve arbitrarily close to P (but not passing through P) connecting points on C on opposite sides of P along which $|\omega(z)|^{\varepsilon}|f(z)| \leq M$ for arbitrarily small positive ε.

The proof is based on consideration of

$$G(z) = [\omega(z)]^{\varepsilon} f(z)$$

which by hypothesis is analytic and free of singularities *inside* C.

For any point a *inside* C a curve C' can be constructed isolating P from the region interior to C and from z (Fig. C.1). Since $|f(z)| \leq M$ on any closed contour D including a and $|\omega(z)| \leq 1$ on D, one has

$$|G(z)| \leq M$$

* The content of this appendix is based substantially on the presentation in E. C. Titchmarsh, *Theory of Functions*, 2nd ed., Oxford: Oxford University Press, 1939. The material is presented by courtesy of The Clarendon Press, Oxford.

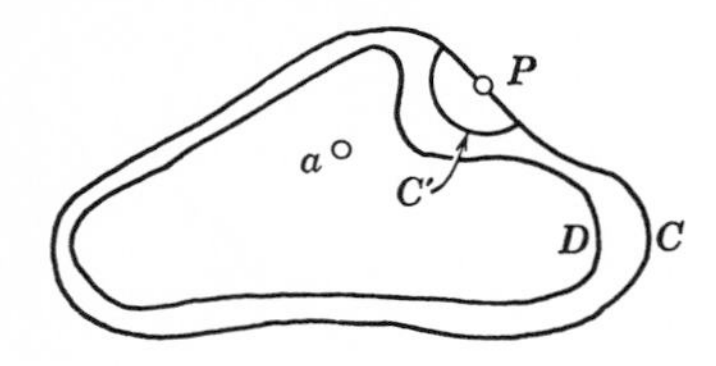

Figure C.1. Diagram for Phragmén–Lindelöf theorem.

on D. Hence, by the maximum modulus theorem,

$$|G(a)| \leq M$$

or

$$|f(a)| \leq M\,|\omega(a)|^{-\varepsilon}$$

and as $\varepsilon \to 0$

$$|f(a)| \leq M$$

The point of the theorem is that a function subject to the theorem is bounded at all interior points of a region by the same value with which it is bounded at all points *but one* of the surrounding contour. It means that the requirement that $|f(z)| \leq M$ at *all* points of the contour in the maximum modulus theorem can be relaxed at a *single point* or by a simple extension of the argument at a *finite* number of points under the conditions of the Phragmén–Lindelöf theorem.

In practice P is usually the point at infinity ($z = \infty$). The theorem can then be stated as

$$|f(z)| \leq M \qquad \text{for} \quad |\arg z| \leq \frac{\pi}{2\alpha}$$

if

(1) $f(z)$ is analytic and free of singularities for all finite values of $|z|$ with $|\arg z| \leq \pi/2\alpha$;
(2) $|f(z)| \leq M$ for finite $|z|$ and $|\arg z| = \pi/2\alpha$;
(3) $|f(z)|e^{-|z|^\beta}$ converges uniformly to K as $|z| \to \infty$ with $\beta < \alpha$.

To prove this let $G(z) = e^{-\varepsilon z^\gamma} f(z)$ with $\beta < \gamma < \alpha$ and $\varepsilon > 0$. Then $[\cos(\gamma\pi/2\alpha) < 1]$ and

$$|G(z)| = e^{-\varepsilon|z|^\gamma \cos(\gamma\pi/2\alpha)}\,|f(z)| \leq |f(z)| \leq M$$

for $|\arg z| = \pi/2\alpha$. Also, on $|z| = R$, $|\arg z| \leq \pi/2\alpha$,

$$|G(z)| \leq e^{-\varepsilon R^\gamma \cos(\gamma\pi/2\alpha)}\,|f(z)| \leq K e^{R^\beta - \varepsilon R^\gamma \cos(\gamma\pi/2\alpha)}$$

and since $\gamma > \beta$, $\lim_{|z|\to\infty} |G(z)| = 0$. Hence, $|G(z)|$ is bounded as $|z| \to \infty$, so by the maximum modulus theorem $|G(z)| \leq M$ and so $|f(z)| \leq M e^{\varepsilon|z|^\gamma}$. As $\varepsilon \to 0$, then $|f(z)| \leq M$.

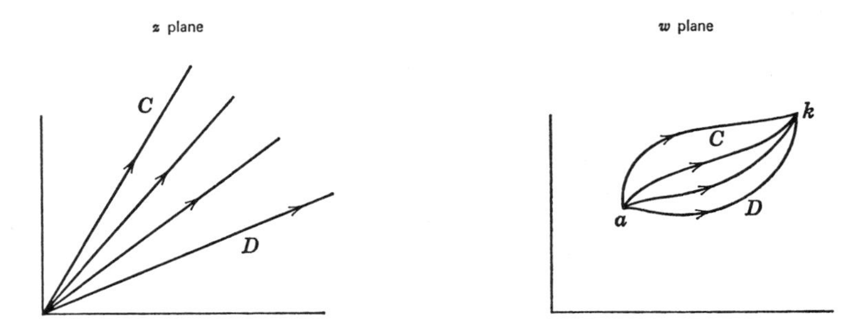

Figure C.2. Common limit along all vertex lines.

A corollary of the Phragmén–Lindelöf theorems is the following: If $f(re^{i\alpha}) \to k \leftarrow f(re^{i\beta})$ as $r \to \infty$ and $f(z)$ is analytic, free of singularities, and bounded for $\alpha \leq \arg z \leq \beta$, then $f(re^{i\theta}) \to k$ for $\alpha \leq \theta \leq \beta$. This may also be seen from the mapping argument illustrated in Figure C.2. Here, $w = f(z)$ and $a = f(0)$, and since the terminus of the images of the two boundary lines are both the identical number k, the images of the vertex lines between cannot terminate except at $w = k$, as was to be shown.

Similarly, if $f(re^{i\alpha}) \to k_1$ and $f(re^{i\beta}) \to k_2$ as $r \to \infty$ and $f(z)$ is bounded, analytic, and free of singularities for $\alpha \leq \theta \leq \beta$, then $f(re^{i\theta}) \to k_1 = k_2$ as $r \to \infty$. To see this consider

$$g(z) = [f(z) - \tfrac{1}{2}(k_1 + k_2)]^2$$

Then

$$g(re^{i\alpha}) \to \left[\frac{k_1 - k_2}{2}\right]^2$$

and

$$g(re^{i\beta}) \to \left[\frac{k_2 - k_1}{2}\right]^2 \qquad \text{as} \quad r \to \infty$$

so that by the preceding result

$$g(re^{i\theta}) \to \left[\frac{k_1 - k_2}{2}\right]^2$$

uniformly in the angle as $r \to \infty$. Hence,

$$[f(z) - \tfrac{1}{2}(k_1 + k_2)]^2 - \left(\frac{k_1 - k_2}{2}\right)^2 = [f(z) - k_1][f(z) - k_2]$$

approaches zero *uniformly* in the angle. Hence, $k_1 = k_2$ and

$$f(re^{i\theta}) \to k_1 = k_2$$

as $r \to \infty$, as was to be shown.

The vanishing of a difference between two functions in a sector can be used to establish an analytical continuation. According to the Phragmén–Lindelöf results this can be established by showing that the relevant conditions are satisfied along the bounding rays of the sector.

Bibliography

Besides the four basic references mentioned in the Preface, the following should be noted:

Bowman, F., *Introduction to Elliptic Functions.* New York: Dover, 1961.

Byerly, W. E., *Fourier Series and Spherical Harmonics.* Boston: Ginn and Co., 1893.

Carrier, G. F., M. Krook, and C. E. Pearson, *Functions of a Complex Variable.* New York: McGraw-Hill, 1966.

Carslaw, H. S., and J. C. Jaeger, *Conduction of Heats in Solids.* Oxford: Oxford University Press, 1948.

Dienes, P., *The Taylor Series.* Oxford: Oxford University Press, 1931.

Ditkin, V. W., and A. P. Prudnikov, *Operational Calculus in Two Variables.* London: Pergamon Press, 1962.

Doetsch, G., *Anleitung zum praktischen Gebrauch der Laplace-Transformation.* Munich: Oldenbourg, 1961.

Franklin, P., *A Treatise on Advanced Calculus.* New York: Wiley, 1940.

Franklin, P., *Methods of Advanced Calculus.* New York: McGraw-Hill, 1944.

Franklin, P., *An Introduction to Fourier Methods and the Laplace Transformation.* New York: Dover, 1949.

Guillemin, E. A., *Synthesis of Passive Networks.* New York: Wiley, 1957.

Hirschman, I. I., and D. V. Widder, *The Convolution Transform.* Princeton: Princeton University Press, 1955.

Hobson, E. W., *The Theory of Spherical and Ellipsoidal Harmonics.* Cambridge: Cambridge University Press, 1931.

Knopp, K., *Theory and Application of Infinite Series.* London: Blackie, 1928.

Knopp, K., *Theory of Functions.* New York: Dover, 1947 (2 vols. of text and 2 vols. of solved problems).

Levinson, N., and R. M. Redheffer, *Complex Variables.* San Francisco: Holden-Day, 1970.

McLachlan, N. W., *Complex Variable Theory and Transform Calculus.* Cambridge: Cambridge University Press, 1953.

Milne-Thomson, L. M., *Theoretical Hydrodynamics.* New York: Macmillan, 1955.

Murphy, G. J., *Basic Automatic Control Theory.* Princeton: D. Van Nostrand, 1966.

Muskhelishvili, N. I., *Singular Integral Equations.* Groningen: Noordhoff, 1946.

Papoulis, A., *The Fourier Integral and Its Applications.* New York: McGraw-Hill, 1962.

Polya, G., and G. Szegö, *Aufgaben und Lehrsätze aus der Analysis.* New York: Dover, 1947 (2 vols. of solved problems).

Smythe, W. R., *Static and Dynamic Electricity.* New York: McGraw-Hill, 1939.

Sneddon, I. N., *Fourier Transforms.* New York: McGraw-Hill, 1951.

Watson, G. N., *A Treatise on the Theory of Bessel Functions.* Cambridge: Cambridge University Press, 1945.

Widder, D. V., *The Laplace Transform.* Princeton: Princeton University Press, 1941.

The following are books of tables:

Byrd, P. F., and M. D. Friedman, *Handbook of Elliptic Integrals for Engineers and Physicists.* Berlin: Springer, 1954.

Kober, H., *Dictionary of Conformal Representations.* New York: Dover, 1957.

Maxfield, J. E., and R. G. Selfridge, *A Table of the Incomplete Elliptic Integral of the Third Kind.* New York: Dover, 1963.

Moon, P., and D. E. Spencer, *Field Theory Handbook.* Berlin: Springer, 1961.

Oberhettinger, F., *Tabellen zur Fourier-Transformation.* Berlin: Springer, 1957.

Index